Pocket Guide
to Biotechnology and Genetic Engineering

Rolf D. Schmid

Pocket Guide
to Biotechnology and Genetic Engineering

142 color plates by Ruth Hammelehle

WILEY-VCH

Prof. Dr. Rolf D. Schmid
Institut für Technische Biochemie
Universität Stuttgart
Allmandring 31
70569 Stuttgart
Germany

This book was carefully produced. Nevertheless, author and publishers do not warrant the information contained therein to be free of errors. Readers are advised to keep in mind that statements, data, illustrations, procedural details or other items may inadvertently be inaccurate.

Bibliographic information published by Die Deutsche Bibliothek

Die Deutsche Bibliothek lists this publication in the Deutsche Nationalbibliografie; detailed bibliographic data is available in the Internet at http://dnb.ddb.de

ISBN 3-527-30895-4

© 2003 WILEY-VCH Verlag GmbH & Co. KGaA, Weinheim

Printed on acid free paper

IV Composition: epline, Kirchheim/Teck
Printing and Bookbinding: Druckhaus Darmstadt GmbH, Darmstadt
Printed in the Federal Republic of Germany

Contents

V

VIII

Preface

Biotechnology, a key technology of the 21st century, is more than other fields an interdisciplinary endeavor. Depending on the particular objective, it requires knowledge in general biology, molecular genetics, and cell biology; in human genetics and molecular medicine; in virology, microbiology, and biochemistry; in the agricultural and food sciences; in enzyme technology, bioprocess engineering, and systems science. And in addition, biocomputing and bioinformatics play an ever-increasing role. Against this background, it is of little surprise that few concise textbooks try to cover the whole field, and important applied aspects such as animal and plant breeding or analytical biotechnology are often missing even from multi-volume monographs.

On the other hand, I have experienced during my own life-long studies, and also when teaching my students, how energizing it is to emerge occasionally from the thousands of details which must be learned, to look at a unifying view.

The Pocket Guide to Biotechnology and Genetic Engineering is an attempt to provide this kind of birds-eye perspective. Admittedly, it is daring to discuss each of this book's topics, ranging from "Beer" to "Tissue Engineering" and "Systems Biology", on a single text page, followed by one page of graphs and tables. After all, monographs, book chapters, reviews, and hundreds of scientific publications are devoted to each single entry covered in this book (many of them are provided in the literature citations). On the other hand, the challenge of surveying each entry in barely more than 4000 characters forces one to concentrate on the essentials and to put them into a wider perspective.

I hope that I have succeeded at least to some extent in this endeavor, and that you will find the clues to return safely from the highly specialized world of science, and its sophisticated terms, to your own evaluation of the opportunities and challenges that modern biotechnology offers to all of us.

This English version is not a simple translation of the original version, which was published in German in December, 2001, but an improved and enlarged second edition: apart from a general update of all data, it contains three new topics (Tissue Engineering, RNA, and Systems Biology).

At this point, my thanks are due to some people who have essentially contributed to this book. Above all, I wish to acknowledge the graphic talent of Ruth Hammelehle, Kirchheim, Germany, who has done a great job in translating scientific language into very clear and beautiful graphs. Marjorie Tiefert, San Ramon, California, has been more than an editor: she has caught and expressed the original spirit of this book. My thanks also to the publisher, in particular to Romy Kirsten. Special thanks are due to the many colleagues in academia and industry who have contributed their time and energy to read through the entries in their areas of expertise and provide me with most useful suggestions and corrections. These were: Max Roehr, University of Vienna; Waander Riethorst, Biochemie GmbH, Kundl; Frank Emde, Heinrich Frings GmbH, Bonn; Peter Duerre, University of Ulm; Edeltraut Mast-Gerlach, Ulf Stahl and Dietrich Knorr, Technical University Berlin; Udo Graefe, Hans-Knoell Institute, Jena; Jochen Berlin, GBF, Braunschweig; Allan Svenson, Novozymes A/S, Copenhagen; Helmut Uhlig, Breisach; Frieder Scheller, University of Potsdam; Bertold Hock, University of Munich-Weihenstephan; Rolf Blaich, Rolf Claus, Helmut Geldermann and Gerd Weber, University of Hohenheim; Hans-Joachim Knackmuss, Dieter Jendrossek, Karl-Heinrich Engesser, Joerg Metzger, Peter Scheurich, Ulrich Eisel, Matthias Reuss, Klaus Mauch, Christoph Syldatk, Michael Thumm, Joseph Altenbuchner, Paul Keller and Ulrich Kull, University of Stuttgart; Thomas von Schell, Stuttgart; Joachim Siedel, Roche AG, Penzberg; Rolf Werner and Kerstin Maier, Boehringer-Ingelheim, Biberach; Frank-Andreas Gunkel, Bayer AG, Wuppertal; Michael Broeker, Chiron Bering GmbH, Marburg; Bernhard Hauer and Uwe Pressler, BASF AG, Ludwigshafen; Frank Zocher, Aventis Pharma, Hoechst; Tilmann Spellig, Schering AG, Bergkamen; Akira Kuninaka, Yamasa Corporation, Chosi; Ian Sutherland, University of Edinburgh; Julia Schueler, Ernst & Young,

Frankfurt. Among the many members of my institute in Stuttgart who have patiently helped me with the manuscript I wish to especially acknowledge Jutta Schmitt, Till Bachmann, Jürgen Pleiss and Daniel Appel.

In spite of all efforts and patient cross-checking, it would be a miracle if no unclearness or errors exist. These are entirely the author's fault. I would be most grateful to all readers who will let me know, via the web address www.itb.uni-stuttgart.de/pocketguide, where this book can be further improved.

Rolf D. Schmid
Stuttgart, New Year 2002/2003

Introduction

This pocket guide is intended for students of biology, biochemistry, and bioprocess engineering who are looking for a first survey of the many areas of modern biotechnology. Because it is written in a modular manner, with a considerable number of literature citations for each entry and a comprehensive index, it can also be used as a starter for deeper studies.

Color plates on 142 subject areas are complemented by a text page on each subject. There are colored column titles on each page to facilitate general orientation.

The guide starts with a short **historical survey**. Because biotechnology has always been an applied science, you will find here the first data on the economic significance of some bioproducts. **Food biotechnology** was the starting point of the whole field and thus is discussed as the next entry. Based on these old skills, the first modern biotechnology processes were developed and enabled the production of **alcohols, acids, and amino acids**, followed by **antibiotics**, **specialty chemicals**, the important group of analytical and technical **enzymes**, and, finally, **baker's yeast and fodder yeasts**. The section on **biotechnology of environmental processes** features the cleaning of water, air, and soil and includes microbial leaching. A large section is devoted to **medical biotechnology**; it includes entries on the preparation of pharmaceutical drugs such as factor VIII and erythro-

poietin, and also on vaccines, antibodies, immunoassays, and biosensors. The immune system is briefly overviewed, followed by features on stem cells and the emerging field of tissue engineering. Another large section is devoted to **agricultural biotechnology**, in particular to animal and plant breeding, cloning, and genetic modification.

The second part of the pocket guide is devoted to the fundamental knowledge and techniques that are at the heart of modern biotechnology. It starts with **microbiological fundamentals**, followed by **process engineering fundamentals** and, with significant volume and depth, the **fundamentals of molecular genetics**. The final part of the book is devoted to **recent trends and safety issues**, followed by **ethics and economic issues**. It includes modern areas such as DNA arrays, proteomics, metabolic engineering, and systems biology and also public concerns, patenting, and international aspects of this economically important area of technology.

I hope that the comprehensive index will assist you to find help and explanations from other entries, should the many technical terms in an area exceed your experience. The literature index is focussed on English language citations only, with recent monographs and reviews predominating. On many subjects discussed in this guide, further information is available in the web and can be retrieved through, e. g., PubMed.

Early developments

History. The origins of what we call biotechnology today probably originated with agriculture and can be traced back to early history. Presumably, since the beginning people have gained experience on the loss of food by microbial spoilage; on food conservation by drying, salting, and sugaring; and on the effects of fermented alcoholic beverages. As the first city cultures developed, we find documents and drawings on the preparation of bread, beer, wine, and cheese and on the tanning of hides using principles of biotechnology. In Europe, starting in the 6th century, the monasteries with their well organized infrastructure developed protocols for the arts of brewing, winemaking, and baking. We owe our strong, alcohol-rich stout beers to the pious understanding of the monks that "Liquida non fragunt ieiunum" (Liquors do not interfere with the chamfering time). Modern biotechnology, however, is a child of microbiology, which developed significantly in the late 19th century. The First and Second World Wars in the first half of the 20th century next probably provided the strongest challenge to microbiologists, chemists, and engineers to establish modern industrial biotechnology, based on products such as organic solvents and antibiotics. During and after this period, many ground-breaking discoveries and developments were made by biochemists, geneticists, and cell biologists and gave rise to molecular biology. At this point, the stage was set for modern biotechnology, based on genetic and cell engineering, to come into being during the 1970s and '80s. With the advent of information technology, finally, modern biotechnologies gave rise to genomics, proteomics and cellomics, which promise to develop into the key technologies of the 21st century, with a host of applications in medicine, food and agriculture, chemistry and environmental protection.
Early pioneers and products. Biotechnology is an applied science – many of its developments are driven by economic motives. In 1864 Louis Pasteur, a French

chemist, used a microscope for the first time to monitor the fermentation of wine vs. lactic acid. Using sterilized media ("pasteurization"), he obtained pure cultures of microorganisms, thus laying the foundation for applied microbiology and expanding this field into the control of pathogenic microorganisms. At the start of the 20th century, it occurred to the German chemist Otto Roehm and to the Japanese scientist Jokichi Takamine that enzymes isolated from animal wastes or from cultures of molds might be useful catalysts in industrial processes. Otto Roehm's idea revolutionized the tanning industry, since tanning up to this time was done using dog excrements. In the field of public health, the introduction of biological sewage treatment around 1900 was a milestone for the prevention of epidemics. During World War I, Carl Neuberg in Germany and Chaim Weizmann, a Russian emigrant to Britain and of Jewish origin, developed large-scale fermentation processes for the preparation of ammunition components (glycerol for nitroglycerol and acetone for Cordite). The Balfour declaration and the ensuing foundation of the state of Israel, whose first president Weizmann became, is thus directly linked to an early success in biotechnology. In the postwar period. 1-Butanol, the second product from Weizmann's Clostridium-based fermentation process, became highly important in the USA as a solvent for car paints. The serendipitous discovery of penicillin by Alexander Fleming (1922), much later turned into a drug by Howard Florey, initiated the large-scale production of penicillin and other antibiotics during World War II. As early as 1950, more than 1000 different antibiotics had been isolated and were increasingly used in medicine, in animal feeds, and in plant protection. Since 1950, the analytical use of enzymes, later of antibodies, began another important field of modern biotechnology. In the shadow of the 1960s' oil crises and the emerging awareness of overpopulation, the conversion of biomass to bioenergy (ethanol, methane) and of single-cell protein from petroleum or methanol was developed.

Biotechnology in early Egyptian drawings

1 beer brewing
2 bread baking
3 hide tanning

early history	sugar-containing juices are fermented to various alcoholic beverages
	sour milk and sourdough products are prepared by lactic acid and yeast fermentation
	hides are bated to leather employing reagents such as animal feces
1650	France: Orléans procedure for the preparation of vinegar from ethanol
~ 1680	The Netherlands: Anthony van Leuwenhoek observes bacteria through a microscope
1856	France: Louis Pasteur separates brewers yeasts from lactic acid bacteria
~ 1890	France, Germany: Louis Pasteur, Robert Koch develop the first vaccines
1900	Japan: Jokichi Takamine uses α-amylase for starch degradation
1908	Germany: Otto Roehm uses pancreatic trypsin in detergents and for leather bating
1916	UK: Chaim Weizmann develops a fermentation process for acetone, n-butanol
since 1920	citric acid is industrially produced by surface fermentation using *Aspergillus niger*
1928/29	UK: Alexander Fleming discovers penicillin
1943	USA: Selman Waksman discovers streptomycin
since 1949	USA: microbial transformation of steroids on industrial scale
since 1957	Japan: glutamic acid is industrially produced by tank fermentation of *Corynebacterium glutamicum*
since 1960	Denmark: *Bacillus* proteases are used in detergents
since 1965	Denmark: microbial rennet for cheese production
since 1970	USA: high-fructose syrups produced by enzyme technology replace saccharose in softdrinks
1972/73	USA: Stanley Cohen and Francis Boyer develop a procedure for in-vitro recombination of DNA, using plasmid vectors
1975	UK/Switzerland: César Milstein and Georges Koehler prepare monoclonal antibodies using hybridoma cells
since 1977	recombinant proteins can be manufactured by fermentation using bacteria
since 1982	first transgenic plants (herbicide resistance) and animals (knockout)
1985	USA: Kary Mullis discovers the polymerase chain reaction (PCR)
from 1990	USA: the human genome project (HUGO) is initiated
1995	transgenic tomatoes (Flavr Savr) are registered as food in the USA and the UK
since 1995	gene therapy experiments on humans
1996	the yeast genome is completely sequenced
1998	Dolly the sheep is the first cloned animal, a replicate of its mother
1998	over 2 billion basepairs are stored in DNA sequence databases
1999	the Drosophila genome with 1.6 billion bp is completely sequenced in ~4 months
1999	human stem cells can be maintained in culture
1999	the sales of recombinant therapeutic proteins exceed 10 billion US$/yr

Biotechnology today

Genetic Engineering and Cell Technology.
In 1973, Stanley Cohen and Frederick Boyer in San Francisco were the first to cause a designed foreign gene to be expressed in a host organism. After about 10 years the first recombinant drug, human somatotropin, was registered. Since then, more than 50 genetically engineered proteins have been registered as therapeutic agents, including insulin (for diabetics), erythropoietin (for anemic patients), factor VIII (for hemophiliacs) and interferon-β (for multiple sclerosis patients). Many hundred more are under development. Although the new technologies were first applied to medicine, their innovation potential in agriculture and food production soon began to emerge. Thus, transgenic crops were bred that were resistant to herbicides, insects, or viruses. Today, they are predominantly grown in North America. Flowers have been genetically modified to exhibit new colors, vegetables or fruits to show enhanced nutritive properties, and woods to contain less lignin, for improved paper production. In the chemical industry, the number of processes based on biocatalysts is steadily growing, since enzymes and microorganisms can be genetically adapted to process conditions. However, the present focus of biotechnology is on genomics and post-genomics. Methods have been developed that have enabled the rapid sequencing of over 50 microbial genomes, of the first plant and animal genomes, and, in 2001, of the human genome. This information is now widely used to understand the molecular basis of diseases and to develop novel drugs by a target-oriented screening approach. Novel approaches, such as proteomics and structural biology, are contributing to our fundamental understanding of the chemistry of life and disease. Using gene therapy, we attempt to replace malfunctioning with correctly functioning genes. These developments are in step with great advances in cell biology, which focus on the complex interactions of cells in a multicellular organism. Tissue engineering, a practical approach to repairing wounded tissue, and systems biology, a computer-based approach to understanding cell function, are just two applications.

Public acceptance. The sheep Dolly, born in 1998, was the first animal ever cloned from a somatic cell of and thus identical to her mother. The thrust and possible consequences of such developments, e. g., for embryonic manipulations or individual genetic fingerprinting, have led to emotional public discussion. Typical subjects are: at what stage does human life begin? Do we accept the cloning of humans? To which extent can we accept a deterministic view of individual health risks, e. g., by an employer or an insurance company? How will molecular genetics and gene therapy affect the age distribution in our societies? Is it ethical to genetically modify plants and animals at will? To what extent are such manipulations in harmony with the ecosystem and its natural diversity? How will the new biotechnologies influence the relationship of industrialized and developing economies? None of these questions has been completely resolved yet. As we approach another borderline, the functional understanding of the mechanism of the human brain, these questions will become more and more urgent to answer on a global scale.

Markets. Over the past few decades, an increasing percentage of novel drugs either have been recombinant human proteins, or recombinant cellular targets were used in their development. Medical diagnostics will be more and more based on information from the individual genome, as eventually will be pharmacology (pharmacogenomics). In animal and plant breeding, the basis of food production, breeding based on genetic markers plays an ever increasing role, and genetic engineering is more and more facilitated by genome sequencing. The market volume of products obtained by genetic engineering techniques at large has since long overtaken the economic importance of traditional fermentation products such as amino acids or antibiotics and is expected to show further rapid growth.

4

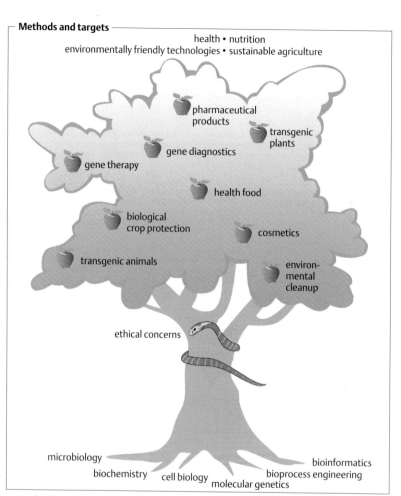

Methods and targets

health • nutrition
environmentally friendly technologies • sustainable agriculture

pharmaceutical products

transgenic plants

gene diagnostics

gene therapy

health food

biological crop protection

cosmetics

transgenic animals

environmental cleanup

ethical concerns

microbiology

biochemistry cell biology

molecular genetics

bioinformatics

bioprocess engineering

Market data of some bioproducts (~ 2000, estimates)

	~ volume	~ value	price/kg
beer	130 000 000 t	330 billion €	2.50 €/kg
ethanol	19 000 000 t	5 billion €	0.25 €/kg
glutamic acid	800 000 t	800 million €	1.00 €/kg
citric acid	700 000 t	700 million €	1.00 €/kg
detergent protease	100 000 t	300 million €	3.00 €/kg
aspartame	10 000 t	50 million €	5.00 €/kg
cephalosporins	5 000 t	2,5 billion €	500.00 €/kg
tetracyclines	5 000 t	250 million €	50.00 €/kg
insulin	8 t	1 billion €	125.00 €/kg
erythropoietin	10 kg	4 billion €	500 million €/kg

Alcoholic beverages

General. Alcoholic beverages have been developed in most if not all human cultures. In Western cultures, these are primarily wine, beer, fermented fruit juices, champagnes, and distilled spirits. The predominant indigenous alcoholic beverage of Asia is rice wine (sake). Regional specialties in other parts of the world are kumys (fermented mare's milk, Mongolian nomads), kwass (from fermented cereals, Russia), pombe (from millet and sorghum, Near Orient) and pulque (from agave juice, Latin America).

Wine is produced from the fermentation of grape must or juice by yeasts. Where the vines are grown, their variety, and the production technology all play key roles in the quality of a wine. The production technology includes harvesting, pressing, must treatment, must fermentation, and cellar treatment. Harvesting depends on the weather and is critical to wine quality. During pressing, the grapes are usually removed from the stems and are squashed into must without destroying the seeds. For white wines, the must is immediately filtered and provides juice, whereas red wine must is traditionally fermented at 20 °C for 6–8 d to dissolve the red anthocyanines occurring in the grape skin in the ethanol developing from fermentation. In modern procedures, the must from red grapes is heated to 40–50 °C; after the addition of pectinases, the anthocyanines from the grape skin are dissolved within 2–4 h, and pressing can ensue. Suitable modifications of the must treatment enables the vintner to obtain (within limits) type-specific musts, even in years of poor grape quality. Depending on national legislation, sugar or acid may be added, acids may be neutralized by adding $CaCO_3$, or fermentation may be stopped by adding SO_2 or potassium pyrosulfite. Such procedures help to harmonize taste, suppress browning of the must (by inhibiting phenol oxidases), protect oxygen-sensitive pigments and aroma components, and suppress the growth of aerobic microorganisms such as acetic acid bacteria, wild yeasts, and molds. Must fermentation now ensues, traditionally in wooden vats, today usually in tanks made from stainless steel or polyester materials. Inoculation may be spontaneous or through the addition of seed cultures of *Saccharomyces cerevisiae* var. *ellpsoideus*. Depending on the type of must, fermentation may take from a few days to several months and can be regulated by the temperature. The residual sugar and ethanol content (7–15 vol%) of the wine can be set either by artificially stopping the fermentation or by adding must ("sweetness reserve") that has been preserved under CO_2 at a pressure of 8 bar. Dry wines contain < 9 g L^{-1} residual sugar. During the ensuing cellar treatment of the wine, chemical, biochemical, biological, and physical processes that are difficult to manage contribute to the maturation and harmonization of the wine. During cellaring, a pH > 3.2 may support the growth of lactic acid bacteria; they convert maleic acid, via malo-lactic fermentation, into much weaker lactic acid and CO_2.

Champagnes are produced from quality wines by adding 1–3 % saccharose and yeast cultures. Fermentation takes place in tanks or, for the more expensive varieties, in the bottle. Champagnes must develop a pressure of at least 3 bar at 20 °C.

Spirits are produced by distillation of sugar extracts from cereals, vegetables, or fruits. Their alcohol content is between 30 and 60 %. Some of the raw materials for spirits are: wine (brandy, cognac, armagnac), sugar cane juice or molasses (arrak, rum), cereals (Korn, whisky), potatoes (vodka), fruit juices (fruit brandy), or agave juice (tequila).

Rice wine (sake). In contrast to wine or beer fermentation, sake is produced by an aerobic solid-stage fermentation. In a first step, soaked and steamed parboiled rice is inoculated with *Aspergillus oryzae* and incubated at ca. 30 °C. At high humidity, this leads within ca. 2 d to the formation of *koji*, a fermented material in which rice starch has been largely depolymerized into sugars. To *koji*, more boiled rice, water, and a yeast starter culture (*moto*) is added to form a mash (*moromi*), which is fermented for 20 d at 25 °C. After filtration and pasteurization *sake* is obtained.

Alcoholic beverages

beer	starch from barley is degraded to sugars by barley amylases. The sugar solution is fermented by yeast to ethanol, in the presence of hop extracts
wine	grape juice is fermented by yeast
champagne	sugar and yeast is added to wine, followed by a second fermentation
cider	apple juice is fermented by yeast
sake	rice starch is depolymerized by amylases from *Aspergillus oryzae*, and sugars are fermented by yeast
whisky	extracts of barley, yeast, rye, or corn are fermented by yeast and distilled
vodka	extracts of potatoes or wheat are fermented by yeast and distilled

Production figures

beer		wine		sake	
total (2001)	1420 million hL	total (1999)	281 million hL		
USA	231 million hL	France	60 million hL		
China	227 million hL	Italy	58 million hL		
Germany	109 million hL	Spain	33 million hL	Japan	9.1 million hL
Brazil	84 million hL	USA	20 million hL	(1997)	
Japan	71 million hL	Argentina	16 million hL		
Russia	63 million hL	Germany	12 million hL		
United Kingdom	56 million hL	Australia	9 million hL		

General production scheme for beer, wine, sake

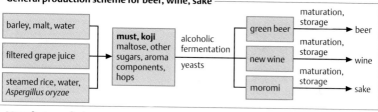

Manufacture of wine

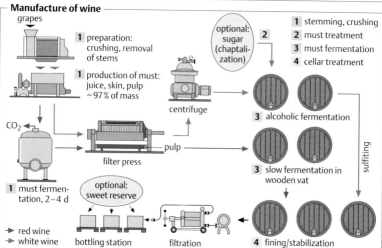

7

Beer

General. Beer is one of the oldest and probably most widely distributed alcoholic beverages of mankind. It is produced by fermentation of malt with yeast *(Saccharomyces cerevisiae)* in the presence of extracts from hops, providing a bitter taste. The world production is about 1.4 billion hl or ca. 140 million t. The largest producers are the USA, China, Germany, Brazil, Japan, Russia, and the UK. The German "purity law" of 1516 requires that only barley (sometimes wheat), yeast, hops, and water may be used for production. In most other countries, a wider variety of raw materials may be used. Modern processing technology allows for the production of low-calorie and light (alcohol-reduced) beers.

Production. By germination and kiln drying, barley is transformed into a storage-stable malt that is rich in carbohydrates and enzymes. When suspended in water, it forms a mash, to which an aqueous extract of hops is added, containing the typical bitter-taste components. By the action of malt enzymes, notably amylases, the resulting wort contains mono- and disaccharides, which can be fermented by yeast. The mash is inoculated with yeast cultures and fermented for some days. Depending on the fermentation process, the alcohol content of beer is between 2 and >18% original wort (dry weight of the extract in g/100g before fermentation). The final maturation is done in a beer cellar at temperatures ca. 0 °C and takes from a few days to several weeks. During this process, turbid components settle and diacetyl, an undesired off-flavor component from the fermentation process, is transformed into acetoin and butane-2,3-diol. In addition, other enzymatic transformations take place which have a strong effect on the aroma of the beer. After filtration, export-type beers are pasteurized.

Types of beer. The many varieties of beer differ according to the time and temperatures used for the main and secondary fermentations, the raw materials and their processing, and the yeast cultures used (e. g., "surface fermenting yeasts" cannot hydrolyze raffinose). Beers are mostly distinguished by the fermentation process used (bottom-fermented beers such as lager, pilsner; and top-fermented beers such as porter, ale and stout), as well as by their wort content: high-gravity beers (>16%), medium-gravity beers (11–14%), draft beers (7–8%), and low-gravity beers (2–5.5%)

Light beers and alcohol-free beers. For the production of light (<1.5% ethanol) or alcohol-free beers (<0.5% ethanol), either fermentation is stopped early, or the alcohol is removed by appropriate processing technology after fermentation (vacuum distillation, membrane-based processes).

Biotechnological innovations. In most countries except Germany, saccharification of starch, degradation of proteins, or filtration steps are enhanced by the addition of microbial enzymes. Addition of the plant protease papain is sometimes used to prevent precipitation of proteins upon storage in the cold ("chill proofing"). Recent developments target improvements such as: 1) breeding and use of transgenic barley, in particular with enhanced activity of thermostable β-glucanases and other enzymes, 2) use of lactic acid bacteria as starter cultures in malting, to minimize microbial contamination of malt by *Fusarium* species and other bacteria, 3) reduction of the lengthy maturation period required to degrade diacetyl, through the addition of bacterial α-acetolactate decarboxylase, an enzyme that transforms α-acetolactate directly into acetoin, 4) the use of recombinant strains of *S. cerevisiae* which express genes such as glucanases or amylases (for better use of raw materials) or α-acetolactate decarboxylase (for faster aroma formation). Using such strains, "low calorie beer" with a very low content of carbohydrates can be brewed. Finally, the processing technology is systematically being improved. Thus, the optimization of beer taste through a week-long storage in the beer cellar can be replaced by a short period of heating to 90 °C followed by 2 hrs of maturation in a bioreactor using immobilized yeasts.

	1 malt	13 storage tank
	2 mill	14 beer filter
	3 water	15 pressure tank
	4 mash tank	16 bottling station
	5 lauter tun	
	6 hops	
	7 wort cooking	
	8 whirlpool	
	9 cooled filter press	
	10 air	
	11 yeast	
	12 fermenter	

	term	process step	details
1	steeping and malting	barley is germinated	8 d at 10–18 °C and pH 7.0. Amylases and proteases are activated
	kilning	barley endosperm is dried	stepwise reduction of water to <3 % and increase of temperature to ~125 °C
	grain removal	barley germ is peeled	animal feed
4	mashing	dried malt is reactivated	in the mash tun, malt amylases depolymerize barley starch to dextrins and maltose, and aroma precursors are formed
5	lautering	turbid polymer material is removed	
6 7	hop and wort	hop extract is added and cooked with wort	after 1½ – 2½ hrs in the cooker, the malt enzymes are inactivated, bitter-taste compounds are dissolved, protein is precipitated and the nutrient solution is sterilized
8	filter	sediments are removed	
9	sparge	wash sediments	
11	pitching	inoculation	with pure culture of *Saccharomyces cerevisiae*
12	main fermentation	fermentation	bottom fermentation: 8–10 d at 5–10 °C ; top fermentation by yeasts and lactic acid bacteria: 2–3 d at 10–25 °C
		yeast is removed	a) decantation or b) removal of surface layer
13		beer matures	several days or weeks in the beer cellar at 0 °C. Polymers sediment, diacetyl is reduced to acetoin
14		filtration, pasteurization	lager beer

Potential of recombinant yeasts

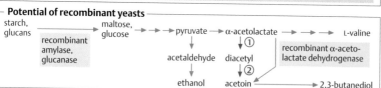

Beer maturation using immobilized yeast

		reaction ①	reaction ②		
from the main fermentation	removal of yeast	heat treatment 10 min at 90 °C	maturation with immobilized yeast in a continuously operated bioreactor, contact time 2 hrs at 15 °C	to conventional process	

9

Fermented food

General. In all human cultures, many foods are modified by microbial fermentation. At first, traditional skills evolved with the aim of preserving foods. Thus, the food value of vegetables can be prolonged by lowering the pH through formation of organic acids (sauerkraut, pickles), digestibility can be enhanced through enzymatic hydrolysis prior to storage (sourdough, sausages, tempeh), taste can be improved (fermented milk products), or the preparation of flavorings (soy sauce, miso). In industrialized countries, about ⅓ of the food is modified by fermentation, usually with defined microbial starter cultures.

Starter cultures are widely available in the food industry for a broad range of fermented foods. They play a crucial role in the manufacture of fermented milk products (such as yoghurts and cheeses), in the production of sourdough breads (starters) and other baked goods (bakers' yeast), and in brewing and wine production (brewers' yeasts). Starter cultures can be classified into single-strain, single-species, and mixed-species cultures. The key criteria for a starter culture are: a fast and reliable start to the fermentation, reliable manufacture of the desired products, and resistance to antibiotics and phages. The production value of starter cultures is on the order of 100 million US$.

Sausages. The types of sausages that can be stored without refrigeration are mostly produced by adding starter cultures of Staphylococci (e. g., *Staphylococcus carnosus*), although Lactobacilli and Penicillium strains are also used. Lactic acid is produced through fermentation from glycogen stored in muscle tissue and lowers the pH to < 5; as a consequence, contamination with acid-sensitive microorganisms is prevented, and muscle protein (isoelectric point: pH 5.3) is gelatinized. Metabolites of the fermentative degradation of fats and proteins play an important role in the flavor of sausages. Starter cultures of salt-stable Staphylococci and Lactobacilli are in use for the production of salted meat and sausages (preservation by the addition of salt, nitrate, or nitrite).

Cheese. In 1999, ca. 15.7 million t of cheese were produced, ca. 6.8 million t in the EU countries and ca. 3.8 million t in the USA alone. There are more than 1000 varieties of European cheeses. The manufacture of cheese is initiated by a spontaneous infection or by the addition of starter cultures to curd (prepared by precipitation from sour milk, via the addition of rennet or recombinant chymosin and the removal of whey). A great variety of microorganisms are used in these processes, often from Penicillium (Camemberts, Rocqueforts), Streptococcus (Emmentals) or Lactococcus (Gouda-type cheese). The traditional crafts vary widely due to differences in the origin of the milk (cow, goat, sheep), the manufacturing process (aerobic, anaerobic, aerobic-anaerobic), and how starter cultures are added (to the surface or by inoculation).

Non-Western fermentation products. Ang-kak is a "red rice" produced in China by the inoculation of moist rice with spores of *Monascus purpureus*. It is used as a flavoring, but also as a digestive therapeutic, due to its content of antibiotics. Kishk is an Asian side-dish prepared by fermentation of sprouted wheat by the addition of sour milk. Miso, a Japanese food flavoring, can be prepared by adding *Aspergillus oryzae* to steamed rice. Soy sauce is a highly aromatic protein hydrolysate that has been prepared for over 1000 years in China. Today, it is manufactured from a mixture of soybeans and wheat. After inoculation with *Aspergillus oryzae* at high humidity at 35 °C, a surface culture forms. It is mixed with an equal volume of salt water (> 13 % salt), and the resulting mash (*moromi*) is fermented for up to a year at room temperature with lactic acid bacteria and yeasts. In Indonesia and Malaysia, the staple food tempeh is manufactured by the fermentation of steamed soybeans and rice with *Rhizopus oligosporus*.

Preparation of starter cultures

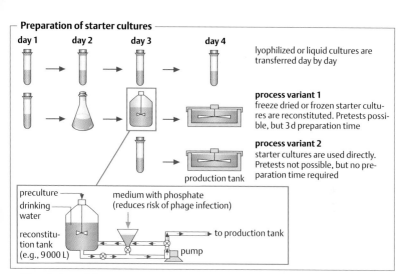

day 1   day 2   day 3   day 4

lyophilized or liquid cultures are transferred day by day

process variant 1
freeze dried or frozen starter cultures are reconstituted. Pretests possible, but 3 d preparation time

process variant 2
starter cultures are used directly. Pretests not possible, but no preparation time required

production tank

preculture
drinking water
reconstitution tank (e.g., 9000 L)

medium with phosphate (reduces risk of phage infection)

to production tank

pump

Manufacture of cheese

raw milk	**starter culture**	**clotting coagulum**	**remove whey**	**salt**
pasteurization	ripening	add rennet, cut, stir, stand		form, press, dry, mature

Manufacture of soy sauce

koji	**moromi**	**finishing**
soy meal, wheat bran, solid phase fermentation with *Aspergillus oryzae*, 72 h, 35 °C, high humidity	add equal volume of salted water; ripening for a long time	filter, pasteurize, conserve

Fermented foods of non-Western cultures

product	country	use	raw material	microorganism
koji	Japan	starting product for soy sauce, miso	soy meal, wheat bran, toasted rice	*Aspergillus oryzae*, *Aspergillus sojae*
shoyu (soy sauce)	Japan	dark flavorant	koji	*Pediococcus* sp.
miso (soy paste)	Japan	light flavorant	koji	*Aspergillus*, *Lactobacilli*
tofu, sufu	Japan, China	coagulated protein	soy beans, soy milk	*Mucor sufu* and others
natto	Japan	spicy flavorant	toasted soybeans in pine wood leaflets	*Bacillus natto*
tempeh	Indonesia	staple food	cooked soybeans in banana leaves	*Rhizopus oligosporus*
Ang-kak	Indonesia, China	flavorant, colorant, therapeutic agent	toasted rice	*Monascus purpureus*
gari	Nigeria	food trimming	*Manihot utilissima* (cassava)	*Geotrichum*, *Corynebacterium*

11

Food and lactic acid fermentation

General. In many cultures, the skills to produce fermented milk products or sauerkraut from fermented cabbage, or to enhance the digestibility of beets as an animal feed by fermentation (silage), have been passed on for hundreds of generations. In 1856, Louis Pasteur laid the foundation for a technological exploitation of these traditions when he discovered the lactic acid bacteria. He found that lactic acid fermentation of food products lowers the pH to ca. 4 which protects against infection by most other microorganisms.

Lactic acid bacteria differ in shape, but can be characterized rather well on the basis of their biochemistry and physiology: they are Gram positive and facultative anaerobic bacteria: though they do not contain heme proteins, e. g., catalase, they are able to grow in the presence of O_2. They cleave lactose to glucose and galactose and metabolize these sugars to lactic acid. "Homofermentative" Lactobacilli such as *Streptococcus pyogenes*, *Lactobacillus casei* and *Lactococcus lactis* form two mole equivalents of lactic acid per mole glucose, "heterofermentative" Lactobacilli such as *Leuconostoc mesenteroides* and *Lactobacillus brevis* just one mole equivalent. How much L-(+) lactic acid (usually 50–90%), D-(−) lactic acid or D,L-lactic acid is formed depends on the occurrence of a species-specific lactate racemase. In addition to lactic acid, fermented milk products contain partially hydrolyzed proteins, no lactose, and a benign microflora; they are therefore considered valuable in the human diet.

Fermented milk products. In Europe, the most important products are sour milk and sour cream, yoghurt, kefir, and buttermilk (fat content < 1%). These can be formed by the spontaneous infection of untreated milk. In commercial dairies, they are manufactured from pasteurized milk by adding starter cultures. The ensuing fermentation results in the formation of lactic acid and an acidification to a pH of 4–5. Yoghurt products with > 95% L-(+) lactic acid, produced, for example, by adding *Lactobacillus acidophilus* starter cultures, possibly supplemented with strictly anaerobic *L. bifidus* (which has been found in the intestinal flora of breast-fed babies), is considered to be especially digestible and to stimulate the immune system. Another feature of milk fermentation is the formation of flavorings from the action of proteases and lipases contained in the starter culture. Microbial strains of particular importance for this effect are *Streptococci*, *Lactobacilli*, *Leuconostoc* and, in some cases, also yeasts.

Lactofermented vegetables, fruits, and juices. Important examples are sauerkraut and pickled cucumbers. Sauerkraut is produced in quantities of several million t/y. The manufacturing process usually relies on the spontaneous infection of finely cut stripes of cabbage in large wooden vats (up to 100 t). The microbial flora thus formed are highly diverse and contain lactic acid and other bacteria, yeasts, and molds. The use of starter cultures is being studied. Other examples of lactofermented vegetables are: borscht (fermented redbeets; Russia and Poland) and kimchi (fermented Chinese cabbage or radishes; Korea). Lactofermented vegetable juices are storage stable, rich in vitamins and minerals, and very digestible. Examples are tomato and carrot juice.

Sourdough. In contrast to wheat flour, rye flour swells significantly only at pH values < 4.3 – a prerequisite for the formation of elastic, digestible crusts of wheat and rye, cracked grain, and whole-grain breads. For this reason, rye flour is turned into sourdough of pH ca. 4.2 *via* a process based on the combined action of lactic acid bacteria and bakers' yeast.

Silage is a fermented winter feed for cattle. Usually sugar beets are used. They are collected in silos or heaps so as to exclude air, leading to lactic acid fermentation. If the fermentation process is incomplete and lactic acid does not form in sufficient quantities to lower the pH to < 5, Clostridia may form, contaminating the feed. Most silages contain the psychotropic pathogen *Listeria monocytogenes*, which can propagate in freezers and thus may lead to the contamination of food products such as soft cheeses, ground meat, and coleslaw, even if these are stored in the cold.

Lactic acid fermentation

homofermentative Lactobacilli		heterofermentative Lactobacilli
Lactococcus lactis *Streptococcus pyogenes* *Lactobacillus delbrueckii* *L. helveticus*	*L. salivarius* *L. casei* and others	*Leuconostoc mesenteroides* *L. lactis* *Lactobacillus brevis* and others

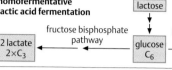
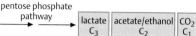

homofermentative lactic acid fermentation

heterofermentative lactic acid fermentation

| 2 lactate $2 \times C_3$ | ← fructose bisphosphate pathway ← | lactose ↓ glucose C_6 | → pentose phosphate pathway → | lactate C_3 | acetate/ethanol C_2 | CO_2 C_1 |

Sourdough

	wheat	rye
% of agricultural area used for production (world)	~33	<3
main product of meal	yeast dough	sourdough
factor responsible for retaining gas on baking	gluten network	pentosanes, proteins formed from acidification
specific volume of bread (volume/weight)	~3.5	~2.0

starter culture 0.5 kg

6 h, 27 °C → initial sour 3.4 kg

9 h, 30 °C → basic sour 25.5 kg

3 h, 28 °C → full sour 81 kg

sourdough 160 kg

cook, form, bake → **sourdough bread**

rye meal — water

Sauerkraut, yoghurt and silage

sauerkraut

| cut cabbage | 2 – 2.5 % salt |

spontaneous inoculation

vat up to 100 t, 18 – 20 °C, protected from air

if lactic acid >1.5 %: filtration, pasteurization, finishing

yoghurt

milk

homogenized, pasteurized, with additives, e.g., fruits

batch fermentation	**stirred yoghurts**
add starter cultures*, fill containers, incubate, cool to 4 °C, distribute	fill tanks, add starter cultures*, incubate, cool to 4 °C, fill containers, distribute

*e.g., *Lactobacillus bulgaricus*, *L. acidophilus*, *Streptococcus thermophilus*

silage

development of microbial population during the fermentation process and acid production

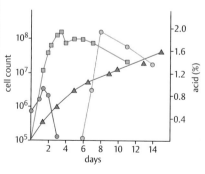

—●— aerobic bacteria
—■— heterofermentative lactic acid bacteria
—○— homofermentative lactic acid bacteria
—▲— acid formation

Ethanol

General. Ethanol is an important industrial solvent, a starting material for the synthesis of organic chemicals, and an energy source ("bioalcohol"). In 1997, about 23 billion L of ethanol (18.5 million t) were produced; 30% was synthesized chemically, and 70% by fermentation. The production of bioethanol is at present economically competitive only if very low prices for glucose (or biomass) are obtainable or if oil prices rise or are regulated by the government. Thus, bioethanol constitutes a reserve technology for energy production. Ethanol has been used in Brazil and the USA since ca. 1975 as a partial or complete replacement for automotive gasoline.

Organisms and biosynthesis. The most important organism used in producing ethanol is bakers' yeast (*Saccharomyces cerevisiae*). By glycolysis, it forms two mole equivalents of ethanol per mole of glucose. *Zymomonas mobilis*, a bacterium isolated from agaves, achieves the same molar yield, but the synthesis is based on the ketodeoxyphosphogluconate (KDPG) metabolic pathway. Since neither organism synthesizes polysaccharide-degrading enzymes such as amylases, saccharose or starch hydrolysates are the C-source of choice. If polysaccharides are used instead, enzymatic saccharification prior to fermentation is required. Alternatively, recombinant yeasts have been prepared which express the appropriate depolymerases and thus can degrade starch, hemicelluloses, or cellulose. Another potential process is based on *Thermoanaerobacter ethanolicus*, a thermophilic anaerobe with a temperature optimum of ca. 70 °C, which metabolizes polysaccharides or sugars over a wide pH range (4.5–9.5).

Fermentation and recovery. The production of bioethanol is usually carried out in the fed-batch mode using large-scale bioreactors of up to 500 m³ capacity. The organism of choice is *Saccharomyces cerevisiae* which, even under not completely sterile (aseptic) conditions, is less sensitive to Lactobacillus infections than *Zy-momonas mobilis*. For cultivation, a sugar medium enriched with N-sources and minerals is used. After an aerobic growth phase, aeration is stopped, and after ca. 20 h ethanol production reaches 90% of the theoretical maximum. High glucose concentrations lead to inhibition of the process by catabolite repression; this is prevented by a continuous or semi-continuous feeding of sugar (fed-batch). Since ethanol concentrations > 8% (which are usually reached after 72 h) inhibit yeast metabolism, the ethanol-containing broth is removed. Ethanol is usually isolated by azeotropic distillation, leading to 95% ethanol, which can be used in cars. Absolute ethanol can be prepared by extractive distillation, molecular sieves, or by membrane technology (pervaporation). In several processes, e.g., the Melle-Boinot process, the cells obtained by a separator during broth removal are recycled and serve as a preculture for the next batch; this procedure reduces the overall fermentation time. Continuous ethanol fermentation methods have been developed, and are occasionally under operation. Experiments with cell reactors, using yeast or bacteria immobilized on a carrier material, have been developed to the pilot scale.

Economic considerations. For industrial ethanol fermentation processes, sugarcane juice (Brazil) and cornstarch hydrolyzates (USA) are mostly used. At present, ca. 700 fermentation plants produce bioethanol. In Brazil, a "Proalcool" program was initiated in 1975, aimed at reducing oil imports. In over 100 fermentation plants, ca. 12 billion L/y (2001) of ethanol are produced from sugarcane molasses using simple technology (batch fermentations, distillation). In the USA, automotive gasoline enriched with 10% ethanol ("gasohol") has been produced since 1975 (2001: ca. 8 billion L/y) from cornstarch. The technology is more demanding (e. g., fed batch, pervaporation). In both countries, fermentation solids (cell mass and solid media components) are sold as animal feeds. Japan has developed several demonstration plants for a bioethanol process based on immobilized yeast.

Ethanol

C_2H_6O

M_R 46,07 | b.p. 78.32 °C | Industrial production:

D. 0.79367 (15 °C) | CAS 64-17-5 | mainly addition of water to ethylene using catalysts

H_3C-CH_2-OH

Production from molasses

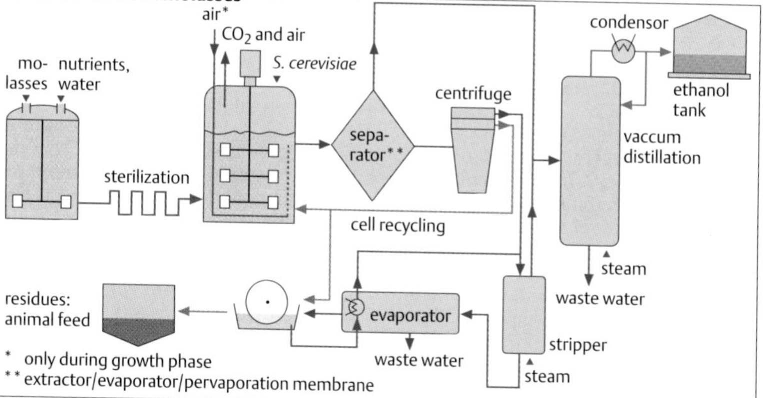

* only during growth phase

** extractor/evaporator/pervaporation membrane

Fermentation and recovery

nutrients
sugarcane juice or beet molasses or saccharified starch, minerals

▼

cell reactor
immobilized cells, continuous process

bioreactor
up to 500 m³ batch or fed-batch process

▼

ethanol separation
distillation or pervaporation

▼

waste material
animal feed

Ethanol

1
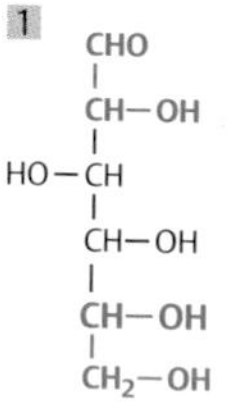

2
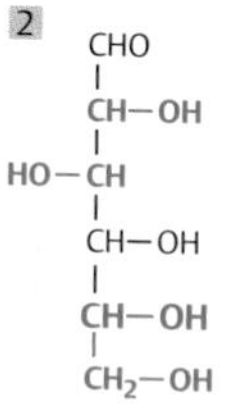

origin of the C atoms of ethanol during metabolism of glucose with

1 *Saccharomyces cerevisiae* — glycolysis

2 *Zymomonas mobilis* — KDPG pathway

organism	system	glucose addition [gL⁻¹]	dilution rate [h⁻¹]	cell concentration [gL⁻¹]	ethanol concentration [gL⁻¹]	maximum productivity [gL⁻¹]
1 *Saccharomyces cerevisiae*	no cell recycling	100	0.17	12	41	7.0
1 *S. cerevisiae*	cell recycling	100	0.08	50	43	29
2 *S. cerevisiae*	cell recycling	150	0.53	48	60.5	32
1 *S. cerevisiae*	cell recycling vacuum (6.7 kPa)	334	0.23	124	110–160	82
3 *Zymomonas mobilis*	cell recycling	100	2.7	38	44.5	120

1 ATCC 4126 2 NRLL Y-132 3 ATCC 10988

1-Butanol, acetone

General. 1-Butanol (world production ca. 1.2 million t/y) is an important solvent for automobile paints and a base chemical for ester formation (e. g., for butyl cellulose). Before the advent of petrochemistry, it was widely used for producing synthetic rubber via buta-1,4-diene. Acetone (world production ca. 3 million t/y) is also used as a solvent. During world war I, it came in great demand for the production of cordite, an explosive used by the British navy. Both compounds are presently produced exclusively from petrochemical raw materials, but before ca. 1950 they were mostly obtained by fermentation using *Clostridium* bacteria and starch or molasses as a carbon source, an industrial process pioneered in 1915 by the Russian/British chemist Chaim Weizmann (who became the first president of Israel). Due to progress in molecular genetics and process technology, the production of either solvent by fermentation might eventually become economically attractive again and is thus being investigated as a reserve technology.

Organisms and biosynthesis. Among the few anaerobic bacteria that can form acetone and 1-butanol, the genus *Clostridium* is the most important. During fermentation, a shift from the formation of butyric and acetic acid ("acetogenesis") to the formation of butanol occurs at the end of cell growth and is accompanied by a decrease in pH to values < 5.0. The composition of the product mixture varies from species to species. The best-studied organism is *Clostridium acetobutylicum*, which also shows the highest tolerance to the cell-toxic solvents formed. It can form up to 38 g of 1-butanol and acetone from 100 g glucose in a ratio of 3:1. Many Clostridia synthesize amylases, amyloglucosidases, and other extracellular hydrolases and thus are capable of metabolizing inexpensive carbon sources such as starch. Another attractive economic option is the use of lactose (whey). The enzymes participating in the biosynthesis of both solvents have been well studied and their genes have been cloned. Pyruvate is formed from glucose by glycolysis. In the presence of pyruvate/ferredoxin oxidoreductase, pyruvate undergoes oxidative decarboxylation to acetyl CoA, which is further reduced to several C_2-, C_3-, or C_4-metabolites using mainly NADH from glycolysis for reduction. A hydrogenase that is also present transfers some of the electrons to protons with the formation of hydrogen. The regulation of these enzymes is being intensively studied with the goal of influencing the yield and the composition of solvents (*metabolic engineering*). The genome of *C. acetobutylicum* has been completely sequenced, and the organism is thus quite promising for cloning work, since shuttle vectors for *Escherichia coli* and *Bacillus subtilis*, specific phages and transposons, are available. The yield of solvent can already be significantly enhanced by using suitable gene constructs for the transformation of wild-type and production strains.

Fermentation and recovery. For more than 40 years, the production of acetone and 1-butanol has been carried out on an industrial scale using *C. acetobutylicum* and batch fermenters of $> 100\,m^3$ volume. Substrate costs in this process were ca. 60 %, energy costs for product distillation ca. 12 %. At present, the batch process based on cornstarch or molasses as the carbon source, which has been used in the USA and South Africa for over 40 years, is not competitive with petrochemical-based synthetic routes. Decisive parameters for renewed use of the fermentation route are the yield of product from raw material (kg solvent kg^{-1} sugar) and the productivity of the process (g solvent $L^{-1}\,h^{-1}$). Modern process developments thus aim at 1) a two-stage process with cell recycling, 2) continuous fermentation processes, 3) using immobilized microorganisms, and 4) improved recovery of the solvents from the fermentation medium by membrane processes (pervaporation, reverse osmosis). Thus, in addition to the development of solvent-tolerant production strains, the enhancement of yields by genetic and metabolic engineering and optimization of process engineering (in particular of downstream processing) are considered essential to revive this historically important fermentation process.

1-Butanol

C$_4$H$_{10}$O

m.w.	74.12
d	0.813
m. p.	−90 °C
b. p.	117–118 °C
CAS	71-36-3

$$H_3C-CH_2-CH_2-CH_2-OH$$

chemical synthesis of 1-butanol:
hydroformylation of propene and hydrogenation

Acetone

C$_3$H$_6$O

m.w.	58.08
d	0.7908
m. p.	−95 °C
b. p.	56 °C
CAS	67-64-1

$$H_3C-CO-CH_3$$

chemical synthesis of acetone:
catalytic dehydrogenation of 2-propanol,
direct oxidation of propene, cleavage of
cumol hydroperoxide

Biosynthesis

1 pyruvate: ferredoxin oxidoreductase
2 hydrogenase
3 thiolase
4 CoA-transferase/acetoacetate decarboxylase
5 aldehyde-/alcohol dehydrogenase
6 two dehydrogenases, crotonase
7 butyraldehyde- and butanol dehydrogenase

the genes coding for all enzymes 1 – 7 have been cloned

Fermentation and recovery

spore suspension
suspension in soil

▼

shake flask culture
anaerobic, 24 hrs at 37 °C

▼

15 m^3 bioreactor
under CO$_2$ layer 18 hrs at 37 °C

▼

700 m^3 bioreactor
6 % molasses 50–60 hrs at 37 °C

▼

continuous distillation
followed by fractionated distillation

~ 38 kg 1-butanol/ acetone (3:1) 100 kg^{-1} glucose

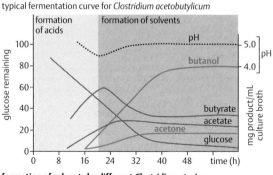

typical fermentation curve for *Clostridium acetobutylicum*

formation of solvents by different *Clostridium* strains

organism	relative yield (%)			
	1-butanol	acetone	2-propanol	ethanol
Clostridium acetobutylicum	30.2	14.0	–	5.0
C. beijerinckii	67.9	6.0	–	–
C. puniceum	75.6	16.8	–	–
C. tetanomorphum	47.1	–	–	42.7
C. butyricum	17.0	–	7.0	–

Acetic acid / vinegar

General. Vinegar is used in many cultures for the acidification and preservation of vegetables, salads, rice, and other food products. Its consumption in these foods and in refreshing drinks is documented back to antiquity. It was and still is produced from fermented fruit juices, e. g., wine. In the 18th century, an "immobilization procedure" was developed in France whereby diluted wine was trickled over twiglets contaminated with acetic acid bacteria. Louis Pasteur succeeded in 1868 in defining selective growth conditions for acetic acid bacteria, thus laying a foundation for the modern technological production of vinegar. These days, vinegar is produced from ethanol by fermentation using strains of Acetobacter. If wine is used as a base material, the product is wine vinegar (a 6 % solution of acetic acid in water, pH ca. 4.8); if rectified ethanol is used, the concentration of vinegar is 5 %. Annual production in the USA alone is ca. 750 million L or 750 000 t. Glacial acetic acid (99.7 %) is an important base chemical. It is produced from ethylene by catalytic oxidation and has a pK_a of 5.6. In the USA, calcium magnesium acetate (m.p. – 7.7 °C) produced from cornstarch has been proposed as an antifreeze ("Nicer De-Icer").

Organisms and biosynthesis. Only a few species of Gluconobacter and Acetobacter can oxidize acetic acid to ethanol by "subterminal oxidation". Taxonomic classification of these species is complicated due to a rapidly changing phenotype during growth and is usually carried out by typing the 16S RNA, more recently also by analysis of plasmid profiles. Oxidation of ethanol proceeds via a sequential reaction of alcohol dehydrogenase (ADH) and aldehyde dehydrogenase (ALDH), which are both membrane-bound enzymes that contain pyrolloquinoline quinone (PAA) as prosthetic groups. ADH contains an additional heme C residue. They transfer the electrons generated by ethanol oxidation to a membrane-bound terminal oxidase via ubiquinone. During growth, these bacteria metabolize glucose to pyruvate both through glycolysis and through the KDPG pathway and further through the citric acid cycle. Both strains are extremely sensitive to lack of O_2. An interruption of the oxygen supply for even a few minutes results in a significant decrease in ethanol oxidation. If ethanol is depleted, acetic acid in the presence of O_2 is further oxidized to CO_2.

Fermentation and recovery. For the technical production of acetic acid, *Acetobacter* sp. is employed. This microorganism is cultivated in a mash of aqueous wine or rectified ethanol, other nutrients and $> 60 \, g \, L^{-1}$ acetic acid under strong aeration, to prevent further oxidation of acetic acid. The process is carried out in a repeated fed-batch mode: once the ethanol concentration has decreased to about 0.2 % (ethanol sensor), a certain amount of the fermenter broth is removed and replaced with fresh mash. Since very homogenous aeration is necessary (0.1 vvm, 0.1 volumes of air per fermenter volume min^{-1}), a highly efficient rotor/stator stirrer with self-priming aerator is used ("Frings Aerator"). Acid formation begins rapidly, with the formation of heat. which is removed by heat exchangers. The average productivity of a 100 m^3 bioreactor using this process is ca. 1.6 g acetic acid $L^{-1} h^{-1}$. With special starter cultures and suitable monitoring and control, this process leads to a ca. 17.5 % vinegar solution in 50–70 h. A more concentrated solution (up to 21 %), as required in the canning industry, can be obtained if fermentation is continued for 45–55 h. Once a concentration of ca. 20 % acetic acid has been reached, the acetic acid bacteria die and fermentation comes to an end. The raw vinegar is filtered and purified by a membrane process, pasteurized, and diluted to a 5–6 % vinegar solution which can be marketed. About 70 % of the world supply of acetic acid is produced in > 700 bioreactor systems of this design (Frings Acetator). Other process variants, e. g., continuous fermentation with cell recycling or the use of immobilized acetic acid bacteria in an airlift bioreactor, sometimes show higher productivity (up to $>100 \, g \, L^{-1} h^{-1}$)

Acetic acid

C$_2$H$_4$O$_2$
M$_R$ 60.05
Sdp.   117.9 °C
pK$_a$ 4.76 (25 °C)   chemical synthesis:
CAS   64-19-7   addition of O$_2$ to ethylene
                               or of CO to methanol

$$H_3C-COOH$$

Biosynthesis by *Acetobacter sp.*

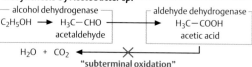

alcohol dehydrogenase

C$_2$H$_5$OH $\rightarrow$ H$_3$C$-$CHO $\longrightarrow$
                acetaldehyde

aldehyde dehydrogenase

H$_3$C$-$COOH
acetic acid

H$_2$O + CO$_2$ $\longleftarrow$ ✗ ⎯⎯⎯
"subterminal oxidation"

the membrane-bound, PQQ-dependent dehydrogenases transfer the electrons generated by the oxidation of ethanol *via* ubiquinone to a membrane-bound terminal oxidase

pyrroloquinoline quinone (PQQ)

Fermentation and recovery

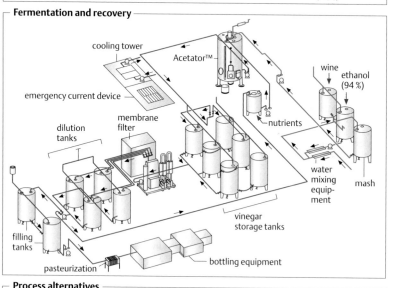

cooling tower
Acetator™
emergency current device
membrane filter
dilution tanks
nutrients
wine
ethanol (94 %)
water mixing equipment
mash
vinegar storage tanks
filling tanks
pasteurization
bottling equipment

Process alternatives

	maximum vinegar production [%]	productivity [L/m^3 d]	remarks
standard procedure (repeated batch)	15	35–50	simple process scheme
one-step high-percentage procedure	18.5	30–50	high vinegar concentration, low storage and transportation cost
two-step high-percentage procedure	>20	30–50	high vinegar concentration, low storage and transportation cost
continuous process	>10	up to 60	high vinegar concentration, low storage and transportation cost
immobilized acetic acid bacteria (experimental)	<9	–	fluid bed- or airlift reactors, up to 460 d

Citric acid

General. Citric acid was first isolated in 1822 from lemon juice by Scheele, who also established its composition. Many fruits form large quantities of citric acid. In 1934, Hans Krebs discovered that citric acid is a central compound in aerobic metabolism (citric acid cycle); for example, in the metabolism of an adult human, 1.5 kg citric acid are formed daily as an intermediary product. Citric acid is a strong tribasic acid. The pK_a-values of its three dissociation steps are 3.13, 4.78, and 6.43 (25 °C). A 1 % solution of citric acid in water has a pH of 2.2. With its three carboxyl and one hydroxyl group, citric acid is also an excellent complexing agent for di- and trivalent cations. Citric acid is exclusively produced by fermentation. The annual production is about 700 000 t (1997), with a market value of ca. 700 million US$. It is used as an acidulant and preservative in the food industry, as a complexing agent in metal treatment, and as a water softener in detergents.

Organisms and biosynthesis. Some molds such as *Aspergillus niger* secrete large quantities of citric acid during and after the late logarithmic growth phase, provided there is an excess of glucose and oxygen. Although the intermediates of the citric acid cycle are usually consumed by the general metabolism, the nearly quantitative conversion of glucose to citric acid by *A. niger* is possible for two reasons: oxaloacetic acid, an intermediate of the citric acid cycle, is replenished via an anaplerotic reaction: the enzyme pyruvate carboxylase, which in this mold is localized in the cytoplasm, catalyzes the addition of CO_2 to pyruvate and thus forms oxaloacetic acid by a shortcut from glycolysis. In addition, citric acid is secreted from the mitochondria, where it is formed, into the cytoplasm by an antiporter which in turn imports maleic acid (a reduction product of oxaloacetic acid) from the cytoplasm.

Manufacture process. *A. niger* is used for the industrial production of citric acid from sugars. A certain part of the production is still carried out by a traditional surface (shallow pan) process: acid-resistant trays in a sterile compartment are filled with the sugar mash and inoculated with spores of *A. niger*. After 5 days, a mycelial mat has formed on the surface that performs the fermentation. A high degree of aeration (up to 10 volumes air/volume broth/min = 10 vvm) is required, mainly to remove the generated heat. After ca. 8 d of fermentation, the mycelium is removed, extracted with hot water, and citric acid is precipitated from the combined liquids The mycelium is removed by filtration, citric acid is precipitated from the filtrate by addition of $Ca(OH)_2$, and recovered from the resulting Ca-citrate by reacting with sulfuric acid. Addition of activated charcoal or ion exchangers allows for crystallization of very pure citric acid. During the above process, > 1 t gypsum/t citric acid is formed, resulting in high wastewater treatment costs. Yields are on the order of 50 g kg^{-1} sugar. However, today most citric acid is produced in stirred or airlift fermenters of 100–500 m³ volume. The stainless steel reactors (harvest solution: pH 2.0) are first sterilized by steam and then filled with an inexpensive carbon source such as starch hydrolysate or sucrose. By a mechanism that is not fully understood, high citrate concentrations are favored in media limited in Mn^{2+} (< 2 µg L^{-1}, obtained through the addition of hexacyanoferrate, Cu^{2+}, or cation exchangers). Formation of the cell mass is usually complete after 48 h at pH 5. Addition of sugar in a fed-batch mode and increase of aeration initiate formation of citric acid, which is excreted into the medium. Yields are in the order of 50%. Recovery is usually carried out as described above. A modern alternative for recovery consists in sequestering citric acid in the fermentation broth with trilauryl amine and extracting the complex with a mixture of alkanes and 1-octanol. Solvents and sequestering agents can be recovered in this process. Some alkane-degrading yeasts are able to form citric acid either from less volatile alkane fractions or from glucose – an interesting starting point for future technologies.

Citric acid

$C_6H_8O_7$
M_R 192.12   CAS 77-92-9
solubility   600 g/L water (20 °C)
acidity pK_{a_1} = 3.128 (25 °C); 1 % solution of citric acid in water: pH 2.2
complexing constants (lgK): Fe^{3+} 12.5, Ca^{2+} 4.68, Cu^{2+} 3.98 (at 20 °C)

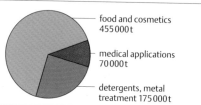

Occurrence and use

occurrence	[g kg⁻¹]
lemons	40–80
grapefruit	12–21
raspberries	10–13
black currants	15–30
strawberries	6–8
tomatoes	2.5

food and cosmetics
455 000 t

medical applications
70 000 t

detergents, metal
treatment 175 000 t

Biosynthesis in *Aspergillus niger*

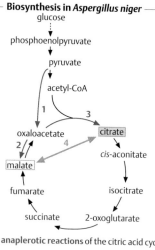

anaplerotic reactions of the citric acid cycle:

1 pyruvate carboxylase (in cytoplasm)
2 malate dehydrogenase (in cytoplasm)
3 citrate synthase (in mitochondria)
4 citrate/malate antiporter (mitochondrial membrane)

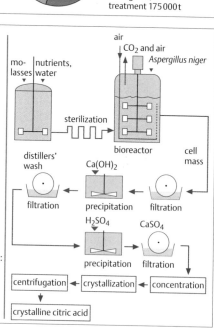

Fermentation and recovery

inoculation material, medium	seed fermenter	bioreactor	filtration	purification	
starch or molasses, N-source, salts	40 m³ bioreactor	400 m³ stainless steel. 8 d at 32 °C. After cell growth pH shift from 5.0 to 2.0	removal of mycelium by vacuum rotary filter	precipitation, recrystallization 1. Ca(OH)₂ 2. H₂SO₄ 3. recrystallization	>200 g/L citric acid after 150 h

Lactic acid, gluconic acid

General. The annual production of lactic acid is ca. 50 000 t, of which 30 000 t are L-lactic acid produced by fermentation. Its main application (85%) is in food and beverages, due to its pleasant acidic taste and its preservative properties. A less pure product is used in the leather and textile industries. A novel application of L-lactic acid is as a chemical building block in the synthesis of biodegradable polyesters (NatureWorks™); present production of polylactide polymers is 4 000 t, but 140 000 t y^{-1} will be produced soon, for a broad range of fiber and packaging applications. Na-D-gluconate, D-gluconic acid, and its δ-lactone are produced at a level of ca. 60 000 t. The δ-lactone is used in the food industry as a mild acidulant. Ca^{2+}- and Fe^{3+}-gluconates are highly soluble and nontoxic and thus are used in medical infusion preparations for the treatment of calcium or iron deficiencies. Sodium gluconate is also a highly alkali-stable complexing agent for calcium and iron; ca. 50% of its production is used as an additive for bottle cleaning and alkaline derusting, in the preparation of concrete, and for the prevention of iron precipitates in textile treatments. The pK_a- of D-gluconic acid is 3.7.

Organisms and biosynthesis. L-lactic acid is produced technically by various Lactobacilli. The choice of organism depends on which carbon source is used. For an exhaustive transformation of the substrate, homofermentative Lactobacilli must be used, since they produce two moles of L-lactic acid per mole of D-glucose during glycolysis. In contrast, D-gluconic acid is the end product of a subterminal oxidation of D-glucose and thus similar to the subterminal oxidation of ethanol to acetic acid. Some molds (*Aspergillus niger*, *Penicillium* species), and also oxidative bacteria, especially Gluconobacter, carry out this reaction. In molds, the responsible flavoenzyme is D-glucose oxidase, which is localized in the fungal cell wall but can also be found in the medium during fermentation. In contrast, Gluconobacter strains use a membrane-bound D-glucose dehydrogenase that contains pyrolloquinoline quinone as a cofactor, similar to the alcohol and aldehyde dehydrogenases of Acetobacter strains.

Fermentation and recovery. Lactic acid can be produced via chemical synthesis or fermentation. Chemical synthesis is done by the addition of H_2O to acrylic acid, or of HCN to acetaldehyde, and leads to racemic lactic acid. In fermentation, the choice of organism depends on the carbon source. *Lactobacillus delbrueckii* or *L. leichmannii* is preferred if dextrose or other sugar solutions are used, and *L. bulgaricus* is used on whey as a carbon source. The fermentation medium contains 12–18% sugar, a nitrogen source, phosphate, and B vitamins. Fermentation, at 45–50 °C under O_2-poor conditions in the presence of $CaCO_3$ as a buffer (to keep the pH constant between 5.5 and 6.0) is complete after 2–6 days, depending on the substrate concentration. After removal of the cell mass, Ca-lactate can be transformed by the addition of H_2SO_4 into the free acid, which can be further purified by ion exchange chromatography. Alternatively, esterification with methanol yields lactic acid methyl ester, which can be purified by distillation. The use of liquid membranes and the direct use of ion exchangers in the fermentation broth, without prior precipitation of the Ca salt, are under development. For polyester applications, a lactide is formed through condensation and purified by vaccum distillation.

D-Gluconic acid is prepared from D-glucose by electrochemical oxidation or by a fermentation process using *Aspergillus niger*. At pH-values > 3, this mold accumulates glucose oxidase in its cell wall, which oxidizes D-glucose to D-glucono-5-lactone, which hydrolyzes spontaneously or faster by enzymatic catalysis (lactonase) to D-gluconic acid. Na- or Ca-gluconate is obtained upon growth of the cell mass at pH 4.5–6.5 (buffered with Na_2CO_3/NaOH or $CaCO_3$) by the addition of 11–25% D-glucose under strong aeration. The salt is obtained from the filtered fermentation broth by concentration and drying. The free acid and the lactone are obtained from the salt by ion-exchange chromatography.

22

D-Lactic acid, L-lactic acid

$C_3H_6O_3$
M_R 90.08
pK_a 3.80 (25 °C)
CAS   50-21-5
     10326-41-7 (R)-form
     79-33-4 (S)-form

chemical synthesis of the racemate:
addition of H_2O to acrylic acid or of HCN to
acetaldehyde, followed by hydrolysis

D-Gluconic acid

$C_6H_{12}O_7$
M_R 196.16
pK_a 3.7 (25 °C)
CAS   526-95-4

Biosynthesis

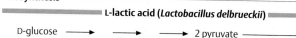

L-lactic acid (*Lactobacillus delbrueckii*)

D-glucose → → → 2 pyruvate → 2 L-lactate
2 NADH

D-gluconic acid (*Aspergillus niger*)

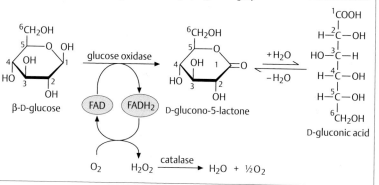

β-D-glucose   glucose oxidase   D-glucono-5-lactone   D-gluconic acid

FAD   FADH₂

O_2 H_2O_2 catalase H_2O + ½ O_2

Fermentation and recovery

L-lactic acid

preculture	bioreactor	recovery	productivity
Lactobacillus	> 100 m³, dextrose or sugar solutions, 50 °C, pH 5.5 – 6.0	precipitation as Ca salt, recrystallization, ion exchange chromatography	$2 – 3\,kg\,m^{-3}\,h^{-1}$

D-gluconic acid

preculture	bioreactor	recovery	productivity
Aspergillus niger	> 100 m³, dextrose or sugar solutions, 33 °C, pH 6.5, 1 vvm	cell separation, concentration, addition of NaOH, drying	$> 13\,kg\,m^{-3}\,h^{-1}$

Amino acids

General. Amino acids have been used for medical purposes, e. g., in infusion preparations, ever since their important metabolic role was discovered in the first half of the 20th century. Some amino acids such as D,L-methionine, L-lysine, or L-threonine serve as additives in animal feed. The findings that L-glutamate has taste-enhancing properties in food and that the dipeptide Aspartame™ is an excellent low-calorie sweetener helped promote the industrial production of amino acids. The 20 proteinogenic amino acids are the building blocks of proteins and peptides. Most higher organisms depend on the uptake of several of these amino acids with their food (essential amino acids). In man and in many of his livestock, these are L-methionine; L-lysine; the aromatic amino acids L-phenylalanine, L-tyrosine, and L-tryptophan; and the hydrophobic amino acids L-valine, L-leucine, and L-isoleucine. Non-proteinogenic amino acids, e. g., having the D-configuration at their Cα atom, occur in natural compounds. They serve as chiral synthons in organic synthesis, e. g., for the manufacture of semi-synthetic antibiotics.

Economic considerations. The annual production of amino acids is > 1 million t/y, their market volume > 2 billion US$. Most commercial producers are located in Asia. The most important product is sodium-L-glutamate (> 1 000 000 t), followed by L-lysine (400 000 t) and D,L-methionine (ca. 200 000 t). L-aspartic acid and L-phenylalanine, starting products for the production of the sweetener Aspartame™, are produced in ca. 10 000 t quantities each. About 65% of industrially produced amino acids are used in food, 30% in animal feeds. Less than 5% are used in high purity and pyrogen-free form in medical therapy, e. g., as additives to infusions, or in cosmetics.

Production. Four different methods are used for manufacturing amino acids: 1) extraction from protein hydrolysates, 2) chemical synthesis, 3) biotransformation of chemical precursors using en-

zyme or cell reactors, and 4) microbial production via fermentation. The extraction of amino acids from protein hydrolysates (the chiral pool) is economically attractive especially for L-cysteine and L-cystine, L-leucine, L-asparagine, L-arginine, and L-tyrosine. Various plant proteins and proteinaceous wastes from the slaughterhouse are used as starting material. After acid hydrolysis, the hydrophobic amino acids L-phenylalanine, L-leucine, and L-isoleucine are separated first by precipitation and ethanol extraction. The water-soluble amino acids are separated in a basic, an acidic, and a neutral fraction by ion-exchange chromatography, followed by crystallization. Chemical synthesis usually leads to racemic amino acids. They can be commercially used if the lack of chirality does not impede function, as is true with D,L-alanine (for the taste finish of fruit juices), but in particular for D,L-methionine as a feed additive. Racemic amino acids are separated into the two enantiomers differing at the Cα-atom by using a biocatalyst. The biocatalyst used is either an isolated enzyme or whole cells containing an appropriate enzyme. Under commercial conditions, it is usually preferable to immobilize the biocatalyst, since this allows for a continuous process and a high operational lifetime of the catalyst. The economic success of such processes is usually based on the simple and inexpensive synthesis of the chemical precursor. Nearly all proteinogenic amino acids can also be manufactured via fermentation processes with selected block mutants, which usually have also been optimized through genetic engineering. Fermentation procedures are often the best alternative for a manufacturing process. With the availability of the complete genome sequence of *Corynebacterium glutamicum*, an important amino acid producer, this approach might gain further importance. Because the biosynthetic pathway of amino acid production is already well understood and the pertinent genes have been cloned, methods such as metabolic engineering may contribute further to optimizing amino acid yields by fermentation.

Industrially important amino acids

amino acid	annual production* [t/y]	value* [US-$/kg]	manufacturing procedure	main application
proteinogenic amino acids				
L-glutamate	>1 000 000	1	fermentation	taste enhancer
L-lysine	200 000	2	fermentation, enzyme reactor	feed additive
D,L-methionine	200 000	2	chemical synthesis	feed additive
L-threonine	10 000	5	fermentation	feed additive
L-aspartic acid	10 000	10	chiral pool, cell reactor	Aspartame™
L-phenylalanine	10 000	10	fermentation, enzyme reactor	Aspartame™, medicine
glycine	10 000	10	chemical synthesis	sweetener
L-arginine	1 000	20	fermentation, chiral pool	medicine, cosmetics
L-tryptophan	1 000	20	fermentation, enzyme reactor	feed additive
all others	3 000		chiral pool, fermentation, enzyme and cell reactors	medicine and other applications
other amino acids (examples)				
D-phenylglycine, D-4-hydroxyphenylglycine			chemical synthesis	precursor for Ampicillin, Amoxicilin
(S)-5-hydroxytryptophan			chemical synthesis	Oxitriptan™, an anti-depressant

*estimates, 2000

Biosynthesis

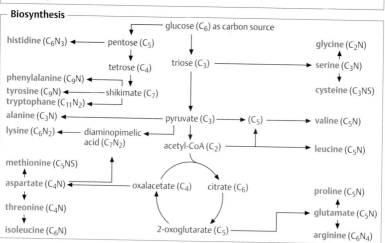

Manufacturing processes

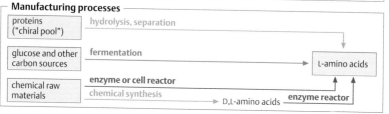

L-Glutamic acid

General. L-glutamic acid is a proteinogenic amino acid. In higher organisms, one of its key function is to act as a neurotransmitter in the brain. The brain is protected from the influx of excess L-glutamic acid by the blood–brain barrier (BBB) and forms L-glutamic acid from L-glutamine, which can cross the BBB. As early as 1908, the flavor-enhancing effect of *Konbu* algae was discovered in Japan and traced to the presence of L-glutamic acid. In 1909, Ajinomoto started to produce this amino acid from acid hydrolysates of wheat gluten and soy protein. In 1957, researchers at Kyowa Hakko discovered that *Corynebacterium glutamicum* secretes L-glutamic acid when grown on sugar-containing media. In subsequent decades, this strain was improved by mutagenesis, and the fermentation technology was optimized, resulting in yields of L-glutamic acid up to 150 g L^{-1}.

Organism and biosynthesis. *C. glutamicum* forms glutamic acid as a byproduct of the citric acid cycle, via isocitrate and 2-oxoglutarate. In wild-type strains, the ratio of glutamate biosynthesis and oxidation of C_2 units via the citric acid cycle is strictly regulated. Industrial mutants, in contrast, exhibit the following properties: 1) they secrete glutamate much better, 2) key enzymes of the biosynthetic pathway are deregulated, and 3) anaplerotic pathways (fill-up reactions) are activated. Further details regarding these include (respectively): 1) mutants with enhanced membrane permeability have been obtained using various measures such as reducing the availability of biotin, oleic acid, or glycerol (using oleic acid- or glycerol-auxotrophic strains), using strains with deformed cell walls, adding penicillin; 2) in production strains, the activity of 2-oxoglutarate dehydrogenase is much lower than that of L-glutamate dehydrogenase (K_m ca. 70 fold, v_{max} ca. 150 fold); 3) carboxylation of phosphoenol pyruvate (PEP) via PEP carboxylase and activation of the glyoxylate cycle lead to enhanced formation of oxaloacetic acid,

the precursor of citric acid, from glycolysis. Since PEP carboxylase requires biotin, this cofactor must be available in sufficient quantities. In addition, several of the enzymes involved in glutamate biosynthesis have been deregulated towards intermediary metabolites, various end products, NH_4^+, and the $NAD^+/NADH$ pool. These classical methods are increasingly being complemented by methods of genetic and metabolic engineering, since the genome of *C. glutamicum* (3.1 million bp) has been completely sequenced. Thus, the effect of multi-copy gene cassettes of glutamate dehydrogenase on glutamate productivity has been investigated.

Fermentation and recovery. Molasses or starch hydrolysates are often used as a carbon source. Under optimized culture conditions, high-performance mutants of *C. glutamicum* convert 50–60 % of these carbon sources to glutamate. Ammonia or ammonium salts are used as a nitrogen source. The biotin content of the medium is optimized, and the pH value is kept between 7 and 8. The oxygen supply is critical (optimal k_d value 3.5×10^{-6} mole $O_2 \text{ atm}^{-1} \text{ min}^{-1} \text{ mL}^{-1}$). Production is carried out in bioreactors of up to 500 m³. Preculture fermenters of increasing volume are used for inoculation. To prevent catabolite repression, a fed-batch mode is preferred for the process, and after ca.14 h (after formation of the cell mass) the glucose content of the medium is kept at 0.5 %, by monitoring the CO_2 content of the exhaust gas. Glutamic acid is isolated from the medium (ca. 150 g L^{-1} after 60 h) after separation of cells via ultrafiltration, followed by ion-exchange and absorption chromatography.

Economic considerations. L-glutamic acid is mainly used as a flavor enhancer in the food industry, often in combination with nucleosides such as IMP or GMP. Commercial production of L-glutamic acid is in excess of 1 million t, with China being the largest producer (ca. 700 000 t). Most other producers are also located in Asian countries. At a price of ca. 1000 US$ t^{-1}, the market volume exceeds 1 billion US$.

L-Glutamic acid

C$_5$H$_9$NO$_4$
M$_R$ 147.13   CAS   56-86-0 (L)-form
m.p.   247–249 °C (decomp.)
solubility   600 g/L water (20 °C)

(2S)-form, L-form
HOOC–...–COOH, NH$_2$

Biosynthesis and high-performance mutants

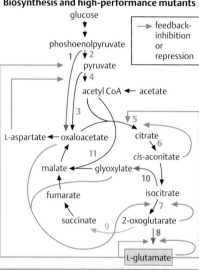

enzyme	gene
1 **phosphoenolpyruvate carboxylase (PEPC)**	*ppc*
2 pyruvate kinase	*pyk*
3 pyruvate carboxylase	*pyc*
4 pyruvate dehydrogenase	*pdh*
5 citrate synthase	*gltA*
6 aconitase	*citB*
7 isocitrate dehydrogenase	*icd*
8 **L-glutamate dehydrogenase (GDH)**	*gdh*
9 α-ketoglutarate dehydrogenase (KDH)	*aceE*
10 isocitrate lyase (ICL)	*aceA*
11 malate-synthetase (MS)	*aceB*

high-performance mutants:
1. higher activities of **PEPC, GDH, ICL** and **MS**
2. reduced activities or block at KDH
3. **PEPC** with reduced feedback inhibition by L-glutamate

Enhancement of secretion

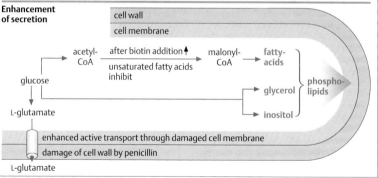

Fermentation and recovery

inoculation, precultures	main fermentation	cell separation	concentration	finishing	
increasing reactor volume	bioreactors up to 500 m³, C-source mostly molasses or starch hydrolysates, N-source mostly NH$_3$, defined biotin content	filterpress or diafiltration	of the filtrate by ultra-filtration		ca. 150 g/L in 40–60h

D,L-Methionine, L-lysine, and L-threonine

General. These three amino acids are mainly used as feed additives. For humans and most domesticated animals, these are essential amino acids, which are not produced by these organisms. Many crops used as food or feeds, e.g., corn, soybeans, oats, barley, wheat, and rice, do not contain enough of these amino acids for healthy nutrition. It is thus recommended that predominantly vegetarian diets be supplemented with these amino acids. In feeds, this deficiency plays an even more important role, since the increased mass during fattening of an animal on wheat or rice reaches the nutritional standard of casein only if L-lysine and L-threonine are added. Similarly, a corn-based feed requires the addition of D,L-methionine, L-lysine, and L-tryptophan. These amino acids are produced industrially by chemical synthesis or fermentation.

D,L-Methionine. The synthesis consists of five steps involving acrolein, methanethiol, and HCN, leading to the intermediary formation of a hydantoin. Since D-methionine is converted to L-methionine by higher animals, racemic D,L-methionine can be used as a feed additive; separation of the enantiomers is not necessary. Small amounts of L-methionine are manufactured by hydrolyzing derivatives of D,L-methionine by enzymes which selectively cleave the L-enantiomer derivative.

L-Lysine. is mainly produced by fermentation using mutants of *Corynebacterium glutamicum*. It is formed via oxaloacetate of the citric acid cycle through condensation of aspartic and pyruvic acid in a multi-step pathway leading through diamino pimelic acid (DAP) as an intermediate. Branches of this pathway also lead to L-threonine and L-methionine, which can suppress the formation of lysine by feedback inhibition. In lysine-overproducing mutant strains, this regulatory process is eliminated due to deregulation or by bypassing strongly regulated enzymes via auxotrophic block mutants (predominantly based on homoserine auxotrophy). Since the complete genome sequence of *C. glutamicum* has been obtained and annotated, and all genes coding for the enzymes involved in this pathway have been cloned, methods of genetic and metabolic engineering based on flux analysis play a most important role in achieving even more potent mutant strains. Today, yields are in the range of $120\ g\,L^{-1}$ in 60 h, and (as for L-glutamate) fed-batch protocols in bioreactors up to 500 m^3 in volume are used. The carbon source is usually sugarcane molasses. The content of biotin in the medium must exceed $30\ \mu g\,L^{-1}$. Recovery includes separation of cells by separators, ion-exchange chromatography, and spray drying. An interesting alternative, though no longer economically viable, is the production of lysine from D,L-α-amino-ε-caprolactam (ACL), an inexpensive intermediate from the chemical synthesis of Nylon™, with acetone-dried cells of *Cryptococcus laurentii* in a cell reactor. After enantioselective hydrolysis, re-racemization of the remaining D-α-amino-ε-caprolactam with a D-aminocaprolactame racemase, using cells of *Achromobacter obae*, leads to nearly quantitative yields of L-lysine.

L-Threonine. The organisms of choice for producing this amino acid are deregulated mutants of *Escherichia coli*. The best strains form $80\ g\,L^{-1}$ in only 30 h. The threonine operon has been cloned and is used for further strain improvement. Recovery is initiated by cell separation, followed by ultrafiltration and then crystallization of the product from the ultrafiltrate.

Economic considerations. Annual production of L-lysine is ca. 400 000 t, of D,L-methionine ca. 200 000 t, and of L-threonine ca. 10 000 t. The preferred process for methionine is chemical synthesis of the racemate. L-Threonine and L-lysine are produced exclusively by fermentation. The price of methionine and lysine is between 2 000 and 4 000 US\$ t^{-1}, leading to a market value of about 1 billion US\$. In the more distant future, the addition of industrially produced amino acids to food and feeds is expected to receive competition from the generation of transgenic crops containing an amino acid composition optimized for nutritional purposes.

D,L-Methionine

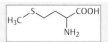

C$_5$H$_{11}$NO$_2$S
m.w.   149.21
m.p.   280–281 °C, dec.
CAS   63-68-3

L-Lysine

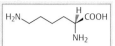

C$_6$H$_{14}$N$_2$O$_2$
m.w.   146.19
m.p.   224 °C, dec.
CAS   56-87-1

L-Threonine

C$_4$H$_9$NO$_3$
m.w.   119.12
m.p.   255–253 °C, dec.
CAS   72-19-5

Amino acids in animal feed

wheat	without additions
	addition of 0.25 % L-lysine
	addition of 0.4 % L-lysine and 0.4 % L-threonine
corn	without additions
	addition of 0.05 % L-lysine
	addition of 0.4 % L-lysine and 0.035 % L-tryptophan
rice	without additions
	addition of 0.1 % L-lysine
	addition of 0.2 % L-lysine and 0.1 % L-threonine
for comparison: casein without any additives	

0   0.5   protein efficiency ratio   2.0   2.5

Deregulated biosynthesis

L-lysine production strain of *Corynebacterium glutamicum*

first block and addition of L-homoserine

aspartic acid → aspartyl phosphate → aspartyl-semialdehyde → L-homoserine → L-threonine

dihydrodipicolinic acid
↓
diaminopimelinic acid
↓
L-lysine

— feedback inhibition
— repression

L-methionine

2-oxo-butyric acid
↓
L-isoleucine

mutants where L-lysine biosynthesis is no longer repressed by L-lysine and L-threonine, obtained through screening in the presence of antimetabolites of the two amino acids

---

Fermentation and recovery of L-lysine

preferred process: fermentation

production strain	30 m^3 bioreactor	500 m^3 bioreactor	cell separation	recovery	
auxotrophic, regulatory and transgenic mutant of *Corynebacterium glutamicum*	dextrose, soybean meal	dextrose, soybean meal, optimized mass transfer	separators or cross flow filtration	ion exchangers, crystallization or direct spray drying	120 g/L L-lysine in 80 hrs

alternative process: cell reactor

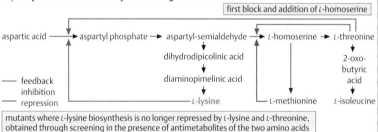

D-α-amino-ε-caprolactam (ACL)   L-α-amino-ε-caprolactam   L-lysine

1 ACL racemase
Achromobacter obae

2 ACL hydrolase
Cryptococcus laurentii

29

Aspartame™, L-phenylalanine, and L-aspartic acid

General. Aspartame (L-α-aspartyl-L-phenylalanine-methylester) is a synthetic low-calorie sweetener that is about 200 times sweeter than sucrose. It is registered as a food additive and produced on a scale of ca. 10 000 t y^{-1}. The starting materials are L-aspartic acid and L-phenylalanine. Chemical synthesis of Aspartame™ requires the use of several protecting groups and thus is not competitive with enzymatic synthesis.

L-Aspartic acid can be isolated by extraction from protein hydrolysates. The preferred synthesis, however, is the addition of ammonia to fumaric acid by *Escherichia coli* cells. Usually, a cell reactor is used, with the bacteria immobilized on κ-carrageenan or polyacrylamide. The productivity of this system is around 140 g L^{-1} h^{-1} and the operational stability (half life of the catalyst) up to 2 y. Using lyophilized cells after induction, yields were up to 166 g L^{-1}. The formation of aspartase in *E. coli* K12 could be increased 30 fold when a plasmid containing the gene coding for aspartase (*aspA*) was expressed. Compared to a cell reactor, fermentation processes are not competitive, even when high-performance mutants are used.

L-Phenylalanine. In the past, industrial production was usually based on enzyme reactors using readily available chemical building blocks. Recently, fermentation processes based on high-performance mutants have become competitive. The availability and price of the synthetic precursors compared to the space-time yields of the fermentation process are the decisive factors in the economic preference. At present, the best results are obtained by the addition of ammonia to *trans*-cinnamic acid, using phenylalanine ammonia lyase from *Rhodotorula glutinis*. In a cell reactor with immobilized microorganism, yields are ca. 50 g L^{-1} at 83% turnover. Cleavage of D,L-5-benzylhydantoin by L-hydantoinase and L-N-carbamoylase from *Flavobacterium ammoniagenes* is also promising. For fermentation processes, high-performance mutants of

E. coli or *Corynebacterium* are mostly used. The biosynthesis of L-phenylalanine proceeds from the precursors erythrose-4-phosphate and phosphoenolpyruvate via the intermediates shikimic acid, chorismic acid, and prephenic acid. This biosynthetic pathway includes branches leading to L-tyrosine and L-tryptophan, and many enzymes on this pathway are highly regulated. Consequently, auxotrophic mutants are mostly used for the production of L-phenylalanine. Most genes of this pathway have been cloned and genetic methods can also be used to produce strains with derepressed production. For example, yields reach 45 g L^{-1} using recombinant strains of *Brevibacterium fermentum*. Fermentation is carried out in a fed-batch mode, the cells are removed by filtration, and product is usually recovered by concentrating the broth by ultrafiltration, followed by adsorption chromatography and crystallization.

Aspartame™. The chemical synthesis of Aspartame™ from its two constituent amino acids requires the use of protecting groups and their later removal. Compared to this multi-step process, the currently used enzymatic production process is much simpler: only the amino group of L-aspartic acid must be protected, and an enzyme is used to catalyze amidation of the α-carboxy group of L-Z aspartic acid (the isomeric L-β-aspartyl-L-phenylalanyl-methylester tastes bitter) with L-phenylalanyl methylester. The regioselectivity of this enzyme even allows racemic phenylalanine methylesters to be used in this reaction. The preferred enzyme for this reaction is immobilized thermolysin from *Bacillus stearothermophilus*, and the solvent is i-amyl alcohol. Thermolysin is quite temperature stable and allows for a process temperature of 70 °C, leading to very high space-time yields of 30 g L^{-1} h^{-1}. The reaction product is of high purity and is further purified by ion-exchange chromatography. Aspartame is just one of a series of natural low-calorie sweeteners which were recently introduced. Other examples are stevioside (a glycosylated diterpene), thaumatin or monellin (glycosylated proteins or peptides, respectively).

L-Phenylalanine	L-Aspartic acid	α-Aspartame

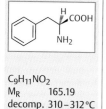

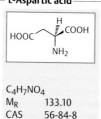

L-Phenylalanine

$C_9H_{11}NO_2$
M_R 165.19
decomp. 310–312 °C
CAS   63-91-2

L-Aspartic acid

$C_4H_7NO_4$
M_R 133.10
CAS   56-84-8

α-Aspartame

$C_{14}H_{18}N_2O_5$
M_R 294.31
ADI*   40 mg/kg
CAS   22389-47-0

*acceptable daily intake

Production of L-phenylalanine

preferred process: fermentation		
high-performance mutants of *E. coli* or Corynebacterium	C-source: glucose Recovery: membrane processes, ion exchange chromatography, crystallization	> 40 g L⁻¹ in 60 h

alternatively: enzyme reactor		
addition of ammonia to *trans*-cinnamic acid	carrier-bound L-phenylalanine-ammonia lyase from *Rhodotorula glutinis*	~ 50 g L⁻¹

> 40 g L^{-1} in 60 h and ~ 50 g L^{-1}

Production of L-aspartic acid

multi-stage cell reactor	recovery	
immobilized *E. coli* cells; 1.2 M ammonium fumarate pH 8.5, 37 °C	with H_2SO_4 at 15 °C to pH 2.8: aspartic acid precipitates	90–95 % yield half-life of *E. coli* cells: 2 yr

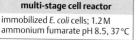

aspartase (immobilized *E. coli*)

Production of α-Aspartame

L-Z-aspartic acid   +   D,L-phenylalanine-OMe ◄────────

immobilized thermolysin 60 °C, 2 h   racemization

                         D-phenylalanine-OMe

L-Z-α-aspartyl-L-phenylalanine-OMe

chemical hydrogenation   ──→ ZH

L-α-aspartyl-L-phenylalanine-OMe α-Aspartame

──→ enzyme catalysis
──→ chemical steps
Z = benzyloxycarbonyl

sweetener	chemical constitution	relative sweetness
saccharose	disaccharide	1
cyclamate	synthetic cyclohexyl sulfamide, Na salt	40
α-Aspartame	dipeptide methylester	200
stevioside	glycosylated diterpene	300
saccharin	synthetic 2-sulfobenzoic acid imide, Na salt	450
thaumatin	nonglycosylated protein from 208 amino acids	2500
monellin	nonglycosylated protein, two peptide chains of 44 and 50 amino acids	2500

Amino acids *via* enzymatic transformation

General. As was shown before for several examples (L-lysine, L-aspartic acid, L-phenylalanine), chiral amino acids can be prepared by the enzymatic transformation of racemic building blocks. The advantage of enzyme processes over fermentation procedures rests in the possibility of preparing non-proteinogenic and even non-natural amino acids. In most enzymatic transformations of this kind, hydrolases and racemic precursors are used. Examples are esterases, aminoacylases, amidases, and hydantoinases/carbamoylases. The disadvantage of this process lies in the need to reracemize the "wrong" enantiomer in a coupled reaction, in order to feed it again into the reaction. Consequently, addition reactions based on lyases and redox reactions based on oxidoreductases are being intensively studied, since they lead to only one enantiomer.

Enantioselective hydrolysis. To date, enzyme reactors based on aminoacylases and hydantoinases are the most advanced and, in some cases, have already been industrialized. In the aminoacylase reaction, carrier-bound enzymes (e. g., from *Aspergillus oryzae* or *Bacillus thermoglucosidius*) hydrolyse racemic N-acyl amino acids. Only the L-enantiomer is hydrolyzed. The N-acyl-D-amino acid remains in the reaction mixture, from which the L-amino acid is separated by crystallization. After the wrong enantiomer has been reracemized, often in a thermic reaction, it is combined with new racemic precursor and fed into the reactor again. Using this technology, several 100 t y^{-1} of L-methionine, L-tyrosine, L-proline, and L-valine are produced for clinical use (mainly infusions). Although D-amino acids can also be produced by this procedure, hydantoinases, which can be isolated in many variations from microorganisms, offer the better choice for the preparation of non-proteinogenic and non-natural amino acids. In this process, racemic hydantoins are cleaved with hydantoinases of the desired specificity. N-carbamoylated amino acids are formed as the primary product and can be hydrolysed to the desired chiral amino acid. The "wrong" hydantoin is then racemized at pH 8.5 and becomes available for a new hydrolysis cycle. This method is used for the industrial manufacture of R-phenylglycine and R-4-hydroxyphenyl glycine, the side chains of Ampicillin and Amoxicillin (semisynthetic penicillins).

Enantioselective addition reactions. Oxynitrilases occur mainly in plants. They exhibit R or S selectivity and so do not lead to formation of a byproduct of undesired chirality. Both types of oxynitrilases have been cloned and expressed in *Escherichia coli* or yeasts and thus are readily available. Examples are R-oxynitrilase from Manihot and the S-oxynitrilase from almonds. The crystal structure of both enzymes has been solved; and researchers are now using protein engineering to enhance their substrate specificity for their eventual use in various industrial applications.

Enantioselective redox reactions. The example of the stereoselective synthesis of L-leucine from synthetic α-oxo caproic acid makes it evident that reductive amination of the oxo compound by L-leucine dehydrogenase from *Bacillus sp.* requires not only NH_3, but also NADH. Since NADH is expensive, its regeneration is imperative for achieving an economic process. An elegant procedure is based on the use of formate dehydrogenase from *Candida boidinii*, since the reaction product CO_2 evaporates from the mixture and thus shifts the reaction equilibrium towards L-leucine. If polyethylene glycol (PEG) is bound to NADH, the cofactor remains functional and is retained in an enzyme membrane reactor. With this technology, up to 6×10^5 mole equivalents of product could be obtained per mole of NADH-PEG consumed. Alternatively, NADH regeneration can also be achieved in a by-pass by chinone hydride acceptors such as methylene blue or, with reduced yields, by electrochemical procedures. Most enzymes used in this reaction have been cloned and modified via protein engineering to increase their stability and specificity (e. g., for the regeneration of NADPH instead of NADH).

Major reaction types

reaction type	enzyme type	comments
hydrolysis of racemic precursors	hydrolases	preferred, simple reaction, but expensive, since the "wrong" enantiomer must be reracemized
addition of HCN or ammonia to carbonyl compounds	lyases	simple, quantitative reaction, but limited choice of enzymes
reductive amination of α-oxo carboxylic acid	dehydrogenases	only one enantiomer is formed, but auxiliary enzymes and cofactors render the process expensive

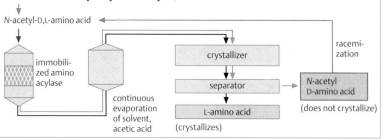

1 hydantoinases
2 carbamoylases
3 esterases (R^1 = OR, R^2 = H)
4 amino acylases (R^1 = OH, R^2 = Acyl)
5 amidases (R^1 = NH$_2$, R^2 = H)
6 oxynitrilases (in our example an S-specific enzyme)

→ enzyme catalysis   →⟶ chemical steps

Enantioselective hydrolysis of N-acyl-D,L-amino acids

N-acetyl-D,L-amino acid

immobilized amino acylase

continuous evaporation of solvent, acetic acid

crystallizer

separator

racemization

N-acetyl D-amino acid
(does not crystallize)

L-amino acid
(crystallizes)

Reductive amination of α-oxocarbonic acids

the broad substrate specificity of leucine dehydrogenase allows the reductive amination of pseudosubstrates to non-proteinogenic amino acids, e.g., t-leucine

leucine dehydrogenase from Bacillus sphaericus	relative activity	K_m
α-oxoisocaproate	100	0.31
α-oxoisovaleriate	126	1.4
α-oxovaleriate	76	1.7
α-oxobutyrate	57	7.7
α-oxocaproate	46	7.0

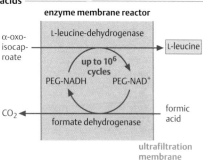

enzyme membrane reactor

L-leucine-dehydrogenase

α-oxo-isocaproate → L-leucine

up to 10^6 cycles

PEG-NADH   PEG-NAD$^+$

CO$_2$ ← formate dehydrogenase ← formic acid

ultrafiltration membrane

PEG-NADH is a synthetic derivative of NADH (M_R ~ 3 000). It is accepted by many dehydrogenases as a cofactor and retained by ultrafiltration membranes

Antibiotics: occurrence, applications, mechanism of action

General. In 1928, the British microbiologist Alexander Fleming noticed that a fungal infection on his agar plate inhibited the growth of Staphylococci. Only a decade later, Howard Florey, an Australian working in the UK, succeeded in the isolation, purification, and structural identification of the mixture of fungal antibiotics responsible for Fleming's observation. Successes in animal experiments and in treating a patient severely affected by a Streptococci infection led to the initiation of a large-scale project by the British/US allies in World War II. By 1945, penicillin could be prepared in kg amounts. In 1947, Selman Waksman, a Russian-American microbiologist, detected in cultures of *Streptomyces griseus* streptomycin, a new antibiotic which, other than penicillin, was also effective against Gram-negative microorganisms. In the following years, systematic screening led to the discovery of very many new antibiotics, and processes for their manufacture in the industrial scale became established. However, the uncritical use of antibiotics against trivial diseases and as a feed component in mass animal production soon led to the development of antibiotic resistance in microorganisms, which is now a major concern in hospitals (nosocomial infections). In an attempt to counteract this phenomenon, novel types of antibiotics are being developed and their use is being restricted to specific applications. Screening methods have recently been complemented by genetic methods such as cell fusion of different antibiotic producers and shuffling of gene clusters responsible for the biosynthesis of antibiotics.

Occurrence. More than 25 000 microbial products are known, and ca. 50 % of them display antibiotic activity under suitable conditions. In addition, ca. 4000 antibiotics have been isolated from higher organisms such as lichens, plants, and animals. Actinomycetes by far outnumber all other organisms in their capacity to synthesize antibiotics.

Applications. Only ca. 200 antibiotics are produced industrially. They are mostly semisynthetic compounds, in which a biologically active lead structure is modified chemically. β-Lactam antibiotics (penicillins and cephalosporins) make up about half of a world market of ca. 24 billion US$. Most antibiotics are manufactured as antimicrobial agents for chemotherapy. They can be classified as broad-spectrum antibiotics, affecting a wide range of pathogens (e. g., cephalosporins, tetracyclines), and selective antibiotics, used for highly special therapies (e. g., rifampicin against tuberculosis, amphotericin B against fungal infections). Antitumor antibiotics are valuable cytostatic agents, but also exhibit high toxicity; an example is adriamycin. Several antibiotics are used for plant protection. They are often effective in lower concentrations than herbicides and show low toxicity against mammals. Examples are blasticidin S and kasugamycin. Only a few antibiotics are used in food preservation, such as pimaricin which is sometimes used as an antifungal agent in cheeses. Feed antibiotics lead to a better mast performance and faster growth in mass animal production, but are usually restricted for feed use (e. g., monensin in broiler production), to prevent clinical cross resistance. In molecular biology, antibiotics serve as a research tool for the selective inhibition of various cell functions.

Mechanism of action. Antibiotics can affect: 1) biosynthesis of the components and function of the genetic machinery of the cell, 2) biosynthesis of cell components, 3) biosynthesis and function of proteins, 4) biosynthesis and function of the cytoplasmic membrane or, as in Gram-negative bacteria, the outer cell membrane, and 5) biosynthesis of the cell wall. The interactions of individual antibiotics are diverse. As a consequence of their short generation time and their rapid adaptation to changing environments, microorganisms usually become quickly resistant to antibiotics, resulting in a permanent race between the development of new antibiotics and the occurrence of new resistant strains.

Occurrence

taxonomic group	relative number (%)
Actinomycetes	50
other bacteria	10
fungi	20
lichens	1
algae	2
plants	15
animals	2

~25 000 compounds from nature

Systemic antibiotics (2001)

type	value (billion US $)
cephalosporins	6.7
penicillins	4.6
chinolones (synthetic)	4.6
macrolides	4.3
tetracyclines	0.7
aminoglycosides	0.6
peptide antibiotics, glycopeptides	0.5
other	2.2
total	24,2

Classification by chemical structure

1	**carbohydrate antibiotics** aminoglycosides	streptomycin (medicine), kasugamycin (rice fungicide)
2	**macrocyclic lactones** macrolides polyene antibiotics ansamycines	erythromycin (medicine) pimaricin (cheese production) rifamycin (against tuberculosis)
3	**chinones and related antibiotics** tetracyclines anthracyclines	tetracycline, chlorotetrycycline (medicine, feed antibiotic) doxorubicin (cancer therapy)
4	**amino acid and peptide antibiotics** amino acid derivatives β-lactam antibiotics peptide antibiotics chromopeptides glycopeptides	cyclosporin (organ transplantation) phosphinothricin (plant protection) penicillins, cephalosporins (medicine) bacitracin (medicine), virginiamycin (feed antibiotic) actinomycin (cancer therapy), bleomycin (cancer therapy), vancomycin (medicine), avoparcine (cattle feed antibiotic)
5	**N-heterocyclic compounds** nucleoside antibiotics	polyoxins, blasticidin S (fungicides for plant protection)
6	**O-heterocyclic compounds** polyether antibiotics	monensin (chicken feed)
7	**alicyclic compounds** cycloalkane derivatives	cycloheximide (leaf fungicide)
8	**aromatic antibiotics** benzene derivatives	chloramphenicol (medicine) griseofulvin (fungicide)

Antibiotics – point of attack

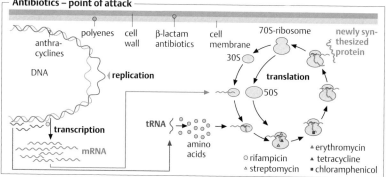

35

Antibiotics: industrial production, resistance

Screening. Traditionally, inhibition of microorganisms' growth in the presence of a producer or its culture filtrate is used to test for an antibiotic. If an interesting biological activity is detected, the antibiotic is enriched and purified from the culture broth, and its structure is elucidated. If this procedure is used today, an already known antibiotic is usually rediscovered. Thus, to increase the number of "hits", many new screening procedures have been developed. For example, other biochemical or biological assays may replace growth inhibition tests using microorganisms, culture filtrates of putative antibiotic-forming strains may be analyzed by chemical procedures, or target-oriented methods may be used (e. g., the inhibition of a microbial enzyme involved in pathogenesis is monitored in a highly parallel microtiter plate assay against whole or fractionated extracts of (micro)organisms.

Strain improvement. If a promising new antibiotic is identified, its yield must be optimized at an early stage, since antibiotics are secondary metabolites and thus formed in rather low concentrations (several $mg L^{-1}$ culture or less). Strain improvement usually follows empirical rules dominated by labor-intense repetitions of mutation and selection. Occasional back-crossing with the wild-type strain may enhance the robustness of production strains. Based on such methods, yields of economically important antibiotics were increased 10^3–10^6 fold relative to the original isolate. Genetic engineering techniques, e. g., increasing the number of gene copies for key enzymes, have led to further improvements in yield.

Fermentation and recovery. The chemical constitution of most antibiotics is complex. They usually contain several stereo centers, rendering chemical synthesis complicated. In fact, the syntheses of many antibiotics have been brilliant examples of chemical synthesis but not applicable to industrial production. As a result, most antibiotics are produced by fermentation in bioreactors. Inexpensive carbon and nitrogen sources such as molasses, dextrose, whey, soybean meal, and corn steep liquor are used. Since most antibiotic producers are subject to catabolite repression, the carbon source is often added in a fed-batch mode. Because antibiotics are secondary metabolites, their formation starts only after cell growth has reached its stationary phase. Antibiotics are usually extracellular products, and often they are only moderately soluble in water. For their isolation at the end of the fermentation process, cells are removed and the culture broth is extracted with organic solvents (the two steps can be integrated into one). The raw antibiotic can be further purified by various chromatographic procedures or by crystallization. The manufacture of antibiotics is subject to strict regulations and safety controls following certified rules, e. g., GMP and ISO 9000.

Resistance. The increase of antibiotic-resistant microorganisms is a major problem in modern medicine. Cross-resistant microorganisms (i. e., microorganisms that are resistant to several or many antibiotics) are on the increase among Salmonella, *Escherichia coli*, Streptococci, Staphylococci, and *Mycobacterium tuberculosis* (which causes tuberculosis). The most important mechanisms underlying microbial resistance to antibiotics are 1) enzymatic modification of the antibiotic itself (e. g., by hydrolysis or acylation), 2) interference with the import or a fast re-exportation of the antibiotic (e. g., by antiporters or changed membrane permeability), or 3) modification of the antibiotic target (e. g., through modification of the binding site of the antibiotic at the ribosome or the translation machinery). Resistance is either encoded in the microbial genome or in mobile genetic elements such as plasmids or transposons, which may give rise to horizontal transfer. Phages are another vehicle for transferring resistance among different strains. There are calls to reduce the use of antibiotics in agriculture (where the largest quantities are used) to limit the spread of resistance.

Screening

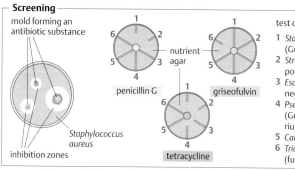

mold forming an antibiotic substance

Staphylococcus aureus

inhibition zones

nutrient agar

penicillin G

griseofulvin

tetracycline

test organisms:

1 Staphylococcus aureus (Gram-positive bacterium)
2 Streptococcus sp. (Gram-positive bacterium)
3 Escherichia coli (Gram-negative bacterium)
4 Pseudomonas aeruginosa (Gram-negative bacterium)
5 Candida albicans (yeast)
6 Trichophyton rubrum (fungus)

Strain improvement

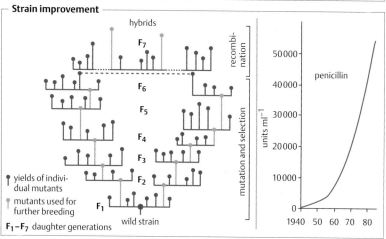

hybrids

F7

recombination

F6

F5

F4

F3

F2

F1

mutation and selection

yields of individual mutants

mutants used for further breeding

F1–F7 daughter generations

wild strain

penicillin

units ml^{-1}

50000
40000
30000
20000
10000
0

1940 50 60 70 80

Mechanisms of resistance

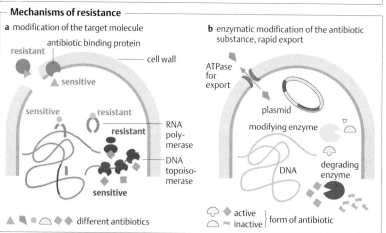

a modification of the target molecule

resistant

antibiotic binding protein

cell wall

sensitive

sensitive

resistant

resistant

sensitive

RNA polymerase

DNA topoisomerase

different antibiotics

b enzymatic modification of the antibiotic substance, rapid export

ATPase for export

plasmid

modifying enzyme

DNA

degrading enzyme

active
inactive } form of antibiotic

β-Lactam antibiotics: structure, biosynthesis, and mechanism of action

General. β-Lactam antibiotics (*Penams*: penicillins. *Cephems*: cephalosporins) are the most important antibiotics both in value and quantity, due to their high efficiency, low toxicity, and wide variety of ways to turn them into semi-synthetic lactam derivatives. Annual production is in the range of some 10^4 tons ; cephalosporins constitute about one third, penicillins about one fifth of the total antibiotic market. They are exclusively used in clinical applications. In contrast, penicillins are also used in veterinary medicine. The most important primary β-lactams are penicillin G and cephalosporin C. They serve as starting materials for the manufacture of semi-synthetic penicillins and cephalosporins. Important criteria for the efficacy of β-lactam antibiotics are: acid stability (for oral delivery) and stability against β-lactamases, plasmid-coded enzymes which are a key factor responsible for microbial resistance to β-lactam antibiotics.

Penicillins. *Penicillium chrysogenum* forms isopenicillin N as the primary product. It contains a nonproteinogenic amino acid in its side chain (L-α-aminoadipic acid). If acids, e.g., phenylacetic acid, are added to the culture medium in the late logarithmic stage, a fungal N-transacetylase catalyzes the incorporation of these acids into "biosynthetic penicillins" which exhibit different pharmacological properties. Penicillin G and 6-aminopenicillanic acid (6-APA), the intermediate obtained by hydrolysis of the acid side chain, are the most important intermediates for making semisynthetic penicillins.

Cephalosporins. Following observations by Giuseppe Brotzu, an Italian physician, cephalosporin C 1953 was the first to be discovered as a new class of β-lactam antibiotics. It is formed by *Acremonium chrysogenum* (former name: *Cephalosporium acremonium*). *A. chrysogenum* does not contain an N-transacetylase; in consequence, it is impossible to prepare biosynthetic cephalosporin derivatives from added acid precursors. Most semisynthetic cephalosporins are therefore produced from the synthetic intermediates 7-amino cephalosporanic acid (7-ACA). Cephalosporins of the 2nd and 3rd generations show an excellent, broad efficacy towards a wide range of Gram-positive and Gram-negative pathogens at low human toxicity.

Biosynthesis. *P. chrysogenum* contains a cluster of three genes that code for the biosynthesis of isopenicillin N. The building blocks L-α-aminoadipic acid, L-valine, and L-cysteine are condensed by a single synthetase to a tripeptide, which is transformed by a second synthetase into bicyclic isopenicillin N. An acyl-CoA:isopenicillin-N-acyl transferase coded by the same gene cluster allows for the exchange of the L-amino acid side chain against a wide range of other acyl groups. The biosynthetic pathway is distributed among several cell compartments. Cephalosporins are formed in *A. chrysogenum* from intermediary isopenicillin N, after epimerization of the amino acid side chain, via an oxidative ring expansion catalyzed by expandase, followed by reactions on various substituents of the lactame and the thiazine ring. The O-acyltransferase of the fungus allows modification reactions at the 3-acetoxymethyl position, but the lack of an N-acyltransferase prevents reactions at the 7-amide bond. In contrast to Penicillium, the genes for the biosynthesis of cephalosporin C are located on two chromosomes. All the genes mentioned above have been cloned and experiments are under way to modify product formation via genetic and protein engineering.

Mechanism of action. β-Lactam antibiotics prevent the formation of peptide cross links during biosynthesis of the bacterial cell wall (murein). Since murein is a major component of the cell wall of Gram-positive microorganisms, it seems clear why the first β-lactam antibiotics where specific towards this group of bacteria. Since higher animals and humans do not contain mureins, the side effects of β-lactam antibiotics are limited to interference with gastrointestinal microorganisms and to the occasional development of an allergy.

Penicillins

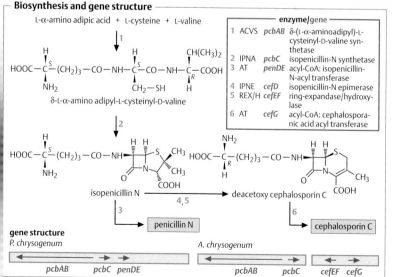

R	name	properties
$HOOC-\overset{(S)}{CH}(NH_2)-(CH_2)_3-$	isopenicillin N	
$H_5C_6-CH_2-$	penicillin G	acid labile, sensitive to β-lactamase
$H_5C_6-\overset{(R)}{CH}(NH_2)-$	Ampicillin	acid stable, sensitive to β-lactamase, active against gram-negative pathogens
$HO-\underset{}{\bigcirc}-\overset{(R)}{CH}(NH_2)-$	Amoxicillin	acid stable, sensitive to β-lactamase, broad-band antibiotic, very well resorbed

R—C—NH structure with: acyl residue; β-lactam ring; thiazolidine ring; 6β-aminopenicillanic acid (6APA); positions 1 10, CH₃ 9, COOH 8, 7, 6, 5, 4, 3, 2, S

Cephalosporins

R^1/R^2	name	properties
$HOOC-\overset{(R)}{CH}(NH_2)-(CH_2)_3-$ $CH_2-O-CO-CH_3$	cephalosporin C	acid labile, sensitive to β-lactamase
$H_5C_6-\overset{(R)}{CH}(NH_2)-$ Cl	Cefaclor	acid stable, stable against β-lactamase, broad-band antibiotic
(thiazole ring) H_2N—$\underset{S}{\overset{N}{\square}}$—$\overset{\|}{C}=N-OCH_3$ (syn) $CH_2-O-CO-CH_3$	Cefotaxim	acid stable, stable against β-lactamase, very broad efficacy

R^1—C—NH structure with: acyl residue; β-lactam ring; dihydrothiazine ring; positions 8, 7, 1, 2, 5, 4, R^2, COOH; 7β-amino cephalosporanic acid (7-ACA, $R^2 = CH_2-O-CO-CH_3$); 7β-amino desacetoxy-cephalosporanic acid (7-ADCA, $R^2 = CH_3$)

Biosynthesis and gene structure

L-α-amino adipic acid + L-cysteine + L-valine

		enzyme/gene
1	ACVS *pcbAB*	δ-(L-α-aminoadipyl)-L-cysteinyl-D-valine synthetase
2	IPNA *pcbC*	isopenicillin-N synthetase
3	AT *penDE*	acyl-CoA: isopenicillin-N-acyl transferase
4	IPNE *cefD*	isopenicillin-N epimerase
5	REX/H *cefEF*	ring-expandase/hydroxylase
6	AT *cefG*	acyl-CoA: cephalosporanic acid acyl transferase

$HOOC-\overset{H}{\underset{NH_2}{C}}-(CH_2)_3-CO-NH-\overset{H}{\underset{CH_2-SH}{C}}-CO-NH-\overset{CH(CH_3)_2}{\underset{H}{\overset{|}{C}}^{R}}-COOH$

δ-L-α-amino adipyl-L-cysteinyl-D-valine

isopenicillin N → (3) → penicillin N

isopenicillin N → (4, 5) → deacetoxy cephalosporin C → (6) → cephalosporin C

gene structure

P. chrysogenum

| *pcbAB* | *pcbC* | *penDE* |

A. chrysogenum

| *pcbAB* | *pcbC* | *cefEF* *cefG* |

β-Lactam antibiotics: manufacture.

Manufacture of penicillins. The penam ring of the penicillins contains three stereocenters, and only one of the nine possible isomers [3(S):5(R):6(R)] is biologically active. As a result, a fermentation process is economically advantageous, relative to chemical synthesis. High-performance strains of *Penicillium chrysogenum* are used in fermentation. A range of aromatic or aliphatic precursors may be added to the culture medium, leading to the formation of "biosynthetic penicillins"; only penicillins G and V are produced in large scale, using phenylacetic acid or phenoxyacetic acid, respectively, as the precursors. Both penicillin G and V are also used in the manufacture of 6-aminopenicillanic acid (6-APA), the key intermediate for preparation of semisynthetic penicillins and some cephalosporins.

Penicillin G and 6-APA. For the industrial manufacture of penicillin G, high-performance strains of *P. chrysogenum* are grown in bioreactors of up to 200 m³ volume. To prevent catabolite repression, sugar is added via a fed-batch fermentation process. The oxygen supply is critical and requires careful optimization of the stirrer and the aeration system, since the fungal mycelium is both viscous and sensitive to shear. The complex nutrient broth may contain lactose as a C-source and corn steep liquor as an N-source. Formation of penicillin occurs mainly after ca. 40 h, when fungal growth has been completed. During the production phase of ca. 100 h, phenylacetic acid is fed to the medium, and penicillin G is secreted by the fungus. When fermentation is complete, the mycelium can be separated by filtration or centrifugation, and the filtrate is extracted within minutes with amyl or butyl acetate at 0–3 °C and pH 2.5–3.0 using a two-stage countercurrent extractor. Alternatively, both steps can be combined by using a direct extraction procedure with two extraction decanters working in countercurrent mode. After re-extraction, using aqueous ammonia or carbonate, the crude antibiotic (> 3 tons for a 110 m³ bioreactor) is purified by crystallization. Subsequent gentle hydrolysis of the amide bond leads to 6-APA, and phenylacetic acid that can be reused for fermentation. 6-APA is a key intermediate for the preparation of most semisynthetic penicillins and some cephalosporins. Over the past few decades, enzymatic hydrolysis of penicillin G, using immobilized penicillin G amidase from *E. coli*, has largely replaced chemical hydrolysis. It is carried out at ca. 30 °C and pH 7.5–8.0, usually in batch mode. The high stability of the enzyme allows repetition of this step up to 1000 times before the enzyme must be replaced. Precipitation of 6-APA, filtration, and washing leads to a very pure product, which is further processed to semisynthetic penicillins or, through (chemical) ring expansion, to 7-ADCA, the base intermediate for some cephalosporins.

Cephalosporins and 7-ACA. The fermentation procedure leading to cephalosporin C, using *Acremonium chrysogenum*, resembles that of penicillin G manufacture, but yields are lower. *A. chrysogenum* does not contain a N-transacetylase and thus cannot form biosynthetic cephalosporins. 7-Aminocephalosporanic acid (7-ACA) is prepared by hydrolysis of cephalosporin C. In the enzymatic procedure, which is more and more preferred for reasons of its favorable ecobalance, the D-α-aminoadipyl side chain is oxidatively deaminated into α-ketoadipyl-7-ACA by the action of an immobilized D-amino acid oxidase. Spontaneous decarboxylation leads to glutaryl-7-ACA, from which the glutaryl side chain is cleaved off by an immobilized glutaryl-7-ACA-acylase. Cephalosporin C amidases, that can hydrolyze the D-α-aminoadipyl side chain in one step, have been reported, but their large-scale application is not yet established. Genetically engineered strains of *P. chrysogenum* have been shown to produce cephalosporin C if expandase/hydroxylase, cloned from *A. chrysogenum* or from *Streptomyces clavuligerus*, is expressed under the control of the β-tubulin promoter in the presence of adipic acid.

Synthetic routes

Penicillium chrysogenum
fermentation medium, phenylacetic acid

Acremoneum chrysogenum
fermentation medium

penicillin G

cephalosporin C

→ fermentation
→ chemical steps
→ enzyme catalysis

immobilized
penicillin G amidase

immobilized
enzymes

6-aminopenicillanic acid (6-APA)

7-amino-deacetoxy cephalosporanic acid (**7-ADCA**)

7-aminocephalosporanic acid
(**7-ACA**)

semi-synthetic penicillins

semi-synthetic cephalosporins

Process steps

penam series	cephem series
inoculation material	**inoculation material**
spore suspension of *P. chrysogenum*	spore suspension of *A. chrysogenum*
prefermenter, production fermenter	**prefermenter, production fermenter**
lactose, corn steep liquor, $CaCO_3$; glucose in fed-batch mode, good O_2 supply; addition of precursors, e.g., D-phenyl acetic acid, leads to biosynthetic penicillins; $>30\,g\,L^{-1}$ after 120 h	fed-batch procedure, good O_2 supply; ~ 120 h; fatty acids, e.g., methyl oleate, increase yield $>17\,g\,L^{-1}$ after 150 h
recovery	**recovery**
1. separation of mycelium 2. extraction of medium with acetyl-, butyl- or pentyl ester in a continuous countercurrent process	1. separation of mycelium 2. ion exchange and precipitation steps, XAD absorber materials

enzymatic hydrolysis to 6-APA	ring expansion to 7-ADCA	enzymatic hydrolysis to 7-ACA
immobilized penicillin-G-amidase from *Escherichia coli*, mostly discontinuous operation at 37 °C	4-step chemical procedure, ~ 70 % yield or enzymatic process	immobilized D-amino acid oxidase and glutaryl 7-ACA acylase $>90\%$ yield

semi-synthetic penicillins	semi-synthetic cephalosporins
chemical addition of acyl side-chains, mostly by Schotten-Baumann reaction	chemical addition of acyl side-chains, mostly by Schotten-Baumann reaction

Amino acid and peptide antibiotics

General. β-Lactam antibiotics are the most important therapeutic antibiotics in human medicine and are discussed above. Among the > 5000 antibiotics originating from secondary amino acid metabolism, some have found practical applications in medical therapy, the treatment of wounds, and agriculture. These include cycloserin and phosphinothricine, cyclic peptide antibiotics such as gramicidin and bacitracin, chelating peptides (bleomycin), chromopeptides (actinomycin), and depsipeptides (virginiamycin). Valinomycin selectively binds K^+ ions. Many of these antibiotics were isolated from Streptomyces strains, but some from other Gram-positive microorganisms such as Streptococci and Bacilli.

Amino acid derivatives. D-Cycloserine is synthesized by *Streptomyces orchidaceus*. An analoge of D-alanine, a component of the bacterial cell wall, it inhibits alanine racemase, a key enzyme in murein biosynthesis. Due to its excellent efficacy against *Mycobacterium tuberculosis*, it was for a long time, in combination with rifampicin, the antibiotic of choice for treating tuberculosis. Alanyl-alanyl phosphinothricine, first isolated from *Streptomyces hygroscopicus*, is an analog of L-glutamic acid and inhibits the glutamine synthetase of plants. Phosphinothricine (Glyphosate®, Basta®) is industrially manufactured by chemical synthesis. If acetyl transferase from *S. hygroscopicus* is cloned and expressed in an agricultural plant, the plant becomes resistant to phosphinothricine while weeds remain susceptible.

Peptide antibiotics can be synthesized either on the ribosome or through non-ribosomal biosynthesis. Ribosomal peptide antibiotics are usually linear peptides but can undergo post-translational modifications, e. g., through the epimerization of L- to D-amino acids, resulting in non-proteinogenic amino acid components. An example is the antibiotic nisin, produced by Lactobacilli, which is found in essentially all fermented milk products. It lyses the cytoplasmic membrane of bacteria and thus helps preserve milk products. Non-ribosomal peptide synthesis takes place on the thio template of a soluble multi-enzyme complex that resembles the fatty acid synthase of eukaryotes. In this system, rather short linear peptides are produced, which can be further transformed into cyclic peptides (e. g., the lantibiotics). They rarely contain > 20 building blocks and frequently contain unusual amino acids or additional structural elements. Due to toxicity, their use is limited to external applications, e. g., in the treatment of wounds or burns. Bacitracin is also used as a feed additive. Cyclosporin is synthesized by *Tolypocladium inflatum* and is the immunosupressant of choice in solid organ and bone marrow transplantation. As it inhibits T-cell activation, it is also used in some chronic inflammatory diseases such as nephrotic syndrome, refractory Crohn's disease, ulcerative colitis, rheumatoid arthritis and others. The colistins (polymycines) produced by *Bacillus polymyxa* are important reserve antibiotics against infections by Gram-negative microorganisms. The bleomycins produced by *Streptomyces verticillus* are among the most important anti-tumor antibiotics. Their 1:1 complex with Fe^{3+}, in the presence of triplet oxygen, acts like a DNase and leads to the rupture of single DNA strands. Actinomycin, a peptide derivative of phenoxazinone, is formed by various Streptomyces strains. It intercalates into 5′-TGCA-3′ palindrome sequences of DNA, thus blocking translation. The resulting cytotoxicity effects was used for some time in tumor treatments. The amino acid building blocks of the depsipeptides are alternately linked via ester and amide bonds. They are mainly active against Gram-positive bacteria. Virginiamycin, produced by *Streptomyces virginiae*, is used in large quantities in pig and calf fattening. The siderochromes are peptide antibiotics that contain or bind Fe and contain hydroxamic acid groups. They are sometimes used for treatment of iron-storage diseases.

Amino acid derivatives

C$_5$H$_{12}$NO$_4$P
M$_R$ 181.13
CAS 35597-44-5
(S)-Form

phosphinothricin

C$_3$H$_6$N$_2$O$_2$
M$_R$ 102.09
CAS 68-41-7

cycloserine

Cyclic peptide antibiotics

```
12       11      10
L-Asn ◄ D-Asp ◄ L-His
6↓      7       8      9
L-Lys ► D-Orn ► L-Ile   D-Phe ····NH₂
5↑      4       3
L-Ile ◄ D-Glu ◄ L-Leu
        bacitracin
                        L-Cys 2
                        L-Ile 1  CH₃
                              CH₃
```

bacitracin

C$_{66}$H$_{103}$N$_{17}$O$_{16}$S
very soluble in water
M$_R$ 1422.71
CAS 1405-87-4

```
MeLeu → MeVal → MeBmt → Abu-Sar
MeLeu                          ↓
D-Ala ◄ L-Ala ◄ MeLeu ◄ L-Val ◄ MeLeu
            cyclosporin A
```

C$_{62}$H$_{111}$N$_{11}$O$_{12}$
M$_R$ 1202.63   CAS 59865-13-3
Abu   L-aminobutyric acid
Sar   sarcosine
MeBmt   butenyl dimethyl threonine
MeLeu   methyl leucine

antibiotic	production	amount and value, USA		applications
phosphinothricine	chemical synthesis			herbicide
bacitracin A	*Bacillus licheniformis*	4 t, 100 million US $ > 200 t, 20 million US $		wound healing feed antibiotic
polymyxin	*Bacillus polymyxa*	10 kg		medicine
gramicidin	*Bacillus* sp.	400 g	25 million US $	medicine
bleomycin	*Streptomyces verticillus*	2 kg		cancer therapy
cyclosporin	*Tolypocladium inflatum*	3 t, 130 million US $		organ transplantation
virginiamycin	*Streptomyces virginiae*	70 t, 12–15 million US $		pig fattening
valinomycin	*Streptomyces fulvisimus*	–		ionophore

Fermentation of bacitracin

prefermenter
1–3 m^3, 6 h at 37 °C

bioreactor
~ 100 m^3, 30 h at 37 °C; saccharose, soy meal, salts

recovery
pharmaceuticals: extraction with 1-butanol; ion exchanger
animal feed: spray drying of fermentation broth

9 g L^{-1} in 30 h

Biosynthesis of gramicidin S (*Bacillus brevis*)

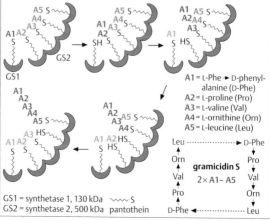

A1 = L-Phe → D-phenyl-alanine (D-Phe)
A2 = L-proline (Pro)
A3 = L-valine (Val)
A4 = L-ornithine (Orn)
A5 = L-leucine (Leu)

```
Leu ·············► D-Phe
↑                    ↓
Orn   gramicidin S   Pro
↑     2 × A1– A5     ↓
Val                  Val
↑                    ↓
Pro                  Orn
↑                    ↓
D-Phe ◄············· Leu
```

GS1 = synthetase 1, 130 kDa
GS2 = synthetase 2, 500 kDa   ∿∿ S pantothein

Glycopeptide, polyether, and nucleoside antibiotics

General. This category includes the very important glycopeptides vancomycin, which is indispensable for the treatment of methicillin-resistant *Staphylococcus aureus* strains (MRSA) ("first-line treatment"), and its analog avoparcin, which is used as a feed additive. Other examples are the glycoside lincomycin, which is highly effective against Gram-positive enterobacteria, and the chicken feed antibiotic monensin, which shows prophylactic efficacy against protozoa. Although nucleoside antibiotics occur naturally, only synthetic analogs are used in therapy, e.g., the guanosine analog acyclovir for the therapy of viral meningitis.

Vancomycin and avoparcin. Vancomycin is produced by *Amycolatopsis orientalis*, an Actinomyces strain. It is used against penicillin-resistant Enterococci, e.g., in septic endocarditis, and for patients who are allergic to penicillin. Since it is nephrotoxic and has been or is sometimes used in combination with other nephrotoxic antibiotics, such as aminoglycosides and cyclosporin, thorough monitoring of nephrotoxic side-effects is absolutely required. The effect of vancomycin, like that of β-lactam antibiotics, relies on inhibition of bacterial cell wall biosynthesis (binding to UDP muramyl pentapeptide). Resistant strains form a modified cell wall glycopeptide that does not react with vancomycin. It is assumed that this type of resistance is horizontally transferred via transposons between humans and domestic animals. The feed antibiotic avoparcin, produced by *Streptomyces candidus*, was used for animal feed in amounts about 10-fold higher than vancomycin. In view of the isolation of conjugative transposons that code for resistance to both vancomycin and avoparcin, it is probable that vancomycin resistance can travel via this pathway to the human food chain and to hospitals. It is now banned in the EU as well as in the USA.

Lincomycin, an antibiotic formed by *Streptomyces lincolnensis*, is active against Gram-positive pathogens and is used in veterinary medicine. Similar to chloramphenicol, it binds to the 50S subunit of the ribosome, inhibiting extension of the growing peptide chain. Resistant strains occur frequently. They either produce rRNA modified by methylation or detoxify lincomycin by enzymatic transformation.

Monensin, a polyether antibiotic, is manufactured by fermentation of *Streptomyces cinnamoensis*. Biosynthesis proceeds via polyketides, using acetate, propionate, and butyrate as building blocks. Monensin incorporates as a ionophore into membranes and causes osmolysis of cells through the influx of Na^+ ions. This mechanism of action affects not only bacteria and fungi, but also protozoa, e.g., *Eimeria* sp. and *Toxoplasma* sp., which occur during mass production of domestic animals. Although its toxicity is high in humans and horses (but not in chicken and cattle), it has become one of the most important broad-spectrum antibiotics in chicken feeds and is well tolerated if dosed appropriately. It is registered both in the European Union and in the USA. Its market share for this application is 80%, counting also salinomycin which is structurally related. It is produced in quantities of several thousand tons. Care must be taken not to poison other farm animals such as horses, or farm workers when using it.

Nucleoside antibiotics have found only limited applications. The cytosine analog blasticidin S from *Streptomyces griseochromogenes* is used as a fungistatic agent for the treatment of blight in rice agriculture. It prevents the binding of aminoacyl tRNA to the ribosome. It is quite toxic to plants as well as to fish, animals, and man, affecting the mucous membrane, skin, and lungs. A purine derivative important in medical therapy is acyclovir, a synthetic guanosine derivative. It is active against Herpes virus and can be used against viral encephalitis. It is obtained by chemical synthesis.

Glycopeptide, glycoside, and nucleoside antibiotics

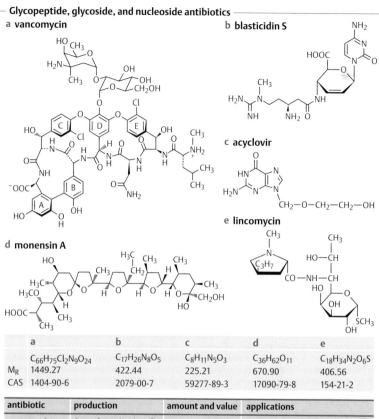

a vancomycin

b blasticidin S

c acyclovir

d monensin A

e lincomycin

	a	b	c	d	e
	$C_{66}H_{75}Cl_2N_9O_{24}$	$C_{17}H_{26}N_8O_5$	$C_8H_{11}N_5O_3$	$C_{36}H_{62}O_{11}$	$C_{18}H_{34}N_2O_6S$
M_R	1449.27	422.44	225.21	670.90	406.56
CAS	1404-90-6	2079-00-7	59277-89-3	17090-79-8	154-21-2

antibiotic	production	amount and value	applications
vancomycin	*Amycolatopsis orientalis*		human infections
avoparcin	*Streptomyces candidus*		feed antibiotic
monensin	*S. cinnamonensis*	>3 000 t, > 200 million US $	chicken feed, active against protozoa
blasticidin S	*S. griseochromogenes*		plant protection (blight fungicide)
acyclovir	chemical synthesis		antiviral agent
lincomycin	*S. lincolnensis*		veterinary medicine

Antibiotics as growth promotors in animal feed: the problem of resistance

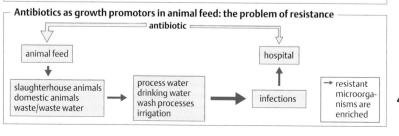

antibiotic

animal feed

hospital

slaughterhouse animals
domestic animals
waste/waste water

process water
drinking water
wash processes
irrigation

infections

→ resistant microorganisms are enriched

45

Aminoglycoside antibiotics

General. The discovery of streptomycin by Selman Waksman (1943) was a milestone in the development of antibiotics. Streptomycin permitted, for the first time in history, a treatment of tuberculosis, caused by *Mycobacterium tuberculosum*. Today, the nephrotoxic properties of streptomycin (which are typical of aminoglycoside antibiotics) have led to its replacement by isonicotinic acid hydrazide and rifampicin (formerly by cycloserine). The lead structure of most aminoglycoside antibiotics is an aminocyclitol ring, i. e., 2-deoxy streptamine, which is linked by glycosidic bonds to other amino sugars. Aminoglycoside antibiotics exhibit broad biological efficacy and are also active against many Gram-negative pathogens. As a result, they are the antibiotics of choice for severe infections and thus hold a firm place in human therapy, in spite of their high toxicity and the formation of resistant strains. They are also used in plant protection. Their sales in the world market are ca. 600 million US$. The most important compounds for human therapy are the gentamicins, the neomycins, tobramycin, kanamycin, and semi-synthetic products such as sisomicin and amikacin. Streptomycin is occasionally still used. Spectinomycin has been successfully used against penicillin-resistant *Neisseria gonorrhoe* infections. Kasugamycin is an important agricultural antibiotic for combating rice blight, and hygromycin is used in veterinary medicine.

Biosynthesis. Aminoglycoside antibiotics are mainly formed by prokaryotic microorganisms of the genera Streptomyces and Micromonospora. The multi-step biosynthesis (streptomycin: 24 steps) always starts from D-glucose and usually leads, on the scaffold of nucleotide sugars, to an amincyclitol unit which is glycosidically linked to other unusual sugars (amino sugars, C-branched sugars). The 33 proteins required for streptomycin biosynthesis are located on a single gene cluster of the *Streptomyces griseus* genome. They comprise 30–40 bp and have mostly been cloned.

Production. As with other industrially manufactured antibiotics, the production strains have undergone significant strain improvement by series of mutation, selection, and back-crossing. Using such procedures, the antibiotic yield of the wild-type strain, which is on the order of some mg L^{-1}, has been raised more than 1000 fold (streptomycin: > 10 g L^{-1} after 120 h fermentation). Industrial production is carried out in large bioreactors, using glucose, starch, or dextrin as the carbon source and soymeal as the nitrogen source. Aminoglycosides that tend to bind to the mycelium, such as gentamicin, are released by acidification to pH 2.0. After separation of the cell mass by filtration or centrifugation and concentration of the fermentation broth, the antibiotic is purified by several cycles of ion-exchange chromatography and crystallization.

Mechanism of action and resistance. Aminoglycoside antibiotics bind to the 30S subunit of the ribosome and interfere with translation, eventually resulting in the inhibition of protein biosynthesis. Some of them also bind specifically to type-I introns of RNA. A major disadvantage is that they lead quickly to the formation of resistant phenotypes, which can be coded on plasmids or on the chromosome. Resistant strains are able to block key hydroxyl groups via acetylation, phosphorylation, or adenylation, thus preventing the binding of the aminoglycoside to their ribosomes.

Semisynthetic aminoglycosides of high potency can be obtained by chemical derivatization, preferentially at amino groups. Examples are sisomicin, amikacin, and tobramycin. Experiments to achieve semisynthetic aminoglycosides through pathway engineering (combinatorial biosynthesis), however, have led to only minor success so far, although in many cases all genes of a pathway have been cloned and suitable expression cassettes have been prepared. It is assumed that many of the enzymes in such a pathway are subject to complex individual regulation, rendering pathway interchanges difficult.

Aminoglycoside antibiotics

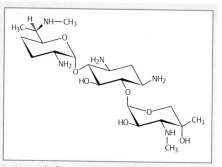

streptomycin M_R 581.58
$C_{21}H_{39}N_7O_{12}$ CAS   57-92-1

gentamycin C_1 M_R 477.60
$C_{21}H_{43}N_5O_7$ CAS   25876-10-2

antibiotic	production	sales 2001 (million US $)	applications
streptomycin	*Streptomyces griseus*		broad-band antibiotic *Mycobacterium tuberculosum*
gentamycin	*Micromonospora purpurea*		broad-band antibiotic
tobramycin	*S. tenebrarius*	~ 600	broad-band antibiotic
amikacin*	*S. kanamyceticus*		broad-band antibiotic
netilimicin*	*M. purpurea*		broad-band antibiotic
neomycins	*S. fradiae*		skin infections
kasugamycin	*S. kasugaensis*		rice blight
validamycin	*S. hygroscopicus*		rice agriculture
*chemically modified			

Biosynthesis of streptomycin

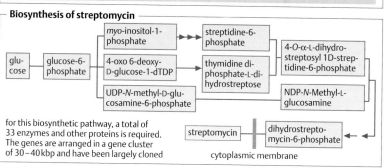

for this biosynthetic pathway, a total of 33 enzymes and other proteins is required. The genes are arranged in a gene cluster of 30–40 kbp and have been largely cloned

Manufacture

preculture	fermentation	removal of cells	purification	
inoculation and growth	up to 150 m³ bioreactor, fed-batch mode, 28–30 °C, 70–120 h, glucose or dextrins, soy meal, 1–3 g L⁻¹ NaCl, trace elements, 0.5–1 vvm O₂	for mycelium-bound antibiotic, acidify to pH 2	ion exchangers, crystallization and recrystallization	>10 g L⁻¹ after 120 h

Tetracyclines, Chinones, Chinolones, and other aromatic antibiotics

General. Due to their broad range of activity, the tetracyclines are an important class of antibiotics. They are used in medicine and agriculture. Chinolones are synthetic analogs of nalidixic acid; they are also active against a broad range of pathogens. They are prepared by chemical synthesis and constitute one of the major classes of medical antibiotics after the lactames (market value > 4 billion US$, 2000).

Tetracyclines. First described in 1945 as metabolites of *Streptomyces aureofaciens,* tetracyclines have become very important antibiotics due to their low toxicity and broad activity spectrum (market value 2000: > 600 million US$) They are active against Gram-positive and Gram-negative bacteria, Rickettsia, Mycoplasma, Leptospira, Spirochaeta, and some larger viruses. Unfortunately, tetracyclines are widely used in some countries as feed additives for chicken and pig fattening, which has led to the development of resistant strains. Mechanisms of resistance that have been observed often are: reduced penetration of the antibiotic through the outer membrane of Gram-negative cells (altered porines) and plasmid-coded synthesis of so-called tet-proteins, which support rapid export of tetracyclines from the bacterial cell. Tetracyclines inhibit protein biosynthesis by binding to the 70S ribosome. They are formed exclusively by Streptomyces strains; the primary product is usually oxytetracycline. Biosynthesis from glucose requires over 70 individual steps and passes through polyketide intermediates. For industrial manufacturing, optimized production strains of several Streptomyces species are cultivated in large bioreactors. Yields may reach > 25 g L^{-1} if the O_2 supply is optimized and the phosphate content of the medium is well controlled. For isolation, the cell mass is separated, and the broth is extracted using n-butyl acetate. Ion-exchange chromatography is used for purification.

Anthracyclines. Anthracycline glycosides such as doxorubicin (adriamycin) inhibit DNA replication through intercalation and inhibition of topoisomerases. They are used in chemotherapy for tumor treatments. They are produced by fermentation.

Chinolones. Nalidixic acid is a side-product of the chemical synthesis of chloroquinone, an anti-malaria agent. Its bactericidal effect was discovered in 1962 and shown in 1977 to be due to the inhibition of bacterial topoisomerase (gyrase A subunit). Since the structure and function of bacterial and human topoisomerases differ quite significantly, chinolones exhibit low toxicity to humans. On the other hand, they show a wide activity spectrum against Gram-positive and -negative bacteria, Mycobacteria, Clamydia, and anaerobic microorganisms. Microbial resistance develops only slowly and is not plasmid-coded. If developed, it is due to modifications of the gyrase subunit or to reduced membrane permeation. Among the > 5000 chinolone derivatives, which are exclusively produced by chemical synthesis, some have found wide application. Ciprofloxacin (Ciprobay®) is a chinolone antibiotic active against *Bacillus anthracis*.

Chloramphenicol was isolated in 1950 from cultures of *Streptomyces venezuelae* but today is completely produced by chemical synthesis. Its biosynthetic pathway branches from the biosynthesis of aromatic amino acids at the level of chorismic acid. It acts through binding to the 50S subunit of 70S ribosomes, blocking peptidyl transferase. Chloramphenicol was the first broad-spectrum antibiotic. It is effective against a wide range of Gram-positive and -negative bacteria, Actinomycetes, Rickettsia, and large viruses. Since it damages the bone marrow, it is used as a reserve antibiotic, in particular for the treatment of typhus, shigellosis, and Rickettsia-based infections.

Griseofulvin, a benzophenone derivative, is a fungistatic antibiotic that blocks mitosis through binding to the spindle apparatus of fungi, which is detected by the formation of short, curly hyphae. It is produced by fermentation. Applications include the treatment of dermatomycosis and agricultural use as a leaf fungicide against blights.

Chinoid and aromatic antibiotics

antibiotic	production	application
(chloro-)tetracycline oxytetracycline	*Streptomyces aureofaciens* *Streptomyces rimosus*	broad-spectrum antibiotics for human therapy and animal feed
anthracycline	*S. peucetius*	cancer therapy
chinolones	chemical synthesis	broad-spectrum antibiotics for human therapy and animal feed
chloramphenicol	chemical synthesis	broad-spectrum antibiotic
griseofulvin	*Penicillium griseofulvium*	fungal infections on skin or leaves (blight)

tetracyclines

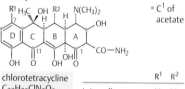

- C^1 of acetate

chlorotetracycline
$C_{22}H_{23}ClN_2O_8$
M_R 478,89
CAS 57-62-5

	R^1	R^2
tetracyline	H	H
chlorotetracycline	Cl	H
oxytetracycline	H	OH

doxorubicin

- C^1 of acetate

$C_{27}H_{29}NO_{11}$
M_R 543,53
m.p. 204–205 °C
CAS 23214-92-8

nalidixic acid and ciprofloxacin, two chinolone antibiotics

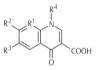

	R^1	R^2	R^3	R^4
nalidixic acid	N	CH_3	H	CH_3
ciprofloxacin	CH	$C_4H_9N_2$	F	C_3H_5

chloramphenicol

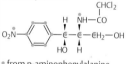

- from p-aminophenylalanine

$C_{11}H_{12}Cl_2N_2O_5$ CAS 56-75-7
M_R 323,13

griseofulvin

- C^1 of acetate

$C_{17}H_{17}ClO_6$ CAS 126-07-8
M_R 352,77

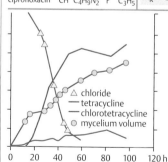

△ chloride
— tetracycline
— chlorotetracycline
○ mycelium volume

oxytetracycline biosynthesis cluster
~ 30 kbp
polyketide synthesis
resistance genes

ori
Streptomyces rimosus
linear chromosome, ~ 8 Mbp,
GC-content ~ 71 %

unstable telomer region, ~ 550 kbp

Fermentation and isolation of chlorotetracycline

preculture	bioreactor	isolation	
high-performance strains *Streptomyces aureofaciens*	~ 150 m³, saccharose, corn steep liquor, salts, 1 vvm air; 28 °C, pH 5,8–6,0; 60–65 h	separation of mycelium through filter press or separator, multi-stage extraction with n-butyl acetate, purification by ion-exchange chromatography	~ 10 g/L after 140 h

Macrolide antibiotics

General. This important group of antibiotics includes the macrolide, the polyene, the macrotetrolide, and the ansamycine antibiotics. They all share a macrocyclic lactone or lactam ring that is formed from a long-chain polyhydroxy fatty acid and a terminal hydroxyl or amino group. The ring can be glycosylated with unusual sugars (macrolides), it may contain conjugated double bonds (polyenes) or an aromatic chromophore (ansamycines), or it may be built as a polylactone (macrotetrolides). Most polyketides are isolated from Streptomyces strains. They are used in human therapy, and also for food protection and in animal feeds. The market value of macrolides used in medicine is in the range of 4.3 billion US$ (2000) or about ⅕ of the total antibiotic production.

Macrolide antibiotics are lipophilic, often basic, compounds. The key structural element is a 10–60-membered macrocyclic lactone ring, which is formed by a multi-enzyme complex resembling fatty acid synthase, through condensation of an acyl-CoA starter unit with malonyl-CoA or its methyl and ethyl homologs. Polyketides are the hypothetical intermediates of this reaction, and they are further modified by unusual sugars containing, for example, amino groups, C-methyl branches, and deoxy groups. Macrolide antibiotics exhibit low toxicity and thus are often used in pediatric medicine. They preferentially inhibit Gram-positive microorganisms, binding to their 50S ribosomal subunits and interrupting translocation of the growing peptide chain. Formation of resistant pathogens is frequently observed and mostly due to methylation of the ribosomal 23S RNA. Erythromycin and spiramycin are preferred antibiotics against bacterial infections of the respiratory tract. Tylosine, a related macrolide, was a valuable feed antibiotic for pig fattening, due to its activity against mycoplasms. Since this application led to the development of cross-resistant strains, tylosine has been banned in most Northern European countries of the European Union.

Polyene antibiotics are formed preferentially by Streptomyces strains. They are constituted from 26- to 38-membered macrocyclic lactone rings with 3–7 conjugated double bonds and may contain additional building blocks such as amino sugars in glycosidic linkage. Several polyene antibiotics are used as fungistatic agents, e. g., amphotericin B or nystatin for medical therapy of *Candida albicans* and pimaricin as a food preservative in cheeses. Polyene antibiotics function by complexing with fungal membrane sterols such as ergosterol. As a result, they are not effective against bacteria. Since they are nephro- and hepatotoxic, their use is limited to severe infections. They are too labile to be used as fungicides in agriculture.

Ansamycins are macrolactam antibiotics containing an aromatic chromophore. Their most important representative is rifamycin, a product of *Nocardia mediterranei*. It is highly active against Gram-positive bacteria and Mycobacteria. Its biosynthesis involves the assembly of a polyketide by addition of acetate and propionate chain extension units to a unique starter unit, 3-amino-5-hydroxybenzoic acid (AHBA) which is formed by a pathway similar to shikimic acid formation. Rifampicin, a semi-synthetic derivative of rifamycin, is presently the most important agent for the treatment of tuberculosis (pathogen: *Mycobacterium tuberculosum*) and is also applied in the treatment of leprosy or Hansen's disease (*Mycobacterium leprae*). It inhibits the bacterial DNA-dependent RNA polymerase by binding to its β-subunit. Since rifampicin does not bind to human RNA polymerase, it is nontoxic to humans. In resistant strains, the RNA polymerase is modified through mutations.

Fermentation and purification. The commercially important macrolide antibiotics are produced industrially by high-performance strains in large bioreactors. As an example, yields of erythromycin by *Saccharopolyspora erythraea* (former name: *Streptomyces erythreus*) are in the range of $7 \, g \, L^{-1}$ after 72 h. The extracellular products are isolated, after removing the cell mass by filtration or separators, via solvent extraction and further purified by chromatography.

Polyketide antibiotics

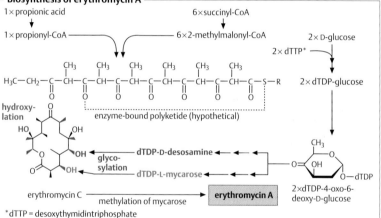

b rifampicin: R = CH=N—N⟋ ⟍N—CH₃

c rifamycin: R = H

a erythromycin

d amphotericin B

a $C_{37}H_{67}NO_{13}$, $M_R = 733.94$, CAS 114-07-8

b $C_{43}H_{58}N_4O_{12}$, $M_R = 822.95$, CAS 13292-46-1

c $C_{37}H_{47}NO_{12}$, $M_R = 697.78$, CAS 6998-60-3

d $C_{47}H_{73}NO_{17}$, $M_R = 924.09$, CAS 1397-89-31

Biosynthesis of erythromycin A

1 × propionic acid
↓
1 × propionyl-CoA ⟶ 6 × 2-methylmalonyl-CoA

6 × succinyl-CoA
↓

2 × D-glucose
2 × dTTP*
↓
2 × dTDP-glucose

$$H_3C-CH_2-\underset{\underset{O}{\|}}{C}-\underset{\underset{CH_3}{|}}{CH}-\underset{\underset{O}{\|}}{C}-\underset{\underset{CH_3}{|}}{CH}-\underset{\underset{O}{\|}}{C}-\underset{\underset{CH_3}{|}}{CH}-\underset{\underset{O}{\|}}{C}-\underset{\underset{CH_3}{|}}{CH}-\underset{\underset{O}{\|}}{C}-\underset{\underset{CH_3}{|}}{CH}-\underset{\underset{O}{\|}}{C}-\underset{\underset{CH_3}{|}}{CH}-\underset{\underset{O}{\|}}{C}-S-R$$

enzyme-bound polyketide (hypothetical)

hydroxylation

glycosylation ⟵ dTDP-D-desosamine ⟵ ⟵ ⟵

dTDP-L-mycarose ⟵ ⟵ ⟵

2×dTDP-4-oxo-6-deoxy-D-glucose

erythromycin C ⟶ methylation of mycarose ⟶ **erythromycin A**

*dTTP = desoxythymidintriphosphate

Manufacture

erythromycin	rifamycin/rifampicin
fermenter	**fermenter**
high-performance strains of *Saccharopolyspora erythraea*, bioreactor > 120 m³ in fed-batch mode, glucose, soy meal, trace elements, 0.2–0.5% propanol; 33°C, 70–120h	high-performance strains of *Nocardia mediterranei*, bioreactor > 120 m³ fed-batch mode; glucose, soy meal, trace elements, 0.2–0.5% propanol; 33°C, 70–120 h

removal of cells by filtration or separators

purification

counter current extraction with acetic acid butyrate, chromatography, recrystallization

~7 g L⁻¹ erythromycin after 72h	~7 g L⁻¹ rifamycin after 72h

4 step Mannich synthesis to rifampicin

51

Antibiotics

New pathways to antibiotics

General. Although antibiotic therapy has been a success story, and new antibiotics are isolated year after year, infections that were believed extinct do reappear and prove more difficult to treat. The re-emergence of tuberculosis is just one of these problems, and even treating infections from Staphylococci or Streptococci today may be difficult. At the root of this problem lies bacterial antibiotic resistance against one or several antibiotics ("cross-resistance"), which constitutes a major medical and scientific challenge. Antibiotic resistance is often coded on plasmids or transposons and can be horizontally transferred among different microorganisms, e.g., by phage infections. Another critical problem is the relatively small number of antibiotics that are effective against pathogenic fungi and yeasts that are nontoxic to man. Since the metabolisms of eukaryotic organisms are closely related, antibiotics that affect fungi are usually toxic against man as well. As a consequence, there are good reasons to look for new leads in the discovery of antibiotics.

New screening procedures. If standard screening procedures, e.g., bioassays, are used for the discovery of new antibiotics, nine out of ten hits turn out to provide structures that have already been described. As a result, a range of unconventional procedures have been developed in order to discover new lead structures. The most important examples are: 1) precursor-directed biosynthesis, where the addition of synthetic precursors to the fermentation leads to biosynthetic antibiotics, 2) screening for antibiotics in hitherto neglected genera such as Myxobacteria, rare Actinomycetales, lichens, or sponges, 3) modifications of the screening procedure, relying on novel assays, 4) searching for active intermediates in antibiotic biosynthesis, 5) searching for new intermediates by reverse genetics, 6) recombination of related gene clusters through cell fusion, 7) recombination of related gene clusters by combination of related gene clusters in vitro (gene shuffling), and 8) searching for pathogen-specific target using genomic information and the tool of bioinformatics.

Reverse genetics. After enough amino acid or nucleotide sequences are known for enzymes that catalyze typical steps in antibiotic biosynthesis, consensus sequences of these enzymes can be used to screen for equivalent activities in the genomic DNA of related strains. This procedure was successfully used to screen for unusual polyketide structures among macrolide antibiotics.

Combinatorial biosynthesis. Modular biosynthetic pathways based on the repetitive incorporation of C-2, C-3, and C-units into a growing polyketide chain are especially suited for combinatorial genetic experiments. The genes required for macrolide synthesis are usually organized as a few modules along a single operon that can be transferred into plasmids, leading to attractive possibilities using gene shuffling. With this procedure, new polyketides having antimicrobial activity were isolated which had never before been isolated from nature.

New targets from genome analysis. The number of pathogenic microorganisms whose genome has been fully sequenced is growing year by year. Thus, the genomes of *Hemophilus influenzae* (1.83 Mbp; chronic bronchitis), *Helicobacter pylori* (1.67 Mbp; ulcers), *Borrelia burgdorferi* (0.91 Mbp; Lyme disease), *Mycobacterium tuberculosum* (4.41 Mbp; tuberculosis), *Treponema pallidum* (1.14 Mbp; syphilis) and *Chlamydonomas trachomatis* (1.04 Mbp; eye infections) have all been elucidated. It is assumed that metabolic or signal transduction pathways specific for the pathogen can be deciphered from genomic information, leading to pathogen-specific targets which can be used in drug development. As an example, experimenters are attempting to identify lead structures directed against Ni^{2+} metabolism which is a particular feature of *Helicobacter pylori*, since this microorganism buffers highly acidic gastric juice via a nickel-dependant urease.

52

type of antibiotic	producer	precursor fed to the medium
penicillins	*Penicillium chrysogenum*	many aliphatic, alicyclic and aromatic carbonic acids
cephalosporins	*Acremonium chrysogenum*	S-carboxymethyl-L-cysteine
bleomycins	*Streptomyces* sp.	modified amines
bacitracins	*Bacillus licheniformis*	D-allo amino acids
streptomycin	*Streptomyces griseus*	2-deoxy streptidine

Combinatorial biosynthesis of macrolide antibiotics

a formation of the multi-enzyme complex of deoxyerythronolide-B-synthase (DEBS) 1–3

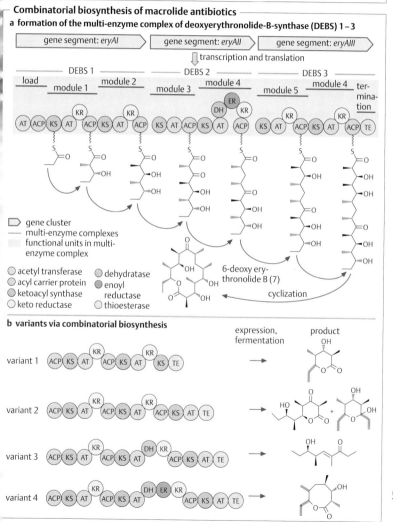

6-deoxy erythronolide B (7)

b variants via combinatorial biosynthesis

expression, fermentation — product

variant 1

variant 2

variant 3

variant 4

53

Vitamins

General. Vitamins are used as additives in medical preparations and animal feed. Their market volume is ca. 3 billion US$ (world). Most vitamins are manufactured by chemical synthesis or by extraction of plant material. Biotechnological processes are used in the syntheses of vitamins B_2, B_{12}, and C.

Vitamin B_2 (riboflavin). The riboflavin derivatives FAD and FMN are cofactors of many redox enzymes. Free riboflavin occurs only in milk. Animals fed with a riboflavin-deficient diet develop dermatitis, growth deficiencies, and damage to the eyes. Biosynthesis proceeds on a complex pathway starting from guanosine triphosphate. Riboflavin can be produced by chemical synthesis, fermentation, or chemoenzymatic synthesis. Approximately 70% of the world production (ca. 4 000 t) is produced by fermentation. In a process using high-performance mutants of the fungus *Ashbya gossypii* in a batch fermenter (carbon source: plant oil, nitrogen source: soymeal) yields reach ca. 15 g L^{-1} in 72 h. Recovery includes cell separation and chromatographic procedures. In a chemo-enzymatic process which is not yet used commercially, the alloxazine ring is synthesized, D-ribose is prepared from D-glucose with mutants of *Bacillus pumilus*, and both components are chemically linked.

Vitamin B_{12} (cyanocobalamin). Vitamin B_{12} is required as an enzyme cofactor in several methylation and isomerization reactions. Vitamin B_{12} deficiencies lead to anemic conditions, in particular to pernicious anemia. Although it is used in medical preparations (mainly for liver protection) and in diets, half the world production of ca. 20 t is added to animal feed. Biosynthesis starts from succinoyl CoA and glycine, which condense to 5-amino 4-oxo valeric acid (δ-amino levulinic acid). In an additional 30 steps, which may vary among different organisms, 5′-deoxyadenosyl cobalamin is formed. For the manufacture of vitamin B_{12}, a fermentation process based on *Propionibacterium shermanii* or *Pseudomonas denitrificans* is used.

Using a medium containing molasses or dextrin as a carbon source, ammonia and cobalt salts, and using 5,6 dimethyl benzimidazol as a precursor compound, yields may reach 150 mg L^{-1} after 120 h. In *Propionibacterium shermanii*, all genes of the vitamin B_{12} biosynthetic pathway have been cloned, and modern production strains have already been optimized through metabolic engineering.

Vitamin C (L-ascorbic acid). Ascorbic acid is often called a "physiological reducing agent" and participates in the reductive elimination of active oxygen species, e. g., oxygen radicals. In addition, it participates as a cofactor in several enzyme reactions. Vitamin C deficiency results in damage to the skin and the blood vessels (scurvy). It is sold as a vitamin preparation and is also added to numerous foods and drinks both as a vitamin and as an antioxidant. Annual production reaches ca. 80 000 t (world). The manufacturing process is based on a chemo-enzymatic synthesis starting from D-glucose. The most important route is the Reichstein-Gruessner synthesis or modifications thereof, which comprise the chemical reduction of glucose to sorbitol and subterminal oxidation of the latter with *Acetobacter suboxydans* to L-sorbose. The overall yield of this process is ca. 66 %. Oxidation is done continuously or batchwise, requiring immobilized cells and strong aeration. Yields are quantitative after 24 h. In the competing *Sonoyama* process, D-glucose is oxidized by *Erwinia sp.* to 2,5-dioxo-D-gluconic acid (94 % yield after 26 h), which is reduced by *Corynebacterium sp.* to 2-oxo-D-gluonic acid (92 % yield after 66 h), and which rearranges to L-ascorbic acid upon addition of acid. The two enzymes responsible for these steps have been cloned and functionally coexpressed in one organism (*Erwinia herbicola*). The Genentech process attempts to produce L-ascorbic acid from D-glucose in a single fermentation step using this recombinant organism. This is followed by acid-catalyzed rearrangement. Unfortunately, the glucose tolerance of the production strain is yet too low, resulting in unsatisfactory space time yields.

Vitamins

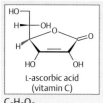

L-ascorbic acid
(vitamin C)

$C_6H_8O_6$
M_R 176.13
CAS 50-81-7

H_3C — 6, 5 — O
NH
H_3C — 8, 9 — N — 1 — 2, O
CH_2
$H—C—OH$
$H—C—OH$
$H—C—OH$
$CH_2—OH$
riboflavin (vitamin B$_2$)

$C_{17}H_{20}N_4O_6$
M_R 376.36
CAS 83-88-5

Manufacturing processes and market volume

vitamin		quantity (t), ~ 2000	manufacturing process	applications
A	β-carotene	30	chemical synthesis	animal nutrition, colorant
B$_1$	thiamine	3000	chemical synthesis	health
B$_2$	riboflavin	4000	fermentation (30%), chemoenzymatic synthesis (50%), chemical synthesis (20%)	health, animal nutrition
B$_6$	pyridoxine	3000	chemical synthesis	health, animal nutrition
B$_{12}$	cyanocobalamine	20	fermentation	health, animal nutrition
C	ascorbic acid	80000	chemo-fermentative synthesis	health, food additive, animal nutrition
D$_2$	calciferol	40000	photochemical synthesis from ergosterol	health
E	α-tocopherol		extraction from plant oil, algae	health

Biotechnological routes to riboflavin

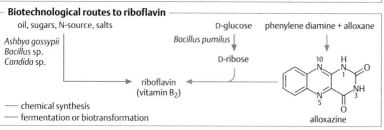

oil, sugars, N-source, salts

Ashbya gossypii
Bacillus sp.
Candida sp.

D-glucose   phenylene diamine + alloxane

Bacillus pumilus

D-ribose

riboflavin (vitamin B$_2$)

—— chemical synthesis
—— fermentation or biotransformation

alloxazine

Synthesis of L-ascorbic acid

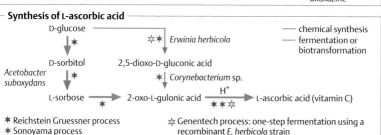

D-glucose

—— chemical synthesis
—— fermentation or biotransformation

☆✶✶ *Erwinia herbicola*

D-sorbitol   2,5-dioxo-D-gluconic acid

Acetobacter suboxydans

✶| *Corynebacterium* sp.

L-sorbose → 2-oxo-L-gulonic acid →$^{H^+}$→ L-ascorbic acid (vitamin C)

✶ Reichstein Gruessner process
✶ Sonoyama process

☆ Genentech process: one-step fermentation using a recombinant *E. herbicola* strain

Nucleosides and nucleotides

General. About 50 years ago, 5′-nucleotides were discovered in Japan as taste-enhancing components in dried mushrooms and dried fish. If even very minor quantities of these compounds (0.0005 – 0.001 %) are added to soups and sauces, their taste is significantly enhanced, and undesired off-flavors, e.g., the metallic taste of canned foods, are suppressed. Addition of Na-L-glutamate enhances this effect. Inosine-5′-monophosphate (IMP), guanosine-5′-monophosphate (GMP) and xanthosine-5′-monophosphate (XMP) exhibit the strongest activity, and the 2′- and 3′-isomers, pyrimidine nucleotides, and nucleosides have no effect. Adenosine-5′-monophosphate, deoxy IMP, and deoxy GMP are less active than 5′-IMP or 5′-GMP. Since 1961, 5′-IMP and 5′-GMP have been produced on an industrial scale, up to several $1\,000\ t\ y^{-1}$. The producers are predominantly located in Asia.

Manufacturing process. Four processes are used in Japan: in 1) enzymatic hydrolysis of RNA, 2) combination of fermentative production of inosine or guanosine, and their chemical phosphorylation, 3) direct production of 5′-IMP by fermentation, and 4) direct production of 5′-XMP by fermentation, and its enzymatic conversion to 5′-GMP.

Enzymatic hydrolysis of RNA. Yeasts show the most favorable RNA/DNA ratio and thus are preferentially used for the production of RNA. If *Candida utilis* is grown on molasses or pulp, using media with a low C/N-ratio, cells contain 10 – 15 % of RNA (dry weight). This amount can be further increased by adding Zn^{2+} and phosphate. The aerobic fermentation is usually carried out continuously on a large scale, using airlift bioreactors. After removal of the cell mass, the RNA is extracted with a hot, alkaline NaCl solution (5 – 20 % NaCl, 100 °C, 8 h), and precipitated with HCl or ethanol. For enzymatic hydrolysis of the RNA, nuclease P_1 preparations from *Penicillium citrinum* are used which do not contain unspecific 5′-nucleotidases or phosphatases. The final isolation of 5′-IMP and 5′-GMP proceeds via adsorption onto activated charcoal, fractionation by ion-exchange chromatography, and crystallization.

Fermentation of 5′-IMP. The classical protocol is based on the formation of inosine by fermentation with *Bacillus* sp. and other Gram-positive microorganisms. Inosine is secreted into the medium and can be precipitated at pH 11. Fermentation of adenine-auxotrophic mutants whose transport properties have been improved by genetic engineering may result in yields of 35 g L^{-1}. Recrystallized inosine is then transformed to 5′-IMP with PCl_3 in trialkylphosphate solvents. The other process is direct formation of 5′-IMP as an extracellular product, using block mutants of *Brevibacterium* (*Corynebacterium*) *ammoniagenes*. The high-performance mutants used for production are no longer repressed by the presence of other nucleosides; they do not degrade 5′-IMP and are not sensitive to the Mn^{2+} content of the medium; and their plasma membrane shows good capacity to excrete 5′-IMP. If several copies of the key enzyme PRPP amidotransferase are cloned into such an organism, yields can enhanced further and may reach 30 g L^{-1}.

Production of 5′-GMP. The preferred processes are 1) the production of guanosine by fermentation, followed by chemical phosphorylation, and 2) production of XMP by fermentation and its enzymatic conversion to GMP. GMP formation via 5-aminoimidazole carboxamide-1-riboside (AICA-riboside), obtained by purine auxotrophic strains of *Bacillus megaterium*, and followed by chemical transformation into 5′-GMP, has also been reported, but is not industrially used at present.

Other nucleotides. Nucleotides such as ATP, cAMP, NAD^+/NADH, $NADP^+$/NADPH, FAD, coenzyme A, and nucleotide sugars are important biochemicals, which have also been used in special biotransformation processes. They are prepared either by fermentation using block mutants or by enzymatic processes starting from chemical precursors. As an example, both NAD^+ and coenzyme A can be prepared by fermentation using block mutants of *Brevibacterium* (*Corynebacterium*) *ammoniagenes* in yields of ca. 2 g L^{-1}.

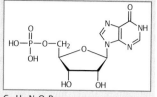

Inosine 5'-monophosphate

$C_{10}H_{13}N_4O_8P$
M_R 348.21
CAS 131-99-7

Guanosine 5'-monophosphate

R = NH_2: guanosine 5'-monophosphate (5'-GMP)
$C_{10}H_{14}N_5O_8P$ M_R 363.22 CAS 85-32-5
R = OH : xanthosine 5'-monophosphate (5'-XMP)

Manufacture and market volume

nucleoside	market volume (t)	manufacturing process	applications
5'-IMP 5'-GMP	2000 1000	enzymatic hydrolysis of yeast RNA fermentation of inosine/guanosine and chemical phosphorylation or direct fermentation of 5'-IMP	flavor enhancer flavor enhancer
inosine	25	fermentation	medical therapy
orotic acid	20	fermentation	liver diseases
adenine, adenosine, ATP	22	fermentation, enzymatic synthesis	medical therapy

DNA-/RNA-content of microorganisms

	bacteria	yeast	fungi	
DNA*	0.37–4.5	0.03–0.52	0.15–3.3	* % of dry cell mass
RNA*	5–25	2.5–15**	0.7–28	** of which 75–80 % is rRNA

Synthetic pathways

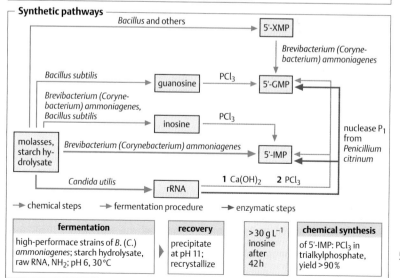

→ chemical steps → fermentation procedure → enzymatic steps

fermentation	recovery		chemical synthesis
high-performace strains of B. (C.) ammoniagenes; starch hydrolysate, raw RNA, NH_2; pH 6, 30 °C	precipitate at pH 11; recrystallize	>30 g L^{-1} inosine after 42 h	of 5'-IMP: PCl_3 in trialkylphosphate, yield >90 %

57

Biosurfactants and biocosmetics

General. Some microorganisms, if exposed to alkanes, plant oils, or even sugar substrates, form surface-active agents which are often termed "biosurfactants". Compared with synthetic surfactants, which are produced in amounts of several million tons, biosurfactants are much more expensive and thus are used only in some limited specialty areas. Due to their good biodegradability, they are being studied for as an environmentally "friendly" means of decontaminating oil-polluted water, soil, and tidal shallows. A few cosmetic cremes also contain biosurfactants and are sometimes called "biocosmetics". This marketing term is not well defined and often includes various kinds of cosmetics to which natural products have been added. Several examples of materials produced by biotechnology and included in biocosmetics are shikonin, a red dye of plant origin used in lipsticks, and hyaluronic acid, a polysaccharide with high water retention which can be produced by microorganisms.

Biosurfactants are formed by several bacteria and yeasts when grown on alkanes or plant oils. Well-studied biosurfactants include bacterial rhamnose- and trehalose-lipids, the lipopeptide surfactin, and the heteropolysaccharide emulsan. Sophorose lipids are a well known example of a yeast biosurfactant. They are formed by *Torulopsis bombicola* in yields of $> 400 \, g \, L^{-1}$ if the organism is grown on triglycerides. The mixture of hydroxy acid glycolipids and their lactones can be separated from the fermentation broth by flotation or solvent extraction. The purified products tend to form micelles, and their CMC (critical micelle concentration, a value indicating surfactant activity) is in a range typical of synthetic nonionic surfactants. Sophorose lipid is added in small quantities to a skin-protecting cosmetic creme sold in Japan. Rhamnolipids, produced by Pseudomonas strains, can be prepared in yields of ca. $100 \, g \, L^{-1}$ and are being tested as an additive to household detergents. Experiments on the performance of biosurfac-

tants in tertiary oil recovery (producing the sugar lipid in-place at the drilling hole) and for cleaning up oil pollution from tidal shallows have led to good results from a technical but not an economical point of view. Other organisms studied are the fungus *Ustilago maydis* which produces cellobiose lipids, and *Rhodococus erythropolis*, which can secrete trehalose tetraesters. Depending on the type of hydrophobic substrate, the chain length of the glycolipids can be modulated in some cases. Emulsan is a heteropolysaccharide (lipopolysaccharide) synthesized by *Acinetobacter calcoaceticus* in the presence of triglycerides. It can be produced by fermentation and isolated from the culture broth by solvent extraction. If added to a suspension of oil in water, it acts as an oil-in-water emulsifier, thus significantly reducing the viscosity of the oil. It can be used in small quantities for improving the flow of oil in pipelines and for cleaning oil trucks and barges. *Bacillus subtilis*, if grown on hydrophobic substrates, forms the acylated heptapeptide surfactin, which exhibits high surfactant activity (a high CMC) but also is hematotoxic to mammals and aquatic organisms. Yields in an optimized fermentation procedure may be as high as $110 \, mg \, L^{-1}$ but are orders of magnitude lower than yields of sophorolipid.

Shikonin is a naphtochinone derivative produced by flowers of *Lithosporum erythreum*, a borage plant. Shikonin possesses wound-healing and antitumor activity and is used as the pigment component of a lipstick series sold in Japan. Industrial production is achieved with plant tissue cultures, and the color of the dye can be modulated by adding transition metals.

Hyaluronic acid, a glycosyl amino glucan of $M_R \, 1-4 \times 10^6 \, Da$, is a component of connective tissue and also occurs in the synovial fluid. It is usually isolated from rooster combs or umbilical cords, but can also be produced, albeit with a somewhat lower M_R, by fermentation with *Streptococcus equi* or *S. zooepidemicus* in yields of ca. $6 \, g \, L^{-1}$ in as little as $20 \, h$. Due to its high capacity for water binding and retention, it is used both in cosmetics and in surgical implants.

Biosurfactants

	organism	structural elements
sophorose lipid	*Torulopsis bombicola*	sophorose, hydroxy fatty acids
cellobiose lipids	*Ustilago maydis*	cellobiose, fatty acids
rhamnose lipids	*Pseudomonas aeruginosa*	rhamnose, β-hydroxydecanoic acid
trehalose lipids	Corynebacteria, Arthrobacter	trehalose, long-chain wax esters
corynomycolates	Corynebacteria, Arthrobacter	mycolic acid esters of mono-, di-, and trisaccharides
emulsan	*Acinetobacter calcoaceticus*	polyanionic heteropolysaccharide, $M_R \sim 10^6$ Da
surfactin	*Bacillus subtilis*	acylated heptapeptide

sophorose lipid

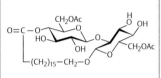

trehalose lipid

$R = CO-CH-CHOH-(CH_2)_m-CH_3$

$m + n = 27 - 31$

organism	fermentation	recovery	
Arthrobacter sp.	C-, N-, P-source, brine water, induction with oil	flotation, chromatography	< 100 g L^{-1} broth

Biosurfactants and marine pollution

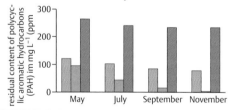

removal of crude oil from tidal shallows (test area 2 m²)

- 10 × contaminated by 1 L crude oil
- 10 × contaminated by 1 L crude oil, 1 g L^{-1} trehalose dicorynomycolate added each time
- 10 × contaminated by 1 L crude oil, 100 mL of a nonionic surfactant added each time

Biocosmetics

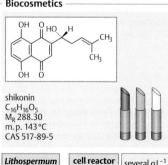

shikonin
$C_{16}H_{16}O_5$
M_R 288.30
m. p. 143 °C
CAS 517-89-5

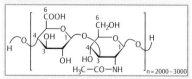

hyaluronic acid

Lithospermum erythreum	cell reactor	several g L^{-1}		*Streptococcus equi*	fermentation	6 g L^{-1}
immobilized	up to several 10 L size	culture medium				

Microbial polysaccharides

General. Several polysaccharides such as starch, cellulose, gum arabic, guar, pectins, alginates and agar have found applications in food manufacture for thickening or stabilization. Xanthan is also used in oil exploration. Although most polysaccharides are isolated from plants or marine algae ("renewable resources"), some are manufactured as extracellular products of microorganisms by fermentation. Because the economics of such fermentation processes compare unfavorably to the preparation of plant or marine polysaccharides, their use is limited to specialty areas. The most important microbial polysaccharides are xanthan and dextran; hyaluronic acid was discussed above.

Xanthan is an acid heteropolysaccharide, formed by the plant pathogenic prokaryote *Xanthomonas campestris*. It is composed of 5 hexose residues as repeating units, and its molecular weight is in the range of $1.5-2 \times 10^6$ Da. The number of pyruvate residues may vary, but this has little influence on the viscosity of the biopolymer. Xanthan viscosity is insensitive to the presence of electrolytes, and the biopolymer exhibits pseudoplastic properties (the viscosity of a xanthan solution decreases reversibly on increasing shear), and thus is very convenient to use in industrial processes. Its most important application is as a thickener in processed food (e.g. salad dressings). The use of xanthan in tertiary oil recovery, for polymer flooding of salt-containing soil formations, was widely studied but, due to the drop in oil prices and improved recovery technology, has not become commercially successful. Xanthan is used in drilling "muds" for oil exploration and development. Fermentation is carried out in a batch mode (C-source: glucose or sucrose; N-source: peptone, ammonium nitrate or urea). The lac operon of *E. coli* has been expressed in *X. campestris*, resulting in strains that can form xanthan from whey, a waste material. This process, however, is not yet commercially competitive. During fermentation, the formation of xanthan is indicated by a steep rise in viscosity to 10 000 cP. A special stirrer configuration is crucial if sufficient O_2 transfer in this viscous medium, as required for high yields, is to be achieved. Isolation of the product from the broth is usually done by precipitation with 2-propanol. At present, ca. $30\,000$ t y^{-1} of xanthan are produced.

Dextran. Dextrans are used as plasma expanders in medicine and, due to their well defined pore size both in the native form and after chemical derivatization, for the purification of proteins. Dextrans are glucans, built from D-glucose through mostly α-1,6-glycosidic linkages. Their molecular mass reaches 5×10^7 Da. They are produced as extracellular products by several microorganisms, e.g., by *Streptococcus mutans* in the oral cavity of humans, leading to dental plaque. For the technical manufacture of dextran, *Leuconostoc mesenteroides* is usually employed. It produces ca. 500 g L^{-1} dextran in 24 h from saccharose, which can be precipitated from the broth by addition of ethanol. If partially hydrolyzed by the addition of acid, dextrans of different molar masses can be isolated by fractionated precipitation. Dextrans of M_R 75 000 Da are used as plasma expanders, of M_R 40 000 Da as antithrombolytic agents in surgery.

Other microbial polysaccharides. *Pseudomonas aeruginosa* and *Azotobacter vinelandii* form microbial alginates whose compositions resemble the algal products. Some fungi form scleroglucan, a polysaccharide from β-1,3-linked glucose residues containing a lateral β-1,6-linked glucose unit. Similar to gellan from *Sphingomonas elodea*, it exhibits pseudoplastic properties similar to xanthan and is used, in small quantities, as a gelling agent in food products. Pullulan, an α-1,4-linked glucan containing ca. 10% of α-1,6-glycosidic bonds, is synthesized by various bacteria. It can be processed to films which are impermeable to O_2 and thus have been investigated for the protection of O_2-labile foods or other materials. However, the high production cost of all these polymers has so far prevented their technical use on a large scale.

Xanthan

degree of polymerization ~10 000

$M^+ = Na, K, \frac{1}{2}Ca$

CAS 11138-66-2

pseudoplastic properties

1.00 %
0.50 %
0.25 %
0.10 %
0.05 %

apparent viscosity [mPa s]

log shear rate [s^{-1}]

0,05 – 1 % dispersion of xanthan in water, 25 °C

Fermentation and recovery

preculture	bioreactor	recovery	
*Xanthomonas campestris**, glucose or saccharose, N-source	>120 m³, 40–80 h, 28 °C, pH 7.0, special stirrer equipment due to high viscosity	pasteurization, precipitation with ethanol or 2-propanol	~30 g L^{-1} in 60 h

*X. campestris is a plant pathogen and must be cultivated under risk group 2 safety conditions

Dextran

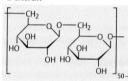

1,6-α-D-glucose, degree of polymerization ~28 000
1,2-, 1,3- and 1,4-linkages occur as well

CAS 9004-54-0

biosynthesis

$(1,6-\alpha-D-glucosyl)_n$ + saccharose

↓ dextran sucrase

$(1,6-\alpha-D-glucosyl)_{n+1}$ + fructose

Fermentation and recovery

bioreactor	recovery	hydrolysis, fractionation	
Leuconostoc mesenteroides 1) growth 2) production phase: saccharose as C-source, 23 °C	precipitation using ethanol, acetone or methanol	dextrans of varying molecular weight	100 g L^{-1} in 24 h yield ~45 % to saccharose

Market data and applications

polysaccharide	production (t)	price (US$/kg^{-1})	microorganism	applications
xanthan	~30 000	10–14	*Xanthomonas campestris*	food additive, oil exploration
dextran, dextran derivatives	2000 600	35–390 400–2800	*Leuconostoc mesenteroides*	plasma expander, food additive, biochemicals
hyaluronic acid	500	2000–100 000	*Streptococcus equi*	surgery, cosmetics

other microbial polysaccharides with some economic importance:
alginate (*Azotobacter*), curdlan (*Agrobacterium*, *Rhizobium*), gellan (*Sphingomonas*), pullulan (*Pullularia*), cellulose (*Acetobacter*), β-glucans (fungal origin)

Biomaterials

General. Biodegradable polymers can be manufactured from natural, chiral hydroxycarbonic acids. Natural polyamides like silk (fibroin, spider's silk) or mussel proteins exhibit unusual properties with interesting technical potential and are the subject of major research projects. Variants of bacteriorhodopsin, formed by *Halobacterium halobium*, are being studied as optical storage materials.

Biodegradable polymers. Polylactides (NatureWorks™) are produced in the > 10 000 t scale from L-lactic acid, and manufacture of copolymers of propane 1,3-diol with terephtalic acid (Sorona™) is shortly to begin. Many microorganisms, e. g., *Ralstonia eutrophia*, store up to 90 % of their cell dry mass in the form of poly[3(R)-hydroxybutyric acid]. The composition of the polymer can be modulated by addition of precursors, depending on the strain used in the experiment. Of particular technical interest are copolymers of 3(R)-hydroxybutyric acid and 3(R)-hydroxyvaleric acid, whose properties resemble those of polypropylene but which are biodegradable and biocompatible. At present, they are produced in small quantities only (Biopol®). Economic progress was recently achieved by expressing the PHB operon, which consists of only three genes, in various plants in high yields. A major focus is on improvements of recovery (e. g., extraction by alkaline hypochlorite solutions, methylene chloride).

Fibroins and spidroins are highly variant polyamides extruded by the spinnerets of silkworms (*Bombyx mori*) and spiders (e. g., *Nephila claviceps*). Their elastic properties and their technical potential are high: the dragline silk of a spider can stretch by ca. 30 % before it breaks. Both types of polymers are formed of repeating polypeptides and have been expressed in *E. coli, Pichia pastoris,* and transgenic animals from synthetic, repetitive gene cassettes whose codon usage had been optimized for the host organism used. Yields of recombinant fibroins and spidroins, however, are still limited to ca. 1 g L^{-1} fermentation broth.

Adhesion protein. Mussels, e. g., *Mytilus edilis*, adhere to smooth surfaces such as crab shells by means of a polyamide that hardens under water; by this trick, they have spread over large distances. The M_R of the precursor protein of this natural adhesive is about 130 000 Da, and the protein is composed mainly of hydrophilic amino acids such as tyrosine, serine, threonine, lysine, and proline. During secretion, the tyrosine and proline residues undergo posttranslational hydroxylation, forming 3- or 4-hydroxy-L-proline and o-hydroxytyrosine (DOPA). Once secreted, these residues, in the presence of oxygen, form chinoid structures, which initiate polymerization of the peptide chains. With an optimized synthetic gene cassette, good expression of the precursor protein was achieved in *E. coli*. The genes for the posttranslational steps involved in adhesive formation, however, cannot be cloned into *E. coli*. As a consequence, the precursor protein is isolated and oxidized by fungal tyrosinase to the polyhydroxylated peptide. L-ascorbic acid is then added to prevent further oxidation. A commercial adhesive based on this material is under development for dental applications.

Bacteriorhodopsin (BR). The archaeobacterium *Halobacterium salinarium* grows in 3–5 M NaCl. During phototrophic growth, the retinal protein BR, which acts as an outward-oriented proton pump, supplies the cell with energy. BR forms 2D crystalline aggregates in the cell membrane ("purple membrane", 75 % protein, 25 % lipids), which are extremely stable and whose isolation is straightforward. The cofactor of BR is retinal. Its light-induced *trans–cis* isomerization with thermal reversibility is the basis for a catalytic photocycle with a turnover of up to 100 s^{-1}. BR is purple in the ground state (λ_{max} 570 nm) and yellow (λ_{max} 410 nm, M intermediate) after deprotonation of the Schiff base formed from retinal and lys^{216}. The speed of this photocycle and thus of the color change can be manipulated by specific mutations. BR and its mutant proteins are thus excellent materials for storing optical information.

Polyhydroxyalkanoic acid

	product	manufacturer
Ralstonia eutropha	copolymers of 3-hydroxybutyric acid and 3-hydroxyvaleric acid	Metabolix (commercial)
Escherichia coli	PHB, cloned operon from *Alcaligenes latus*	ATO-DLO
Pseudomonads	polyester with side-chains ranging from C_3- to C_8- or phenylvaleriate side-chain	not commercial
rapeseed	PHB, cloned operon from *Ralstonia eutropha*	not commercial
Lactobacillus sp.	chiral polylactides from L-lactic acid	Cargill-Dow (commercial)
Klebsiella pathway cloned into E. coli	copolymers of propane 1,3-diol with terephtalate	DuPont, Genencor

Ralstonia eutropha	**bioreactor**	**recovery**	>80 g/L culture medium, >2 g/L × h
heterotrophic production strain	fed-batch, glucose, propionic acid, 30 °C, ~ 40 h	enzymatic break-up of cells	

PHB operon from *R. eutropha*

promoter	PHA synthase	3-ketothiolase	acetyl-CoA reductase

Dragline silk of spiders (*Nephila claviceps*)

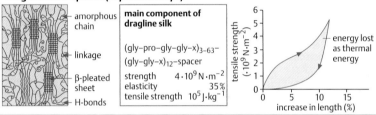

— amorphous chain
— linkage
— β-pleated sheet
— H-bonds

main component of dragline silk

(gly–pro–gly–gly–x)$_{3-63}$
(gly–gly–x)$_{12}$–spacer

strength	$4 \cdot 10^9$ N·m^{-2}
elasticity	35 %
tensile strength	10^5 J·kg^{-1}

energy lost as thermal energy

tensile strength ($\cdot 10^9$ N·m^{-2})
increase in length (%)

Adhesion protein from common mussel (*Mytilus edulis*)

-[ala-lys-(pro or hyp)-ser-(tyr or DOPA)-hyp-hyp-thr- DOPA -lys-]

hyp = 4-hydroxy proline
DOPA = 3,4-dihydroxy-L-phenylalanine

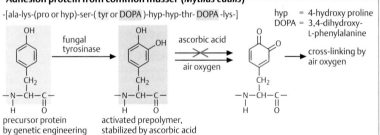

fungal tyrosinase

ascorbic acid
air oxygen

cross-linking by air oxygen

precursor protein by genetic engineering

activated prepolymer, stabilized by ascorbic acid

Bacteriorhodopsin as a proton pump

cytoplasm
membrane
membrane

x-ray structure of bacteriorhodopsin at 0.23 nm resolution
red: retinal
green: Asp^{85}, Asp^{96}, Lys^{216}

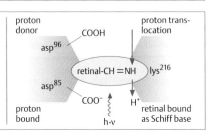

proton donor
proton translocation

asp^{96} — COOH

retinal-CH=NH — lys^{216}

asp^{85}

proton bound

COO$^-$ — H$^+$
h·ν

retinal bound as Schiff base

Biotransformation

General. Biotransformations are key functions of organisms and serve for the biosynthesis or biodegradation of metabolites (anabolism and catabolism) and also for the detoxification of toxic or unnatural (xenobiotic) compounds. Each step is catalyzed by an enzyme. In biotechnology, biotransformation is usually a synonym for biocatalysis, i. e., the transformation of natural or synthetic precursors (educts) into products of higher value. Technical biotransformations are carried out either with microorganisms, mammalian cells, or plant cells in a bioreactor (fermentation) or with isolated enzymes or cells (which can be immobilized on carrier materials). At present, the use of genetically engineered (recombinant) cells or enzymes is increasing rapidly. Whether a given biocatalytic process is termed "biocatalysis", "biotransformation", "fermentation", or "enzyme catalysis" often depends on personal preference, since in all these processes, a single or several enzymes may be required. The use of isolated enzymes can simplify a process, since temperature tolerance is usually better, sterile conditions are not required, and diffusion of educt and product are unhindered. However, using an isolated enzyme may not be an option if the enzyme is expensive to isolate, is unstable, or requires auxiliary enzymes and cofactors.

Microorganisms are used for producing natural metabolites (e. g., glutamic acid) and also for the biotransformation of unnatural substrates (e. g., in the 11β-hydroxylation of synthetic steroid derivatives). Being enzyme-catalyzed reactions, these transformations generally proceed in a regio- and stereoselective manner. Since genes or whole gene cassettes from other organisms can be cloned and expressed in a host microorganism, the possibilities for biotransformations have expanded dramatically (e. g., for the microbial production of indigo). Metabolic engineering and protein engineering, as well as the discovery of novel enzymes and pathways through genome sequencing, will help to further expand the industrial use of biotransformation reactions.

Mammalian cells are used on large scales in industry in the manufacture of pharmaceutical proteins (cell fermenters), but are too expensive for single-step biotransformations. They are being investigated in medicine, e. g., for use as an artificial liver to transform toxic metabolites occurring in blood and bind them to albumen.

Plant cells have been studied for biotransformation, e. g., for the 12-hydroxylation of β-methyl digitoxin to β-methyl digoxin with cell cultures of *Digitalis lanata*. Compared with chemical synthesis or recombinant microorganisms, plant cell cultures have had only limited success in industrial processes.

Enzyme catalysis. Usually, enzymes are used in single-step biotransformation processes. In most processes, isolated enzymes in free or immobilized form are used, but if the enzyme proves too expensive to isolate, an active enzyme in a whole inactivated microorganism may be used (e. g., glucose isomerase in Streptomyces cells). Most industrial examples of enzyme biotransformations involve hydrolases, since they do not require a cofactor and often catalyze regio- and stereoselective reactions. Enzymatic isomerization reactions, as well as addition reactions to double bonds, carbonyl groups, or activated CH bonds have also found industrial applications.

Recombinant metabolic pathways. Important examples are the production of L-ascorbic acid with a mutant of *Erwinia herbicola* and of indigo with a mutant of *E. coli*. For producing indigo, naphthalene dioxygenase from *Pseudomonas* sp. was cloned into *E. coli*, followed by optimization of the novel pathway through metabolic engineering of the host organism. The genes coding for the transformation of glycerol to propane 1,3-diol in *Klebsiella pneumoniae* were engineered into *E. coli*, allowing for the commercial production of this polyester building block from hydrolyzed corn starch. Other reactions of this type include the biodegradation of xenobiotic substances in the environment. Metabolic engineering has become a key method to optimize recombinant organisms in view of its high productivity.

Biotransformations

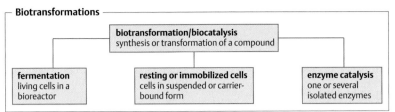

biotransformation/biocatalysis
synthesis or transformation of a compound

| **fermentation** living cells in a bioreactor | **resting or immobilized cells** cells in suspended or carrier-bound form | **enzyme catalysis** one or several isolated enzymes |

Biotransformation reactions (examples)

	organism/enzyme	type	company
D-sorbitol → L-sorbose	*Acetobacter suboxydans*	F	Roche
phenyl-D-lactate → 4-hydroxy-phenyl-D-lactate	*Beauveria gossypii*	F	BASF
fumaric acid → L-aspartic acid	*Escherichia coli*	IC	Tanabe, DSM
D-glucose → D-fructose	glucose isomerase from *Streptomyces* sp.	IC, IE	Novo, Clinton
D,L-acetoxymethoxyphenylethyl-amine → L-phenylethylamine	lipase from *Pseudomonas cepacia*	IE	BASF
tryptophane → indigo	recombinant strain of *E. coli*-Stamm	RIC	Genencor
glucose → propane 1,3-diol	*E. coli* with *Klebsiella* pathway	RIC	DuPont, Genencor

F: fermenation, IC: immobilized cells, IE: immobilized enzyme, RIC: recombinant immobilized cells

Indigo from recombinant *E. coli*

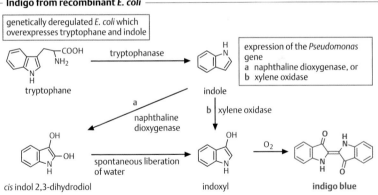

genetically deregulated *E. coli* which overexpresses tryptophane and indole

tryptophanase

tryptophane

indole

expression of the *Pseudomonas* gene
a naphthaline dioxygenase, or
b xylene oxidase

a naphthaline dioxygenase

b xylene oxidase

cis indol 2,3-dihydrodiol

spontaneous liberation of water

indoxyl

O_2

indigo blue

Immobilization of enzymes and cells

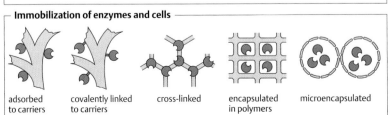

adsorbed to carriers | covalently linked to carriers | cross-linked | encapsulated in polymers | microencapsulated

Steroid biotransformations

General. Steroids are a class of chemicals for which biotransformations have been successfully applied.

Steroids. This large group of compounds comprises over 10 000 natural and synthetic compounds. Many of them are used in pharmacology. Examples are vitamin D (calciferol), anti-inflammatory drugs (corticosteroids), ovulation inhibitors (estrogens and progestins), anti-arhythmics (digitalis glycosides), and diuretics (spironolactone). Several types of biotransformation reactions are used in their industrial manufacture. Important examples are the sidechain degradation of β-sitosterol to androsta-4-ene-3,17-dione (AD), androsta-1,4-diene-3,17-dione (ADD), and the 11β-hydroxylation of cortexolone ("Reichstein S"). The total synthesis of pregnenolone from sugar has been achieved using a recombinant yeast carrying several foreign genes; however, as indigo or ascorbic acid, the economics of this process cannot yet compete with the traditional production process.

Sidechain degradation. For a long time, diosgenin, a natural compound isolated from a Mexican root, was a key substance in the industrial synthesis of steroids. Bile acids from animal gallbladders and stigmasterol, a side product of vitamin E production from soy oil, were other raw materials. Because chemical synthesis of corticosteroids, estrogens or spironolactone from these starting materials requires a large number of steps, sidechain degradation of plant sterols (e. g. β-sitosterol, isolated from rape or soy seeds) eventually became an attractive industrial alternative. Actinomycetales such as Mycobacterium, Nocardia, Arthrobacter, and Corynebacterium have this capacity. They can degrade the sidechain of phytosterols directly to steroid intermediates such as androsta-4-ene-3,17-dione (AD) or androsta-1,4-diene-3,17-dione (ADD), which are starting materials for the synthesis of estrogens and progestins and can also be used to add sidechains chemically, at position 17, which lead to the corticosteroids.

Cholesterol or bile acids have also been used as starting materials for this kind of biotransformation, but with less economic success.

11β-Hydroxylation. Based on microbial screening of many decades, today's culture collections contain microorganisms that can hydroxylate more or less selectively at nearly all positions of the steroid ring. The most important industrial hydroxylation step is the 11β-hydroxylation of Reichstein S, a chemical intermediate in the synthesis of hydrocortisone. *Curvularia lunata*, a mold, is used for this transformation. Because other activities of this organism would lead to hydroxylation at positions 7α and 14α as well, the 17α-acetate ester of Reichstein S is used as a substrate, but not for 11β-hydroxylation. The desired biotransformation step proceeds in this system with high regio- and stereoselectivity, because undesired side-reactions are sterically hindered. Microbial hydroxylation of other positions such as 7α, 9α, 11α or 16α is also used in industry for the manufacture of specialty steroids.

Biosynthesis of pregnenolone from sugar. A recent experimental approach consists in constructing a recombinant yeast strain that can form pregnenolone from sugar. During growth, *Saccharomyces cerevisiae* forms ergosterol as a component of its membrane. In the recombinant strain, oxidative degradation of this mycosterol was first suppressed by switching off the gene for Δ22-desaturation. Cloning and functional expression of three bovine genes involved in steroid metabolism and of a gene for a δ7-reductase from the plant *Arabidopsis thaliana* on chromosomes XIII, XV, and III of the yeast resulted in a strain that forms pregnenolone from D-galactose. Further cloning of a 3β-hydroxysteroid dehydrogenase into this mutant led to the formation of progesterone. Experimentation is under way to complete this pathway by the functional expression of bovine 11β, 17α-, and 21-hydroxylases, leading to the formation of hydrocortisone from D-galactose in one fermentation step. Yields, however, are still far from being economically competitive.

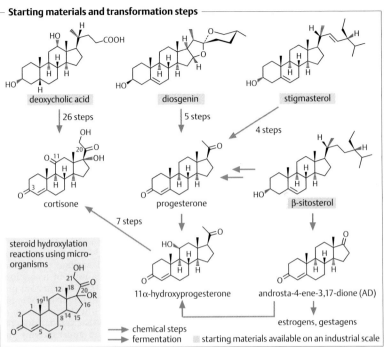

steroid hydroxylation reactions using micro-organisms

chemical steps
fermentation
starting materials available on an industrial scale

11β-Hydroxylation of Reichstein S using *Curvularia lunata*

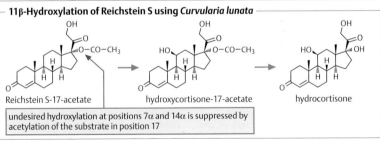

Reichstein S-17-acetate
hydroxycortisone-17-acetate
hydrocortisone

undesired hydroxylation at positions 7α and 14α is suppressed by acetylation of the substrate in position 17

Steroid synthesis using recombinant yeast

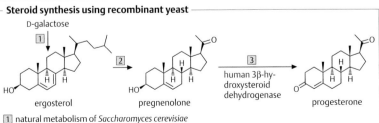

D-galactose

ergosterol
pregnenolone
human 3β-hy-droxysteroid dehydrogenase
progesterone

1 natural metabolism of *Saccharomyces cerevisiae*
2 cloned genes from beef and *Arabidopsis thaliana*
3 cloned human gene

Enzymes

General. Starting about 100 years ago, various enzymes isolated from animals, plants, and microorganisms have been developed into important components used in technical processes and into analytical reagents. Around 1970, immobilized enzymes were developed for industrial biocatalysis (enzyme transformation). Genetic engineering has further increased this potential by providing both pure and genetically modified enzymes.

Enzyme classification. Following international agreements, enzymes are divided into six classes, based on their function. Several thousand enzymes of various functions are known, and for most of them additional variants have been isolated from different organisms. Usually their properties are unfavorable for use in industry. For example, about one third of all enzymes function in the environment of biological membranes and are rather unstable in isolated form. The activity of most oxidoreductases, transferases, ligases, and synthases requires cofactors such as NADH, ATP, or coenzyme A, which are expensive and must be regenerated for economic reasons. Hydrolases and isomerases do not share these disadvantages and thus are preferred enzymes in industrial applications. For analytical and diagnostic applications, however, the high selectivity of enzymes justifies a high prize, leading to the use of all kinds of enzymes.

Manufacture. Production methods vary greatly, depending on the origin of the enzyme (animals, plants, microorganisms), the intended use (degree of purity required), and production scale. Other factors that largely determine the individual isolation and purification protocols are the properties of the desired enzyme (soluble or membrane-bound, stable or labile). If large quantities of an enzyme are required for a technical application (e. g., microbial proteases for use in detergents), extracellular enzymes produced by fermentation are preferred, as are simple steps for downstream processing such as separation of cells, concentration of fermentation broth by precipitation or ultrafiltration, gentle drying in a spray or vortex dryer, or adding stabilizing compounds for finishing. The enzyme preparations obtained by such processes are usually of low purity and are often contaminated with other enzymes (which actually may be useful for the desired application). In contrast, enzyme preparations that are used in therapy (e. g., tPA, DNAse) or for diagnostic purposes must be highly pure. They are often produced inside cells and, because they achieve a higher price in the market, are usually isolated and purified in several chromatographic purification steps that eventually lead to a single enzyme activity. The purity of the enzyme preparation is monitored through 1) determining the specific activity at each purification step, 2) determining the decrease in undesired side-activities, 3) electrophoretic methods. Genetic engineering has revolutionized enzyme production, and many enzymes used in technology today are produced by fermentation procedures based on recombinant host microorganisms. A particular advantage of these processes is that fewer side-activities are present, and a pure enzyme preparation can be obtained in fewer chromatographic steps, producing a reduced amount of waste materials.

Registration. Enzymes are natural compounds and thus can be used in all applications except food production and medical therapy without limitation, if manufactured properly according to GMP procedures ("good manufacturing practice"). Food additives prepared from natural products by means of enzyme technology also are considered natural and need not to be labeled (example: isoglucose). If enzymes are added to food products, the organism used for enzyme production should have GRAS status (generally recognized as safe), which applies to plants, animals, and a restricted choice of microorganisms that have a long history in the production of fermented foods (AMFEP directives). If enzymes from other microorganisms are to be used, complex and expensive registration procedures are required, which are often economically prohibitive. Similar restrictions apply to the use of recombinant enzymes in food processes.

Enzyme classification

EC-number	name	coenzymes	example
1.x.y.z	**oxidoreductases**		
1.1.y.z	react with CH–OH	NAD^+, $NADP^+$, PQQ	alcohol dehydrogenases
1.1.3.z	react with CH–OH	FAD^+	glucose oxidase
1.3.y.z	react with C–H	heme, Fe^{2+}	steroid 11β-hydroxylase
2.x.y.z	**transferases**		
2.4.y.z	transfer glycosyl groups		glucosyl transferase
2.6.1.z	transfer NH_3 to carbonyl groups	pyridoxal phosphate	transaminases
3.x.y.z	**hydrolases**		
3.1.y.z	hydrolyze ester bonds		lipases, esterases
3.2.y.z	hydrolyze glycoside bonds		α-amylase
3.4.y.z	hydrolyze peptide bonds		subtilisin, trypsin
4.x.y.z	**lyases**		
	catalyze elimination reactions with formation of double bonds or add groups to double bonds		
4.3.y.z	add or eliminate NH_3 to/from C=C-double bonds		aspartase
5.x.y.z	**isomerases**		
5.1.y.z	racemize D- and L-amino acids		alanine racemase
5.3.y.z	intramolecular oxidoreductases		xylose (glucose) isomerase
6.x.y.z	**ligases**		
6.2.y.z	C-S coupling	ATP, CoASH	acetyl-CoA synthetase

Preparation of enzymes

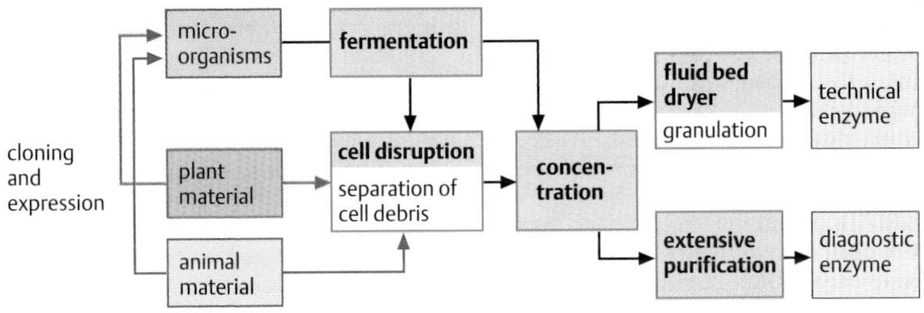

Registration

origin	examples	recommendations
animal organs	**pancreas extracts, rennin, pepsin**	**good manufacturing practice**
plant material	**papain, bromelain**	**good manufacturing practice**
microorganisms		
a) traditionally used for the production of food products	*Bacillus subtilis*, *Aspergillus niger*, *A. oryzae*, *Mucor javanicus*, *Rhizopus* sp., *Saccharomyces cerevisiae*, *Kluveromyces fragilis* and *K. lactis*, *Leuconostoc oenus*, organisms from starter cultures	GRAS = generally recognized as safe: simple tests are enough (recommendations of AMFEP, Association of Microbial Food Enzymes Producers)
b) enzymes from well known microorganisms	*Bacillus stearothermophilus*, *B. lichenformis*, *B. coagulans*, *B. megaterium*, *B. circulans*, *Klebsiella aerogenes*	extensive experiments for registration required

Enzyme catalysis

General. Enzymes are used in chemical syntheses due to their favorable energetics and their regio- and stereoselectivity ("green technology"). Enzymes that do not require coenzymes are preferred. The most important examples are hydrolases, but some isomerases and oxidoreductases with firmly bound coenzymes are also used. Today, many enzymes are available in recombinant form. They can be prepared in large quantities, free of contaminating enzymes of different specificities, and they can be optimized for industrial or diagnostic use by methods such as protein engineering. The discussion here uses the numbering of enzyme nomenclature.

Oxidoreductases. About 980 different types of oxidoreductases have been classified. Oxidases usually contain firmly bound FAD as a cofactor and thus can be used for analytical test strips. Dehydrogenases are very valuable for both analytical and synthetic applications, since their reaction leads to equilibria; consequently, they can be applied either for the oxidation of hydroxyl or the reduction of carbonyl groups. For technical applications, however, their expensive coenzymes, usually NAD(P)$^+$ or NAD(P)H, must be added, or an inexpensive coupled reaction for regenerating these coenzymes must be found. A breakthrough occurred when formate dehydrogenase from *Candida boidinii* and NAD(P)H bound to PEG as a coenzyme were used with dehydrogenases in an enzyme membrane reactor, shifting the equilibrium to complete oxidation of the substrate due to formation of CO_2 as the coupled product. Recently, interest in the technical use of peroxidases, dioxygenases, and P450 monooxygenases has been growing; they all can oxidize nonactivated carbon–hydrogen bonds in a regio- and stereoselective manner. They often contain a firmly bound Fe-S cluster or heme as a cofactor.

Transferases include about 1019 enzyme types. They are not presently used in technical applications.

Hydrolases are the most important group of technical enzymes. They comprise 1002 types of enzymes, among which lipases, esterases, amylases, and proteases have found the greatest technical application. If the water content and water activity of the reaction mixture is thoroughly controlled, they can be used for regio- and stereoselective formation of ester and amide bonds. For example, thermolysin from *Bacillus stearothermophilus* is used for regiospecific esterification of L-aspartate and L-phenylalanine methylester, providing aspartame sweetener; and penicillin amidase from *E. coli* is widely used for synthesis of semisynthetic penicillins from 6-aminopenicillanic acid. Lipase from *Burkholderia cepacia* is used for industrial synthesis of chiral amines from racemic amide precursors, and lipase from *Rhizomucor miehei* is extensively used for manufacturing cocoa butter substitutes. Amino acylase is used for enantioselective hydrolysis of racemic N-acyl amino acids.

Lyases include 341 examples and most are cofactor-independent. Aspartase from *E. coli* is used for manufacturing L-aspartic acid from fumaric acid on a technical scale, using immobilized cells instead of the free enzyme. Acrylonitrile hydratase from *Pseudomonas chloraphis* catalyzes the addition of water to acrylonitrile, forming acrylamide, the monomer of polyacrylamides. Oxynitrilases enable addition of HCN in a stereoselective manner to carbonyl compounds, resulting in chiral D- or L-hydroxy acids after hydrolysis of the nitrile group. Using aldolases, e. g., from rabbit liver, various sugars have been synthesized from suitable building blocks. Pectate and pectin lyase are used in the food industry.

Isomerases are a small group of enzymes, comprising just 147 members. They do not need a coenzyme. Glucose isomerase is used on a large industrial scale for partial isomerization of D-glucose to D-fructose, resulting in isoglucose syrups of high sweetness. The intracellular enzyme is usually used within inactivated, immobilized cells of its producing microorganism, often Streptomyces.

Ligases. All 122 ligases are ATP-dependant. Thus, if they are to be used in a technical process, regeneration of ATP is required. Although laboratory procedures for ATP regeneration have been described, there is no industrial application yet.

Enzymes vs. chemical synthesis

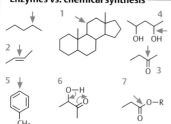

enzymes are usually superior for

1 regio- and stereospecific oxidations at nonactivated C–H bonds
2 stereoselective addition reactions at activated carbon atoms
3 stereoselective reduction of carbonyl groups
4 regioselective oxidation of hydroxy groups
5 position-specific substitution at the aromatic ring
6 position-specific isomerization
7 regio- or stereoselective hydrolysis or formation of esters under mild conditions

Types of enzymes used in biocatalysis

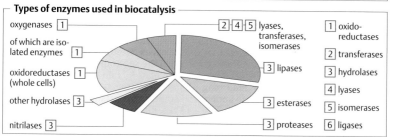

oxygenases [1]

of which are isolated enzymes [1]

oxidoreductases [1] (whole cells)

other hydrolases [3]

nitrilases [3]

[2]-[4]-[5] lyases, transferases, isomerases

[3] lipases

[3] esterases

[3] proteases

[1] oxido-reductases
[2] transferases
[3] hydrolases
[4] lyases
[5] isomerases
[6] ligases

Enzyme processes in industry

	process	enzyme	scale (t/year)	company (examples)
[3]	hydrolysis of penicillin G to 6-amino penicillanic acid	penicillin amidase* from E. coli	40 000	DSM
[3]	hydrolysis of N-acyl-DL-amino acids to L-amino acids	acylase* from Aspergillus sp.	5 000	Tanabe Seiyaku, Degussa, DSM
[4]	addition of NH₃ to fumaric acid yielding L-aspartic acid	aspartase* within E. coli	10 000	Tanabe Seiyaku
[3]	hydrolysis of starch to D-maltose and D-glucose	α-amylase, glucoamylase	100 000	several companies
[5]	isomerization of D-glucose to D-fructose	glucose isomerase* within Streptomyces sp.	100 000	Clinton Corn Products, others
[3]	synthesis of acrylamide from acrylonitrile	nitrile hydratase* within Pseudomonas chloraphis	30 000	Nitto Chemicals
[3]	hydrolysis of racemic phen-ethyl amides to chiral amines	lipase* from Burkholderia cepacia	10 000	BASF
[3]	transesterification of palm oil with stearic acid methyl esters, yielding cocoa butter substitute	lipase* from Rhizomucor miehei	1 000	Unilever
[3]	dehalogenation of 1-chloro propane diol	dehalogenase* from thermophilic organisms	under development	Dow Chemicals

*in immobilized form

Analytical enzymes

General. Enzymes from all six classes are used for analytical or diagnostic purposes. Their main advantage is their high substrate specificity, which allows selective detection of even a minor component in a complex mixture. To guarantee the highest possible specificity, the enzyme preparation must be free of other enzymes that would interfere with the intended application. Thus, analytical and diagnostic enzymes are of considerable purity. Many analytical determinations with enzymes can be carried out conveniently by use of commercial reagent kits and automated laboratory robots or with test strips. Enzymes can also be used as reporter groups indicating binding events, e. g., by linkage to antibodies or DNA fragments; if an excess of enzyme substrate is present in such assays, the signal resulting from a binding event is amplified considerably. The world market for enzyme-based diagnostics at present is ca. 10 billion US$ (2002).

Principles of measurement. Usually, photometric, fluorimetric, or luminometric assays are used for measuring enzyme reactions. If direct measurement of the substrate or product of the primary enzyme reaction is not feasible, auxiliary enzyme cascades are often used. Frequently, they include a NAD(P)H-dependent dehydrogenase as indicator enzyme; NAD(P)H formation or consumption can be monitored easily by photometric measurements at or near 334, 340, or 366 nm. These measurements are usually performed with µL amounts of sample (e. g., blood). Thus, in addition to providing extremely pure enzymes, the development of highly precise pipettes, optical equipment, and other devices adapted to such small volumes (µL-technology) was another milestone in establishing enzyme technology.

Methods of determination. These include basically three procedures: 1) end-point determinations, 2) kinetic methods, and 3) catalytic methods. In *end-point determinations*, the turnover of a substrate or cofactor in a reaction is measured after its completion. Examples are the determination of ethanol with alcohol dehydrogenase or of lactate with lactate dehydrogenase. End-point methods require at least several minutes: reaction velocity increases with enzyme concentration and with a low K_m or high v_{max} value. An example of the use of an auxiliary reaction is the determination of D-glucose with hexokinase, followed by the formation of NADPH in the coupled reaction of glucose-6-phosphate dehydrogenase. With the same reaction, but using a kinetic method, glucose levels can be determined much faster. Kinetic methods do not require that a reaction proceed until completion; rather, its initial velocity is measured, which is linearly proportional to the rate-determining enzymatic step, provided the substrate concentration is low ($< \frac{1}{10}$ of its K_m value). Reaction conditions and measuring time intervals must be kept strictly constant in this type of assay. Thus, kinetic methods are the method of choice with clinical laboratory robots. *Catalytic methods* enable an even lower limit of detection, since the substance being analyzed is used as the rate-limiting factor in a cyclic multienzyme reaction leading to the continuous consumption of a regenerating cofactor. An example is the determination of coenzyme A with the combined reactions of phosphotransacetylase, citrate synthase, and malate dehydrogenase.

Preparation and properties. Analytical enzymes are produced in small quantities, but at high purity. Often, intracellular enzymes are applied, which occur in low concentrations and which must be isolated from disrupted cells in high quality and with little or no side activities, using highly diverse methods of enzyme purification. Today, recombinant enzymes dominate this area, since they can be produced much more easily in larger quantities and without side-activities. Recombinant enzymes can further be optimized for a given purpose by protein engineering. Apart from specificity and purity, another important issue with such enzymes is their stability during shipment and storage. Typically, activity losses of $< 20\%$ per year at temperatures of $4\,°C$ can be tolerated. For this purpose, stabilizers such as sugars or glycerol are often added during the finishing process.

Diagnostic and analytical enzymes

enzyme	EC number	analyte	auxiliary analyte
alcohol dehydrogenase	1.1.1.1	ethanol, other alcohols, aldehydes	NADH
glucose oxidase	1.1.3.4	glucose	dye
pyruvate kinase	2.7.1.40	phosphoenol pyruvate, ADP	
creatine kinase	3.5.3.3	creatine	
citrate lyase	4.1.2.6	citric acid	NAD^+
mannose-6-phosphate isomerase	5.3.1.8	mannose	
succinyl-CoA synthase	6.2.1.4	succinate	

End point determinations

measurement technique	detection limit	
photometry		
– end point determinations	1 –	10 µmolar
– kinetic methods	0,1 –	1 µmolar
– catalytic methods	1 –	10 nmolar
fluorimetry		
– end point determinations	1 –	10 nmolar
– catalytic methods	1 –	10 fmolar
luminometry	1	–1 000 pmolar

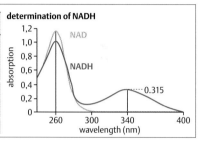

determination of NADH

End point determinations

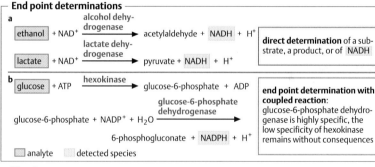

a

ethanol + NAD^+ — alcohol dehydrogenase → acetylaldehyde + NADH + H^+

lactate + NAD^+ — lactate dehydrogenase → pyruvate + NADH + H^+

direct determination of a substrate, a product, or of NADH

b

glucose + ATP — hexokinase → glucose-6-phosphate + ADP

glucose-6-phosphate + $NADP^+$ + H_2O — glucose-6-phosphate dehydrogenase →

6-phosphogluconate + NADPH + H^+

end point determination with coupled reaction: glucose-6-phosphate dehydrogenase is highly specific, the low specificity of hexokinase remains without consequences

☐ analyte ☐ detected species

Kinetic assays

kinetic determination of glucose with hexokinase and glucose-6-phosphate dehydrogenase; formation of NADPH after 10 s in an automated system

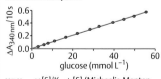

$v = v_{max} \times [S]/K_m + [S]$ (Michaelis-Menten equation): for substrate concentrations $S \gg K_m$ the velocity v of an enzyme reaction is linearly proportional to the substrate concentration

dependence of enzyme dosage on K_m

analyte	enzyme	K_m (µmol L^{-1})	V/K_m (= 1 mL min^{-1})*
ADP	adenylate kinase	1600	1600
glucose	hexokinase	100	100
glycerol	glycerol kinase	50	50
uric acid	uricase	17	17
fumarate	fumarase	1.7	1.7

* given for the following condition: the higher the K_m, the higher dosage of enzyme is required, if 99 % of the substrate should be transformed in a few minutes (V/K_m should be on the order of 1 mL min^{-1})

Enzyme tests

General. Since about 1950, the enzymatic determination of metabolites and the measurement of enzyme activities in body fluids have revolutionized medical diagnosis. A little later, immunoassays and, more recently, DNA assays have become indispensable tools of the physician. In enzyme diagnostics, low-molecular-weight compounds such as lactic acid or glucose are determined in blood or serum by means of appropriate enzymes of high selectivity. Alternatively, suitable indicator reactions are used to determine the concentration of enzymes in blood serum for the diagnosis of destructive processes in individual organs. Enzyme tests have also found entrance in food analysis, in the monitoring of fermentation processes, and in environmental protection. If dehydrogenases are employed, the change in the spectral absorption or fluorescence of NAD(P)$^+$ and NAD(P)H can be used ("optical tests"). Using this principle with suitable dehydrogenases, glucose and ethanol can be measured directly, whereas many other substances, such as fatty acids or glycerol, can be determined quantitatively by coupling two or more different enzymatic reactions. For practical reasons, it is advantageous to use as few auxiliary enzymes as possible. In food analysis, a range of sugars (glucose, galactose, maltose) and acids (citrate, malate) are determined enzymatically. Quasi-continuous determination of glucose is important in most microbial fermentations, and similar monitoring of glucose and lactate is of importance in animal cell culture. The inhibition of an enzyme activity may also be used as an analytical tool: for example, measurement of the inhibition of isolated acetylcholine esterase, a key enzyme in neurotransmission, can be used to indicate the presence of nerve gases or of organophosphate or carbamate pesticides.

Determination of enzyme activities. The measurement of enzyme activities in the serum of a patient is an important tool for the physician. When cells are destroyed during a disease (heart infarction, hepatitis, cirrhosis, pancreatitis, etc.), enzymes are released into the bloodstream. Thus, damage to the cells of various organs (heart, muscle, liver) and their compartments (membranes, cytoplasma, mitochondria) can be monitored by measuring enzyme activity in the blood. An organ-specific distinction (differential diagnosis) is possible, since the cells of different organs often release different enzymes or different enzyme subtypes (isoforms) of an enzyme. Thus, liver diseases can be monitored by measuring transaminases; pancreatic inflammation by measuring α-amylase; and the destruction of heart muscle cells by creatine kinase ("diagnostic key enzymes"). Their determination follows a kinetic regime based on the use of substrates that show little or no cross reactivity with other enzymes.

Laboratory automation. Clinical enzyme assays were originally performed manually, using individual sample preparations, incubation procedures, and determinations using simple photometers with filters of suitable wavelengths. Now, the tremendous increase in clinical enzyme tests has led to almost complete automation in central diagnostic laboratories, where all pipetting, incubation, and measurement steps are carried out by laboratory robots, and barcodes or other procedures are used for sample identification.

Test strips. Occasionally, an assay in a central diagnostic laboratory may be too time-consuming for "point-of-care" use (e. g., daily self-tests by diabetics, urgent tests in hospital emergency rooms). In such situations, diagnostic test strips are often used. They are composed of several solid separation and reaction layers ("dry chemistry"), with which the sample is prepared and the indicator reaction proceeds once a solution (e. g., a drop of blood) is added. The enzymes used here are often oxidases; in the presence of the auxiliary enzyme peroxidase, their primary reaction product, H_2O_2, oxidizes a leuko dye forming a colored compound, the amount of which is proportional to that of the component to be measured in the sample. Test strips can be analyzed semiquantitatively by visual comparison with a graded color chart, or quantitatively with a reflectance photometer.

Frequent enzyme tests (selection)

	typical analyte	relevance
clincial diagnostics	glucose	diabetes
	triglycerides, cholesterol	blood fat, risk of atherosclerosis
	uric acid	gout
	alkaline and acid phosphatase	tumors
	transaminases	liver diseases
	creatine kinase	heart infarction
food analysis	sugars (glucose, maltose)	quality
	acids (citrate, malate)	quality, counterfeits
bioprocess control	glucose, lactate	microbial and cell culture fermentation
environmental monitoring	inhibitors of acetylcholine esterase	presence of organophosphates, carbamates

Differential diagnosis of various organ malfunctions

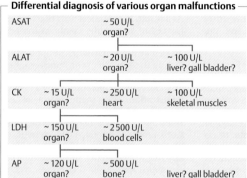

ASAT	~ 50 U/L organ?	
ALAT	~ 20 U/L organ?	~ 100 U/L liver? gall bladder?
CK	~ 15 U/L organ?	~ 250 U/L heart	~ 100 U/L skeletal muscles
LDH	~ 150 U/L organ?	~ 2 500 U/L blood cells
AP	~ 120 U/L organ?	~ 500 U/L bone?	~ 750 U/L liver? gall bladder?
γ-GT	~ 75 U/L liver	~ 15 U/L bone	~ 750 U/L liver? gall bladder?

ASAT aspartate aminotransferase
GLDH glutamate dehydrogenase
ALAT alanine aminotransferase
CHE choline esterase
CK creatine kinase
LDH lactate dehydrogenase
AP alkaline phosphatase
γ-GT γ-glutamyl transpeptidase

key enzymes for the diagnosis of organ diseases

liver, gall bladder: ALAT-GLDH-γ-GT-CHE
heart: CK-LDH (isoenzyme-1)-ASAT
skeletal muscle: CK-GLDH
bone: AP
blood: LDH
pancreas: α-amylase, lipase

Test strip (example: blood sugar)

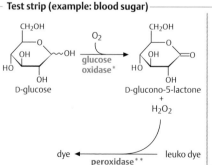

D-glucose → D-glucono-5-lactone + H_2O_2

glucose oxidase*

dye ← peroxidase** ← leuko dye

hand-held photometer for determination of blood sugar

composition of the test strip

addition of blood sample
separation layer
reagent layer
carrier foil

* glucose oxidase from *Aspergillus niger*
** peroxidase from horseradish

75

Enzymes as additives

General. Today, enzymes are used as additives in many areas of technology, e. g., in detergents, in food technology, as well as in paper, textile, and leather processing. Enzymes are also increasingly used for the industrial synthesis of fine chemicals, since they are often superior to chemical catalysts in regio- and stereoselectivity. For the same reason, enzymes are widely used in the analytical determination of medical samples and food products. Many enzymes are available today in a recombinant form. This implies a low price, high purity, and the potential to optimize their properties for the desired use by application of protein engineering techniques. For enzymes to be used as additives, hydrolases are preferred since they do not require the addition of a cofactor.

Purposes. In industry, the use of any enzyme as an additive should eventually pay off. Improvements in product quality, savings in process costs, or environmental benefits are typical examples where this has proved true. For example, proteases in detergents contribute to dissolving protein stains from within fibers - an effect that cannot be had with chemical additives. The enzymatic hydrolysis of starch is superior to acid hydrolysis in terms of the formation of undesired byproducts. By using pectinases in fruit processing, the yield of juice can be significantly enhanced, and process costs during filtration are reduced. Proteases and collagenases permit selective removal of hairs and other skin components during leather treatment, and their first application, about a century ago, dramatically improved the working conditions of tanners, who used to be an outcast profession due to their distasteful working conditions. The coagulation of milk upon addition of microbial or recombinant rennet is significantly less expensive and more hygienic than the classical process, which was based on extracts from calf stomachs. In most of these cases, however, enzyme mixtures are used, partially for cost reasons, partially because side activities are desirable fringe benefit (e. g., due to the presence of starch-degrading α-amylase in detergent proteases). As a result, the preparation of such mixtures is simpler and less demanding than for analytical enzymes. On the other hand, characterization of the resulting enzyme mixtures is often not simple and is subject to a manufacturer's standards. Similarly, the technical or economic benefit of an industrial enzyme becomes apparent only if quite specific determination methods are used, pertinent to a special craft or even a single manufacturer. Such methods may have developed over long periods of time and cannot be easily standardized, since internal quality control may rest on decade- or even century-long standards based on these individual methods. As a result, biochemical tests that are performed in an enzyme manufacturer's laboratory for screening improved enzymes often may not match with the application testing used in the client industry, often rendering the improvement of enzymes as additives very tedious.

Registration. The manufacture of enzymes as additives is regulated by the rules of GMP (good manufacturing practice). Enzymes produced by genetic or protein engineering techniques play a key role in detergents. In smaller market segments, such as food enzymes, recombinant enzymes are the exception rather than the rule. Though there may be a clear advantage for using a recombinant enzyme, the costly registration and real or expected consumers' concerns may form an insurmountable barrier to their use, except in some special cases such as recombinant chymosin.

Enzyme additive costs and markets. The benefit of adding an enzyme preparation obviously relates to its price. Usually, the technical benefit translates into an economic advantage on the order of cents per kg product. As a result, the price for technical enzymes is mostly in the range of 3 – 10 € per kg. In spite of these limitations, the world market for technical enzymes has grown in 2001 to ca. 1.6 billion €, and hydrolases used in detergents, followed by foods and feeds enzymes, have the largest share.

Enzymes as additives in industry

application	enzyme type	organisms (examples)	market size (% of total)	economic advantage
detergents	proteases, cellulases, lipases	*Bacillus licheniformis* *Aspergillus nidulans* *Trichoderma reesei*	40	1
starch hydrolysis	α-amylase	*Bacillus amyloliquefaciens*	5	3, 4
glucose iso-merization	glucose isomerase	*Streptomyces venezuelae*	7	1, 3
beer brewing	amylase	*Bacillus subtilis*	3	3, 4
fruit processing, wine	cellulases, hemicellulases, pectinases	*Aspergillus niger*	5	3, 4, 5, 6
flour, bakery goods	α-amylase, proteases	*Aspergillus oryzae*	8	1, 3
cheese manufacture, aroma	proteases, chymosin, lipases	animal rennin, *Rhizomucor miehei*, *Saccharomyces cerevisiae*	12	2
silage and animal feed	phytases	*Aspergillus niger*	8	3
paper and textiles	α-amylase, lipase	*Bacillus, Humicola*	2	4
leather treatment	proteases	*Aspergillus oryzae*	10	1, 7

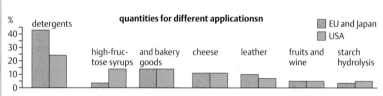

quantities for different applicationsn

%
40 — detergents
30
20 — high-fruc-tose syrups | and bakery goods | cheese | leather | fruits and wine | starch hydrolysis
10
0

■ EU and Japan
□ USA

process/application	enzyme cost per unit quantity (US $)
starch liquefaction	ca. $ 2 per t starch
glucose from starch	$ 3.5 per t starch
isomerization of glucose	$ 6 per t starch
HFS in USA	$ 6–7 per t starch
ethanol	$ 1 per t starch
beer	$ 0.1 per 100 L
bakery goods USA	$ 0.1 per 100 kg flour
bakery goods EU	$ 0.1–0.5 per 100 kg flour
fruit juice	$ 0.1–0.5 per 100 L juice
wine	$ 0.1–0.5 per 100 L wine
stabilization of fruit lemonade by glucose oxidase	$ 0.3–0.8 per 1 000 L
cheese manufacture	$ 0.05 per 100 L milk
detergents	$ 0.05 per kg detergent
leather tanning	$ 1.2–3 per t skin

important goals in application technology

1 higher product quality

2 improved taste

3 better yields

4 reduced process costs

5 better filtration

6 better conservation

7 improved working conditions, reduced environmental load

Detergent enzymes

General. About a century ago, Otto Roehm in Germany was the first to introduce detergents to which pancreatic enzymes had been added, to enhance the removal of proteinaceous stains such as blood, eggs, cocoa, grass stains, etc. When microbial proteases from Bacillus strains became available around 1960, the use of enzymes began to increase rapidly, and at present there is rarely a detergent without enzyme additives. Cloning and protein engineering have resulted in improved proteases that are perfectly adapted to laundry conditions. Recombinant Bacillus proteases are produced on a scale > 10 000 t per year. Other enzymes such as cellulases, lipases, and amylases are used as well in laundry detergents.

Detergents and the laundering process. Laundry detergents usually contain anionic and nonionic, sometimes also cationic, surfactants. Anionic surfactants are inactivated by the Ca^{2+} and Mg^{2+} ions of hard water; as a consequence, detergents contain complexing agents to sequester these ions. Instead of sodium pentasodium triphosphate, which used to be the most important complexing agent, sodium aluminum silicates in the presence of small amounts of "carriers" such as citric acid or phosphonates are preferred today. The traditional bleaching chemicals were sodium hypochlorite or sodium metaborate perhydrate; they have also been mostly replaced by sodium percarbonate or medium-chain organic peroxi-acids. The pH of a detergent solution is about 10; the temperature during laundry washing is 30 – 90 °C and the time required is usually ca. 30 min. As a result, detergent enzymes must be adequately stable against alkali, surfactants, complexing agents, and bleaches. Further, their substrate specificity should be low.

Proteases. At present, serine proteases obtained from Bacillus strains are exclusively used. Strain and enzyme improvement used to be done by mutagenesis and strain improvement, but these methods have been replaced by genetic and protein engineering techniques. Today, Bacillus strains in which the protease gene is integrated into the chromosome are used for production. Using protein design methods, detergent proteases have become more stable to complexing agents and oxidants. For example, the replacement of methionine[222] by alanine led to an enzyme with enhanced properties in dishwashers. During fermentation, production of the protease occurs with growth (constitutive expression, without addition of inducer) and is completed after 72 h or less. After complete removal of the cell mass by separators or filters, the extracellular enzyme is concentrated by precipitation or ultrafiltration, followed by partial purification. Since inhalation of enzyme dust or aerosols in working areas may lead to allergic reactions, the enzyme concentrate is further processed into granules which are encapsulated and dust-free. For this purpose, the concentrate is sprayed on a core particle and then granulated in a rapid mixer in the presence of additives such as salts, waxes, and stabilizers or it is directly sprayed on the core particle in the presence of the additives in a spray dryer. As a final step, each variety is coated with wax and pigment. In several large-scale studies, no allergies from inhalation or skin exposure of such products was found.

Cellulases. Endocellulases having a suitable pH optimum hydrolyze, during the washing of cotton, those cellulose microfibers that protrude from the fabric. As a result, the washed textile feels softer and its color is brighter. In addition, it resists soil pigments better than natural fibers. At present, cellulases from *Humicola insolens* and Bacillus sp. are mainly used.

Lipases. Lipases having an alkaline pH optimum are intended for removing greasy stains (e. g., olive oil) or wax esters (e. g., lipstick stains) which are difficult to remove with chemical surfactants. The most important detergent lipase presently used is from *Thermomyces* sp., pared in recombinant strains of *Aspergillus oryzae*.

Amylases. Amylases loosen starch-containing stains by hydrolytic attack. For this purpose, alkali- and thermo-tolerant amylases have been proposed.

Suitability of various protease types as detergent enzymes

type of protease	serine	cysteine	carboxy	metallo
example	subtilisin	papain	pepsin	thermolysin
activity at pH 10	+	–	–	–
stability at pH 10	(+)	–	–	–
stability at 50°C	+	–	–	+
stability against complexing agents	+	+	+	–
stability against oxidizing agents	+	–	+	+
stability against surfactants	~	~	~	~

Structure

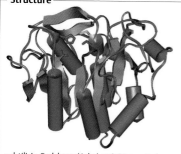

subtilisin Carlsberg (1sbc) at 0.23 nm. Red: catalytic triad (ser, asp, his); green: met^{222}

Transformation vector

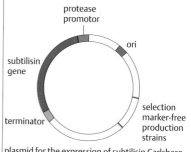

protease promotor

ori

subtilisin gene

selection marker-free production strains

terminator

plasmid for the expression of subtilisin Carlsberg in high-performance strains of *Bacillus lentus*

Fermentation and recovery

preculture

genetically modified high-performance strain of *Bacillus lentus;* 10 m^3, 24 h at 35 °C

▼

bioreactor

120 m^3, 48 h at 35 °C C-source: dextrin; N-source: soy meal, induction by casein

▼

> 15 g L^{-1} subtilisin in 60 h

▼

cell separation and concentration

microfiltration and ultrafiltration or cross-flow filtration, precipitations

▼

1 **layered granulate**	2 **mixer granulate**
core particle (coated salt/sugar granulate), radical scavengers, PEG, TiO$_2$, layered coating in spray dryer	core particle (salt/PEG), formed to granulate in mixer, dried, coated by PEG/TiO$_2$ in spray dryer

1 cm

1 cm

Enzymes for starch hydrolysis

General. Starch is the second most important (after cellulose) polysaccharide produced on earth. Next to glucose, it is the most important carbon source for fermentation. In 1995, ca. 20 million t of starch were produced; 70% came from corn and 20% from potatoes. Only ca. 20% of isolated starch is used directly; 30% is chemically modified, and 50% is saccharified to yield glucose or its oligomers, the dextrins. The preferred method of hydrolyzing starch is with enzymes, which leads to fewer side-reactions than acid hydrolysis.

Starch is a polymer (dp 200–5000) that is composed of linear amylose (poly-α-1,4-D-glucose) and branched amylopectin. Amylose is pseudocrystalline. Its amount in starch can be determined by the iodine-starch reaction. In the amylopectin component, the linear amylose chain is branched at about every 20 glucose residues, in a D-1,6-glycosidic manner. Depending on the origin of the starch, the ratio of amylose to amylopectin may vary; this ratio determines the physical and chemical properties of the starch. Starch is insoluble in cold water. When heated, it dissolves to the extent to which intramolecular hydrogen bridges are destroyed. In the gelatinization range, starch forms a gel by absorption of water, resulting in a strong increase in viscosity. In this state, it can be chemically modified or enzymatically degraded. If gelled starch is cooled, amylose recrystallizes quickly with the formation of intermolecular hydrogen bonds (retrogradation). Starch is an important component or raw material for many basic food materials such as bread and beer. Traditional procedures for their manufacture are being replaced more and more by optimized protocols in which enzymatic processing of starch plays an important role.

Starch-degrading enzymes. Several enzymes are available for degrading starch: α-amylases (synonym: exo-amylase) hydrolyzes starch at α-1,4 positions within the polymer chain; β-amylase (synonym: endo-amylase) hydrolyzes maltose or maltotriose from the non-reducing end; glucoamylase (synonym: γ-amylase, amyloglucosidase) splits maltose to two moles of glucose, but also, if with reduced velocity, the α-1,6-bonds of amylopectin; pullulanases split α-1,6-bonds, preferentially of pullulan, but also of amylopectin; isoamylases also split α-1,6-bonds, but their velocity is higher with amylopectin than with pullulan.

α-Amylases. This aspartyl enzyme occurs in many organisms. It has been crystallized from malt, pancreas, and *Aspergillus oryzae*. The crystal structure has been obtained for the α-amylase of *Bacillus subtilis*. Several α-amylases have been cloned and overexpressed in various host systems. Bacterial α-amylases exhibit a significantly higher temperature optimum (*Bacillus licheniformis*: 78 °C) and are also more alkali-stable than fungal amylases. As a consequence, a wide range of α-amylases of different pH- and temperature stabilities are available for any desired application.

β-Amylase, a sulfhydryl enzyme, is available from wheat malt and also from *Bacillus stearothermophilus*. It is an important enzyme in the preparation of maltose syrups.

Amylases splitting α-1,6-bonds. The most important enzyme of this group is glucoamylase from *Aspergillus niger*. A similar enzyme from *Rhizopus* sp. is also used. Pullulanase is industrially produced by strains of *Klebsiella pneumoniae* and *Bacillus cereus*.

Manufacturing procedures. Some of the above enzymes are still produced by the classical procedure of surface fermentation. This method is gradually being replaced, however, by aerated bioreactors of up to 120 m³. The enzymes are secreted into the culture medium, resulting in a dilute enzyme solution after removal of the cell mass. Since the enzyme product is of relatively low value, the isolation procedures include only a few simple process steps such as ultrafiltration, precipitation, and finishing, in the presence of stabilizing additives.

Composition and properties of various starches

starch from	amylose (%)	amylopectin (%)	gelatinizing range (°C)	swelling capacity (fold)
potato	18–23	77–82	56–66	>1000
wheat	19–25	75–81	52–63	21
corn	21–30	70–79	62–72	24
rice	17–19	81–83	61–78	19

potato   wheat   corn   rice

Structure of amylose and amylopectin

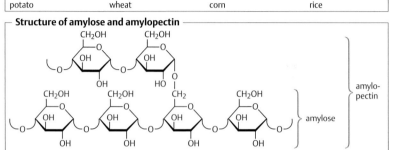

amylopectin

amylose

Enzymatic degradation

glu – glu – glu – glu – glu – glu – glu – glu – glu – glu – glu   **amylose**

glu – glu – glu – glu – glu
         ►|
      glu – glu – glu – glu – glu   **amylo-**
               ►| **pectin**
glu – glu – glu – glu – glu – glu – glu – glu

glu – glu – glu –
       ►|
      – glu – glu – glu –
              ►| **pullulan**
           – glu – glu – glu –

glu – glu **maltose**   glu – glu – glu **maltotriose**   glu = nonreducing end

→ α-amylase (1,4-α-D-glucan-4-glucanohydro-lase, EC 3.2.1.1): splits α-1,4-bonds to maltose and maltotriose

→ β-amylase (1,4-α-D-glucan-galtohydrolase, EC 3.2.1.2): splits maltose units from the non-reducing ends

→ glucoamylase (γ-amylase, (1,4-α-D-glucan-glucohydrolase, EC 3.2.1.3) and pullulanase (amylopectin-6-glucanohydrolase, EC 3.2.1.41): splits α-1,6-bonds

mixture of α-amylase and glucoamylase/pullulanase: splits amylose and amylopectin to D-glucose

Enzymes

enzyme	origin	properties	conditions of application
α-amylase	Bacillus licheniformis	best effect at 60 °C and pH 5.7, requires Ca^{2+}	95–105 °C, pH 6–7, Ca^{2+}
	Aspergillus oryzae	best effect at 50–60 °C and pH 5.0, requires Ca^{2+}	<50 °C, pH >3.5, Ca^{2+}
	barley malt	best effect at 70 °C and pH 5.5	<70 °C, pH >4.5
β-amylase	Bacillus stearothermophillus	best stability at 75 °C, pH 5.0	for maltose syrups
pullulanase	Klebsiella pneumoniae	best effect at 50–55 °C and pH 5.0	55–65 °C, pH 3.5–5
gluco-amylase	Aspergillus niger, Rhizopus sp.	best stability at 55–60 °C and pH 4.0	55–65 °C, pH 3.5–5

Enzymatic starch hydrolysis

General. About half of the starch that is isolated annually (ca. 20 million t y^{-1}) is enzymatically hydrolyzed. About 6 million t are used in the manufacture of isoglucose (or HFS: high fructose syrup). The remainder is partially hydrolyzed to dextrins and maltose syrups, which are used in a wide range of applications, e. g., as a carbon source in fermentations. Corn and wheat, usually from North America, are mostly used as raw materials for starch. Potato and rice starch have less practical importance as raw materials for glucose, since they are produced by intensive agriculture in Europe or Asia and thus cannot compete in price. When Napoleon blocked the access of British merchandise to Europe (the Continental Blockade), sugar began to be produced in Europe from sugar beet saccharose hydrolyzed with mineral acids. This technology leads to colored by-products and today is unattractive.

Enzymatic starch hydrolysis. Starch from corn or wheat is obtained by wet milling. Valuable by-products are corn germ oil, corn and wheat gluten, and feed additives of various compositions. The milled starch is then heated for only a few minutes, in the presence of thermostable bacterial α-amylase, to temperatures between 105 and 140 °C (starch cooking), to swell and gelatinize the starch. After 2–3 h in the presence of Bacillus α-amylase, biotransformation to maltodextrin (dextrose equivalent DE 15–20) is complete. Maltodextrin is a mixture of oligosaccharides with minor quantities of mono-, di-, and trisaccharides. It is an excellent starting material for the ensuing complete saccharification. Maltodextrins are also used as food components of low sweetness, e. g., in baby food, hospital diets, and instant soups.

Enzymatic saccharification can be directed to yield dextrose, glucose, high-maltose, or high-conversion syrups. In this process, mixtures of bacterial α-amylase and glucoamylase from *Aspergillus niger* are the key enzymes. First, maltodextrin obtained from the starch cooker is cooled to ca. 60 °C and the pH is adjusted to ca. 4. Usually, the process is carried out continuously, requiring several tanks in sequence to prevent mixing syrups of different degrees of hydrolysis. This procedure results in glucose syrups with a very high DE value (97–98) after 48–72 h. Addition of pullulanase or isoamylase leads to even higher DE values and reduces the glucoamylase levels. The use of immobilized enzymes has proved of little value, due to diffusion limitation in the highly viscous substrate solution and the formation of reversion products. From syrups with very high DE, pure D-glucose monohydrate (dextrose hydrate) can be isolated by crystallization. Using an appropriate choice of enzymes and saccharification conditions, a wide choice of glucose syrups of different compositions can be tailor-made. Glucose syrups with relatively low and medium DE are used in the manufacture of sweets. High-maltose and high-conversion syrups are often produced from maltodextrin using α-amylase from *Aspergillus niger*; they have high viscosity, but a reduced tendency to form brown colors or to crystallize.

Cyclodextrins. If maltodextrins are processed with the enzyme cyclodextrin transferase, 5-, 6-, or 7-membered cyclodextrin rings are formed. They exhibit good solubility in water, but also contain hydrophobic cavities 0.5–0.75 nm in diameter, which can accommodate hydrophobic guest molecules such as vitamins, fragrances, or drugs. In consequence, cyclodextrins are used to enhance the solubility of such compounds in formulations or to improve their stability. Chiral cyclodextrin derivatives are also used as stationary phases for the separation of mixtures of chiral compounds. The industrial manufacture of cyclodextrins is mostly limited to β-cyclodextrin; suitable cyclodextrin transferases for this process have been isolated from mesophilic and alkalophilic Bacilli. Starch or dextrins are used as a starting material in this process.

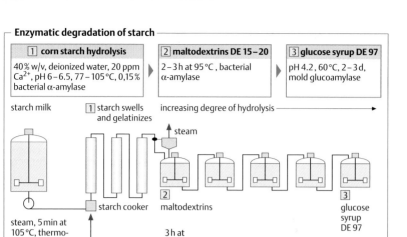

1 corn starch hydrolysis	2 maltodextrins DE 15 – 20	3 glucose syrup DE 97
40% w/v, deionized water, 20 ppm Ca^{2+}, pH 6 – 6.5, 77 – 105 °C, 0,15 % bacterial α-amylase	2 – 3 h at 95 °C , bacterial α-amylase	pH 4.2, 60 °C, 2 – 3 d, mold glucoamylase

starch milk   1 starch swells and gelatinizes   increasing degree of hydrolysis ⟶

steam

starch cooker   2 maltodextrins   3 glucose syrup DE 97

steam, 5 min at 105 °C, thermo-stable α-amylase

3 h at 95 °C   2 – 3 d at 60 °C

Products of enzymatic starch hydrolysis

product	charac-teristics	enzyme used	applications
maltodextrins	DE 15 – 25	α-amylase	food additive with good rheological properties; raw material for sweeteners
maltose syrups	DE 40 – 45	α-amylase, β-amylase	sweetener
high maltose	DE 50 – 55	α-amylase, glucoamylase	sweetener
maltose syrups with high or extra high degrees of saccharification	DE 60 – 70 DE < 80	α-amylase, glucoamylase, pullulanase	sweetener raw material for fermentation
dextrose	DE 97	prolonged hydrolysis	raw material for isoglucose
isoglucose	DE 97	glucose isomerase	sweetener

DE = dextrose equivalent, a measure for the degree of starch hydrolysis

Cyclodextrins

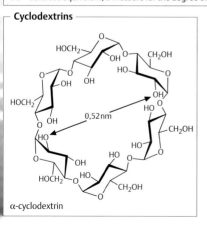

0,52 nm

α-cyclodextrin

Properties

cyclo-dextrins	α	β	γ
number of glucose residues	6	7	8
M_R	972	1135	1297
water solu-bility (in 100 g/mL)	14.5	1.85	23.2
diameter (nm)	≥0.47	≥0.6	≥0.75
CAS	10016-20-3	7585-39-9	17465-86-0

Enzymes and sweeteners

General. Sugar (D-saccharose, sucrose) became an important food additive in Western culture only in the 18th century. It was originally isolated from sugarcane grown in tropical or subtropical countries. In response to Napoleon's continental embargo of Europe, a new technology developed, allowing the production of sugar from sugar beets. Today, corn and wheat starch have become very important raw materials for sugar production: in two enzymatic processes, starch is first hydrolyzed to glucose, followed by isomerization to glucose-fructose syrups (isoglucose). Due to dietetic considerations (reduced intake of calories, cavity prevention), several sugar substitutes have been developed. Some of them are also produced by enzyme technology. The sweetness of sugars or sugar substitutes is evaluated by a sensory panel in comparison to the sweetness of a 10% saccharose solution in water.

Invert sugar. Saccharose is hydrolyzed by acid or by invertase to invert sugar syrup, an equimolar mixture of D-glucose and D-fructose. The sweetness of invert sugar is similar to that of saccharose. Since it does not crystallize, it is often used in the manufacture of candy and soft chocolate fillings. Invertase is obtained from the cell mass of bakers' yeast, *Saccharomyces cerevisiae*. It is located in the cytoplasm and can be obtained in pure form, after breakage of the cell wall in a ball mill, by a small number of downstream processing steps. Invert sugar is produced in an enzyme reactor with immobilized invertase, using 70% saccharose syrup as the starting material. The stability of the enzyme permits the reactor to be used continuously for several days at 55 °C.

Isoglucose. The enzyme glucose isomerase, which can be isolated from Streptomyces and other strains, transforms D-glucose in an equilibrium reaction to D-fructose. At the preferred reaction temperature of 60 °C, a mixture of 42% D-fructose and 55% D-glucose is formed (isoglucose, high-fructose syrup HFS) which is slightly less sweet than saccharose. If glucose is separated from this mixture by chromatography and re-isomerized, isoglucose 55 can be obtained. It contains 55% D-fructose. Its higher sweetness has proven advantageous in the manufacture of soft drinks. The annual production of isoglucose 42 and 55 is ca. 3.5 and ca. 4.5 million t, respectively. 80% of which is produced in North America. Due to EU regulations protecting sugar beet farmers, only ca. 300 000 t of isoglucoses are produced in the EU.

Manufacture. The intracellular enzyme glucose isomerase occurs in several microorganisms. This protein forms a homodimer, and bivalent metals, e.g. Mn^{2+} or Co^{2+}, participate in catalysis. About 1500 t of this enzyme is produced annually, mostly from Streptomyces strains or *Bacillus coagulans*, usually in a fed-batch bioreactor process taking 72 h. To avoid the cost of isolating the intracellular enzyme, isoglucose is often manufactured directly with the immobilized microbial strain, which is inactivated, crosslinked by glutaraldehyde and bound to a suitable carrier material. Within 1 h of contact time between the immobilized microorganism and the glucose feed, the concentrations of substrate and product reach equilibrium. At a process temperature of 60 °C, this is at ca. 42% D-fructose, yielding isoglucose 42. The half life time of the catalyst under these conditions is around 50 d at 60 °C. Thus, continuous operation of the process is possible if a modular process line is used. Due to strain optimization, modern processes no longer depend on the addition of Mn^{2+} or Co^{2+} as a cofactor. The reaction product contains traces of caramelized sugars and is purified by passing it over charcoal columns. Isoglucose is marketed as a syrup.

D-Fructose is a sugar of high sweetness. It can be prepared from invert sugar or from isoglucose by chromatographic procedures, usually by simulated moving bed chromatography. Topinambur (Jerusalem artichoke) tubers contain up to 75% inulin, a fructose polymer. Fructose can be prepared from topinambur in an enzymatic process using inulinase.

D-fructose

HOCH₂, CH₂OH
HO
OH

M_R 180.16
m. p. 106 °C, decomp.

D-glucose

CH₂OH
O
OH OH
HO
OH

M_R 180.16
m. p. 146 °C, decomp.

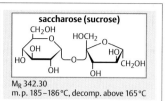

saccharose (sucrose)

CH₂OH HOCH₂
O O
OH HO
HO
OH CH₂OH
OH

M_R 342.30
m. p. 185–186 °C, decomp. above 165 °C

Sugars and sugar substitutes

name	relative sweetness (related to saccharose)	starting material, production process
saccharose	1.00	isolated from sugar cane or sugar beets
glucose	0.5 – 0.8	hydrolysis of starch with α-amylase, glucoamylase
glucose syrups	0.3 – 0.5	hydrolysis of starch with α-amylase, glucoamylase
hydrogenated glucose syrups	0.3 – 0.8	hydrogenation of starch hydrolysates or glucose syrups
isoglucose 42	0.8 – 0.9	enzymatic isomerization of glucose with glucose isomerase
fructose	1.1 – 1.7	• enzymatic hydrolysis of saccharose, or • enzymatic isomerization of glucose, or • enzymatic hydrolysis of inulins followed by chromatographic separation
invert sugar	1.00	hydrolysis of saccharose with invertase
mannitol	0.4 – 0.5	hydrogenation of fructose
sorbitol	0.4 – 0.5	hydrogenation of glucose
xylitol	1.0	hydrogenation of xylose
lactitol	0.3	hydrogenation of lactose
maltitol	ca. 0.9	hydrogenation of maltose
palatinitol, isomaltol	0.45	enzymatic isomerization of saccharose to isomaltulose (palatinose), follwed by hydrogenation yielding a mixture of glucopyranosido sorbitol and glucopyranosido mannitol

the relative sweetness of a sweetener is determined organoleptically by a panel and compared to a 10 % aqueous solution of saccharose. As a result, values can fluctuate around a median

Manufacture

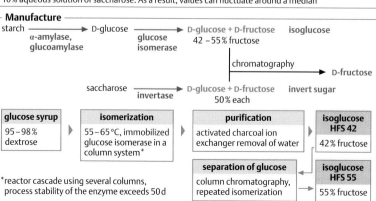

starch ──(α-amylase, glucoamylase)──► D-glucose ──(glucose isomerase)──► D-glucose + D-fructose isoglucose
42 – 55 % fructose

──(chromatography)──► D-fructose

saccharose ──(invertase)──► D-glucose + D-fructose invert sugar
50 % each

glucose syrup	isomerization	purification	isoglucose HFS 42
95 – 98 % dextrose	55 – 65 °C, immobilized glucose isomerase in a column system*	activated charcoal ion exchanger removal of water	42 % fructose

	separation of glucose	isoglucose HFS 55
	column chromatography, repeated isomerization	55 % fructose

*reactor cascade using several columns, process stability of the enzyme exceeds 50 d

Enzymes for the hydrolysis of cellulose and polyoses

General. Cellulases and hemicellulases are mostly used in the food industry for the gentle manufacturing of vegetable and fruit purees. Other applications include pulp- and paper-manufacturing processes. In some detergents, alkaline cellulases are used as softening agents for cellulose fibers (cotton). The use of cellulases and hemicellulases for the enzymatic hydrolysis of biomass, in view of its use for the generation of glucose as a fermentation raw material and of energy (ethanol), has been studied intensely but has not yet become economically attractive.

Cellulose is the shape-determining material of the plant cell wall. It is also the most abundant renewable resource: the annual production of cellulose in the biosphere is estimated to be ca. 20 billion t. Cellulose is built of β-1,4-linked units of D-glucose (average degree of polymerization ca. 10 000), which are ordered via hydrogen bonds into parallel bundles (microfibers).

Polyoses (hemicelluloses) are a heterogeneous group of heteropolymers built from pentoses, hexoses, deoxyhexoses, and hexuronic acids. They constitute ca. 20 % of the cell wall. Xyloglucans (xylans) are directly linked to the cellulose microfibers via hydrogen bonds. They consist of a backbone of β-1,4-linked glucose residues which carry numerous xylose residues as β-1,6-branches, which can reach a considerable length. Arabinogalactans bind to the glycoproteins of the cell wall; their backbone is made up of alternating galactose and arabinose residues linked by β-1,4-bonds. Pentosans consist of alternating residues of xylose and arabinose.

Biodegradation. In nature, celluloses and hemicelluloses are broken down by bacteria and especially by white-rot fungi. These fungi hydrolyze cellulases and hemicellulases in parallel with lignin oxidation, using a number of unusual enzymes. The process starts with mechanical softening of the cell wall via growth of the fungal mycelium. In some organisms, e. g., the Clostridia, a number of cellulose-degrading enzymes exhibiting various activities occur as an ensemble combined on a platform, the cellulosome.

Cellulases can be isolated from many microorganisms. Among the better known bacteria are Cellulomonas and Clostridium strains. Fungal cellulases can be isolated from *Trichoderma reesei, Aspergillus niger, Humicola insolens*, and others. The manufacture of these enzymes in a bioreactor follows the usual pattern for extracellular enzymes. The enzyme preparations obtained after cell separation and fractionated precipitation may still contain side activities from hemicellulases, which may be desirable for the intended application. Cellulases for the saccharification of wood are mainly prepared from the white-rot fungus *Trichoderma reesei*. A mixture of cellulolytic enzymes may be manufactured by surface or submersed fermentation in a bioreactor. If surface cultures are used, the enzyme mixture is obtained by extracting the koji with water. In submersed fermentations, the cells are removed and the broth is fractionated by the addition of ethanol.

Hemicellulases. β-Glucanases are prepared by isolation from *Bacillus subtilis, Penicillium emesonii, Aspergillus niger*, and other strains. Mannanases and galactomannanases are isolated from *Aspergillus niger* or *Trichoderma reesei*. Manufacture follows the standard rules for obtaining extracellular microbial enzymes.

Glucose and xylose. For any production methods that start with cellulose or biomass (wood chips, bagasse, and others), the costs of transporting the substrate to the bioreactor and preparing it for fermentation, as well as the cost of the enzyme, are the key parameters. In substrate preparation, lignin is mostly removed by heat and/or aggressive chemical treatments, which are similar to the those used in preparing pulp. In this process, xyloses are partially degraded. Enzyme costs become lower if thermophilic and recombinant cellulases or hydrocellulases can be used.

Composition of polysaccharides in plant cell walls

compo-nent	structure	degree of polymerization	building block	percentage
cellulose	β-1,4-D-glucose	1 000 – 10 000 microfibers	D-glucose	wood*: 60 % cotton: 90 %
pectins	polygalacturonic acids, rhamnogalacturonic acids, galactans, arabi-nogalactans	100 – 2 000	D-galacturonic acid, D-galacturonic acid methyl esters, L-rhamnose, D-arabinose	10 – 40 %
polyoses (hemi-cellulose)	xylans, xyloglucans, β-1,3- and -1,4-D-glucans, galactomannans, arabinogalactans, glucuronomannans		xylose, glucose, galactose, mannose, arabinose, glucuronic acid	20 %

*wood contains ~ 40 % lignin as second major component

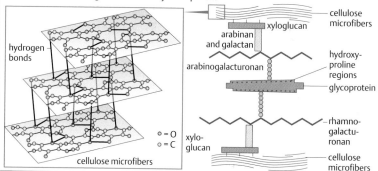

Cellulose and xylan

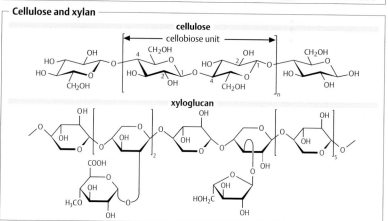

Preparation of cellulase (example)

inoculation culture	surface culture	isolation	
recombinant strains of *Trichoderma reesei* after 100 h	Koji process, 3 – 5 d	extraction with water, precipitation with ethanol	10 000 U after 3 – 5 d

Enzymes in pulp and paper processing

General. Pulp is a fibrous material that consists mainly of cellulose. It is produced by removing most of the lignin and hemicelluloses from wood. Pulp is manufactured into paper, cardboard, and chemical products. In 1996, ca. 150 million t of pulp and ca. 250 million t of paper and cardboard were produced (50 % of the total amount in North America). For producing 1 t of paper, 3.3 t of wood are required; in addition, 0.4 t of petroleum is needed for energy. The wastewater from pulp and paper plants is highly polluted; it contains 1 kg AOX and 55 kg BoD_5 per t of pulp. As a result, improved manufacturing processes are being studied; they should require less material and energy and be less harmful to the environment. The properties of the end products pulp and paper depend both on the type of tree and on the manufacturing process. Besides some hardwoods that grow exceptionally fast (eucalypti, poplars), softwoods are preferred (birches, pines, firs), since their wood consists of longer fibers, which are better to process. Relative to hardwoods, softwoods contain less polyoses (14–17 %) but more lignin 26–32 %). Research based on plant tissue culture and genetic engineering is ongoing in an attempt to reduce the lignin content of softwoods.

Pulp manufacturing. After the trees are cut and their bark peeled, the wood is mechanically cut into chips, which are further transformed into pulp by mechanical, thermomechanical, or chemical means. The dominant chemical process today is the Kraft process (80 % of world production), which is based on the alkaline depolymerization of lignin using $Na_2S/NaOH$ under high pressure at 170 °C. Alternatively, lignin can be removed by sulfonation, treating wood with an excess of sulfite (sulfite pulp). In both processes, a bleaching step using chlorine or chlorine derivatives follows. Enzyme-based process steps are being studied in the context of biopulping and for a milder bleaching of the raw pulp.

Biopulping. Prior to their mechanical degradation, pretreatment of the wood chips with lignin-degrading microorganisms is being studied (biopulping), mostly using white-rot fungi. For example, in the Cartapip™ process of Clariant, wood chips are sprayed with nutrients and spores of the white-rot fungus *Ophiostoma piliferum* and are left for some weeks. This process was investigated to a scale of 100 t; when followed by a Kraft process, yields were higher and the quality of the products was improved.

Enzymatic bleach. The pulp produced both by Kraft and sulfite treatment is of a loose structure and easily accessible to an enzyme. The subsequent bleaching with ClO_2 is necessary to remove the colored by-products of the pulping. The pulping process can be improved and less ClO_2 is required if the pulp is pretreated with xylanases, especially if it comes from hardwoods. Several paper mills in Scandinavia have explored this enzymatic procedure, up to a production scale of 1000 t per day. The amount of ClO_2 could be reduced by 5 kg t^{-1} pulp, compensating for the price of the enzyme; and the wastewater load was reduced by a third. Xylanase complexes from *Clostridium thermocellum* and *Streptomyces roseiscleroticus* were studied in this process, as well as a xylanase preparation from the extreme deep-sea thermophile *Thermotoga maritima* with a temperature optimum of 96 °C; this preparation was obtained by a recombinant process using *E. coli* as a host. Examples of commercially available enzymes are the xylanases from *Trichoderma longibrachiatum* and *T. reesei*. The application of these technologies is, however, still marginal.

Pitch control. Some woods, e.g., pine, contain a rather high amount of triglycerides which, under the drastic conditions of pulp manufacture, tend to form pitches. Microbial lipases have proved valuable for pretreating pine wood chips to prevent pitch formation. Another benefit of this method is that no chlorinated fatty acid material is formed during chlorine bleaching. At present, this is the most important application of any enzyme in pulp and paper processing.

Removal of printers' ink. In recycling printed paper, printers' ink is removed more thoroughly if the paper is pretreated with a mixture of cellulase, xylanase, and lipase.

Processes for pulp and paper manufacture

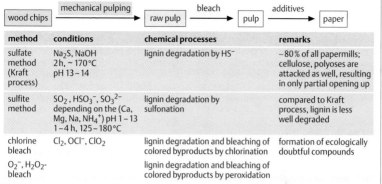

method	conditions	chemical processes	remarks
sulfate method (Kraft process)	Na$_2$S, NaOH 2 h, ~ 170 °C pH 13 – 14	lignin degradation by HS$^-$	~ 80 % of all papermills; cellulose, polyoses are attacked as well, resulting in only partial opening up
sulfite method	SO$_2$, HSO$_3^-$, SO$_3^{2-}$ depending on the (Ca, Mg, Na, NH$_4^+$) pH 1 – 13 1 – 4 h, 125 – 180 °C	lignin degradation by sulfonation	compared to Kraft process, lignin is less well degraded
chlorine bleach	Cl$_2$, OCl$^-$, ClO$_2$	lignin degradation and bleaching of colored byproducts by chlorination	formation of ecologically doubtful compounds
O$_2^-$, H$_2$O$_2$-bleach		lignin degradation and bleaching of colored byproducts by peroxidation	

Biopulping

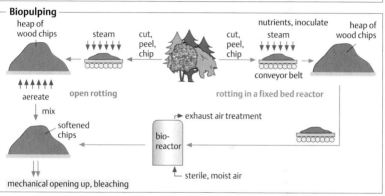

Enzymatic degradation of xylan and pulp bleaching

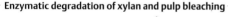

hardwood xylan

softwood xylan

xyl–xyl

xylanases contain:
→ β-xylanase   → α-glucuronidase   → α-arabinosidase   ⇒ esterase   ⇢ β-xylosidase

xyl = xylose, ara = arabinose, MeGlcA = 4-O-methyl-D-glucuronic acid, Ac = acetyl

Paper and cardboard manufacture

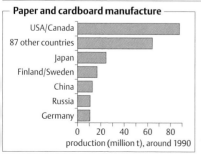

losses in yield trough hydrolysis

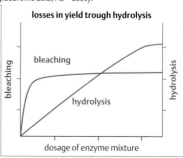

Pectinases

General. Technical pectinases are usually composed of a variety of different hydrolases and lyases. They degrade pectin, thus modulating the texture and viscosity of mashed fruit and vegetables. As a result, they are important enzymes in fruit and vegetable processing and are manufactured on a scale of ca. $1000 \, t \, y^{-1}$ (world).

Pectins are acidic polysaccharides with a M_R $30\,000 - 300\,000$. Their composition varies widely, but their main building block is δ-1,4-linked D-galacturonic acid, which may be esterified with methanol. D-Rhamnose-rich sequences may also occur, carrying sidechains of D-arabinose, D-xylose, and D-galactose. Pectins of high molecular weight which are highly esterified form the water-insoluble middle lamella of the primary cell wall of plants (protopectin, "cell concrete"). Through depolymerization and hydrolysis of the ester bond, protopectin is transformed into shorter-chain pectins which carry an anionic charge and thus can form cross links and gels in the presence of Ca^{2+} ions. This process, which occurs naturally during the ripening of fruits and vegetables, leads to major changes in structural, physical, and chemical properties. Thus, the mild degradation of pectins by pectinase preparations is in demand during the industrial manufacture of purees and fruit juices.

Pectinases include the following enzymes:

1. endopolygalacturonases [EC 3.2.1.15]
2. exopolygalacturonases [EC 3.2.1.67]
3. pectate lyase [EC 4.2.2.2]
   exopectate lyase [EC 4.2.2.9]
4. pectin lyase [EC 4.2.2.10] and pectin esterase [EC 3.1.1.11].

Commercially available pectinases contain different ratios of the above enzymes, leading to the desired property during application.

Technical preparation. Presently, all commercial pectinases are prepared from molds such as Aspergillus or Rhizopus whose enzymes have been cleared for food applications (GRAS status). Most preparations are still prepared by surface fermentation of these molds. For example, the medium may contain wheat bran held in high humidity. After ca. 100 h cultivation, the culture is extracted with buffer solution. Ultrafiltration of the extract leads to an enzyme concentrate which is microfiltered to remove spores. After the addition of stabilizers such as glycerol or sorbitol, the preparation is standardized as to its enzyme activity (which may be quite difficult, because of the presence of a wide range of enzyme activities, e. g., cellulases, hemicellulases, and glycosidases) and sold on the market. Preparations with well defined activities could be prepared by genetic engineering techniques, but their acceptance is still questionable, due to consumers' concerns. In addition to concentrates, spray-dried dry enzyme preparations are also available.

Applications. Pectinase preparations are used 1) for maceration of vegetables and fruits ("macerases"), 2) for treating mashes during preparation of fruit juices or grape juice, to enhance the filterability and yield, 3) in the treatment of grape must, for removing suspended pectin material, and 4) in processing citrus fruits, to prevent the formation of gels. The macerating action of pectinases is often used in the manufacture of fruit and vegetable purees for baby foods, turbid fruit juices, and fruit yoghurts. If pectinase is added during juice manufacture, yields can be improved by 5 – 10 %, and filtration is enhanced by a factor of 1.5 to 5, depending on the type of fruit.

Cellulases and hemicellulases are often used in food processing. Examples include the preparation of starch from corn and the extraction of coffee beans and tea leaves. If added to pectinases, they can further improve vegetable and food processing. Bacterial or fungal glucanases help to open up the malt used in breweries, thus rendering faster mash formation during the beer production process.

Pectin

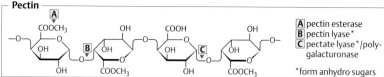

A pectin esterase
B pectin lyase*
C pectate lyase*/poly-
galacturonase

*form anhydro sugars

Juice production (e.g., grape, red currant, apple juice)

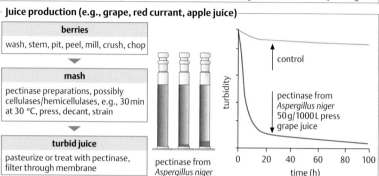

berries

wash, stem, pit, peel, mill, crush, chop

mash

pectinase preparations, possibly cellulases/hemicellulases, e.g., 30 min at 30 °C, press, decant, strain

turbid juice

pasteurize or treat with pectinase, filter through membrane

pectinase from Aspergillus niger

control

pectinase from Aspergillus niger 50 g/1000 L press grape juice

Processing of carrots

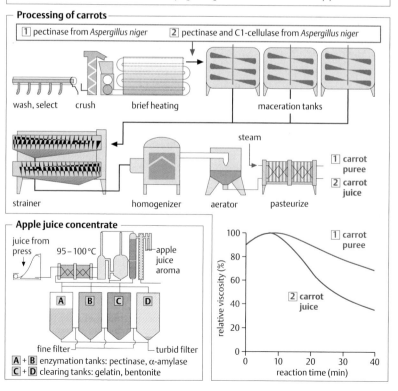

1 pectinase from Aspergillus niger   2 pectinase and C1-cellulase from Aspergillus niger

wash, select   crush   brief heating   maceration tanks

strainer   homogenizer   aerator   pasteurize

steam

1 carrot puree
2 carrot juice

Apple juice concentrate

juice from press   95 – 100 °C   apple juice aroma

A B C D

fine filter   turbid filter

A + B enzymation tanks: pectinase, α-amylase
C + D clearing tanks: gelatin, bentonite

1 carrot puree
2 carrot juice

91

Enzymes and milk products

General. During the manufacture of milk products, proteases, lactases, and lipases are mainly used. A key enzymatic process is the use of highly specific proteases (rennins) for cheese production; ca. 1000 t of enzyme are used for this purpose (1995). Secondly, the aroma of cheese can be influenced by the action of lipases and proteases. Since lactose (milk sugar) is not tolerated well by many adults, lactase (β-galactosidase) -treated milk products have been developed. In some developing countries, milk products are occasionally stabilized against microbial attack by adding lysozyme or H_2O_2/catalase. Finally, whey has been often investigated as a fermentation raw material; it is produced in quantities of ca. 50 million t (world) during the preparation of cottage cheeses and cheese.

Milk is an oil-in-water emulsion with a water content of ca. 90 %. The triglycerides of milk (butterfat) are unusual in that they contain 2–4 % butyric acid. The percentages of lactose and protein are in the range of 3 % each. Casein, the main component of the protein fraction, is a mixture of phosphoproteins with M_R 20–30 kDa. It forms aggregates of ca. 1 million Da, the κ-casein fraction serving as a protective colloid. To become digestible, this colloid must first be degraded. This can occur in various ways: by the presence of Ca^{2+} ions in concentrations > 6 mM, by a decrease of the pH to values < 4.6, and through hydrolysis of a single peptide bond of κ-casein (^{105}phe–^{106}met) with chymosin (rennin), a protease of the upper digestive tract of mammals.

Casein-hydrolyzing proteases. In the stomach of mammals, casein is first hydrolyzed by the combined action of chymosin and acid, becoming susceptible to further proteolytic attack. The related technical procedure for transforming milk into sour milk or cheese by the addition of cow, goat, or sheep rennet is one of the oldest inventions for preserving this highly perishable food. In traditional cheese making, preparations of animal stomachs were used for this purpose. They were, however, rather impure and difficult to standardize. As a consequence, today microbial rennets or, more recently, recombinant chymosin is preferred. Microbial rennet, a protease from *Mucor miehei*, exhibits the same specificity (^{105}phe–^{106}met) as animal chymosin and is widely used. In 1987, calf chymosin was cloned and expressed both in *E. coli* and *Saccharomyces cerevisiae*. Since 1992, the use of recombinant chymosin in cheese making has been allowed by the FDA in the USA, since 1997 also in the EU and in Japan.

Lactose hydrolysis. Lactose (milk sugar) is tolerated well by most mammals only until maturity. Adult humans, with the notable exception of most Northern Caucasians, exhibit a similar intolerance towards lactose. This occurs because the formation of lactase becomes largely reduced by adulthood, resulting in the passage of undigested lactose into the large intestine; there it is fermented by the intestinal flora, leading to excessive gas formation. Lactose intolerance must be clearly distinguished from the rare genetic disease galactosemia. The latter is due to an autosomal recessive gene defect on chromosome 9, leading to a phenotype in which UDP-galactose is not synthesized. This results in an overproduction of galactose metabolites, in particular of toxic galactol. Whereas a galactose-free diet is indispensable for people who have galactosemia, lactose intolerance can be avoided by not ingesting lactose or by enzymatic hydrolysis of lactose in milk products. Milk products pretreated with lactase are available for the manufacture of fermented milk products, e. g., yoghurts, for hydrolyzed lactose syrups used in bakeries, and for animal feeds based on whey. The sweetness of hydrolyzed lactose exceeds that of the native disaccharide.

Cheese aroma. The short- and medium-chain fatty acids occurring in butterfat can be partially hydrolyzed by lipases to a product mixture useful for their aroma in cheese production (enzyme-modified cheeses, EMC). Depending on the chain-length specificity of the lipase used, various aroma notes can be produced.

Composition of milk

	milk (%)	whey (%)
water	~ 88	~ 94
fat	~ 3–4	~ 0.5
protein	~ 3.3	~ 1
casein	~ 2.6	–
lactose	–	~ 4.8

Plasmid

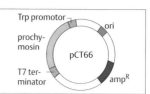

for the expression of chymosin in *E. coli*

Trp promotor — ori
prochymosin
pCT66
T7 terminator — amp^R

Processing of milk

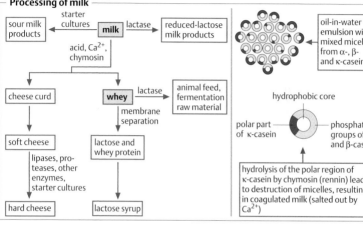

sour milk products ← starter cultures ← **milk** → lactase → reduced-lactose milk products

acid, Ca^{2+}, chymosin

cheese curd — whey → lactase → animal feed, fermentation raw material

soft cheese — whey → membrane separation → lactose and whey protein

lipases, proteases, other enzymes, starter cultures

hard cheese — lactose syrup

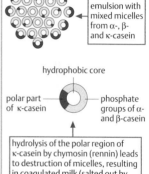

oil-in-water emulsion with mixed micelles from α-, β- and κ-casein

hydrophobic core

polar part of κ-casein — phosphate groups of α- and β-casein

hydrolysis of the polar region of κ-casein by chymosin (rennin) leads to destruction of micelles, resulting in coagulated milk (salted out by Ca^{2+})

Manufacture of chymosin

native	microbial	recombinant
stomachs of young animals	**preculture**	**recombinant microorganism**
cutting, activation at pH < 5	high-yield mutants of *Mucor miehei* or *M. pusillus*	*Escherichia coli*
extraction	**bioreactor**	**bioreactor**
salt water, 14 d	dextrose syrup, soy meal, 30°C, 72 h	maltodextrins, 37°C, 36 h
purification	**purification**	**purification**
ultrafiltration standardization	separation of mycelium, reverse osmosis, precipitation	isolation of inclusion bodies, Tritron-X100/EDTA, urea-/alkali-extract, ion-exchange chromatography, acid treatment
200 U/kg stomach	5 000 U/m^3 in 72 h	20 000 U/m^3 in 36 h

Lactose intolerance and galactosemia

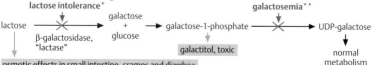

lactose intolerance [*]   **galactosemia** [**]

lactose ⟶ galactose + glucose ⟶ galactose-1-phosphate ⟶ UDP-galactose

β-galactosidase, "lactase"

galactitol, toxic

normal metabolism

osmotic effects in small intestine, cramps and diarrhea

[*] >70 % of adult Bantus, American Blacks, Indians, Chinese, Aboriginees
[**] galactose-1-phosphate-uridyltransferase defect on chromosome 9, frequency 1:100 000

93

Enzymes in baking and meat processing

General. The most important applications of biotechnology for bakery products are the preparation of yeast dough and of sourdough. For sourdough, a combined fermentation of lactic acid bacteria and *Saccharomyces cerevisiae* increases the digestibility of rye meal. Enzymes such as α-amylases, glucoamylases, proteases, and xylosanases are used as well in processing dough and bakery goods. Specific xylanases are used to inhibit the formation of pentosane-based gels, reducing the viscosity and thus improving the processing of rye and wheat flours. α-Amylases, glucoamylases, and proteases can be advantageously used in all sorts of flours since they modulate the properties of starch and of gluten, leading to a desirable viscosity of the dough and an improvement in its fermentability. Maltogenic β-amylase is used as an additive to prevent bread from becoming stale. The industrial use of enzymes in bakeries is estimated at around 1000 t (world). For the processing of sausages, starter cultures are often used. They influence both aroma and preservation. Meat can be tenderized by the addition of proteases; papain is the major protease used for this application.

Meal processing and enzymes. The milling of grain leads to a flour whose composition with respect to starch, pentosanes, and proteins depends strongly on the type of grain (wheat, rye, etc.), but also on the composition of the soil, the climate, and the time of harvesting. The composition can be modulated through the addition of appropriate enzymes. Amylases depolymerize the starch to dextrins (α-amylase), maltose (β-amylase), and finally to glucose. As a consequence, their addition influences the processing, the aroma, and the volume of the dough (because bakers' yeast can ferment only mono- and disaccharides). A maltogenic β-amylase from *Bacillus stearothermophilus* has proven an excellent enzyme additive to prevent the staling of bread. The gluten protein of the grain binds some of the water added during dough production, forming a gel. If proteases are added, this gel is partially destroyed, and the elasticity of the dough is increased – a prerequisite for good retention of the CO_2 that is formed during fermentation. An excess of added protease can lead, however, to a reduction in the required viscoelastic properties and thus to a softening of the overall structure. In practice, α-amylase from malt, from Bacillus, or from mold strains are used. Mold glucoamylase is also used. Since the industrial production of bread requires ever shorter times for dough preparation and fermentation, high enzyme activities are used. For proteases, the preferred enzyme has been isolated from *Aspergillus oryzae*. It loosens the gluten in the desired manner without leading to excessive degradation.

Analytical methods. The field is dominated by traditional methods related to traditional craftsmanship. Thus, gas formation is monitored in gas tubes, and the decrease in viscosity is measured by a falling number method. For the evaluation of protease action, farinographs, extensograms, or alveograms are used, which all give some evidence about the viscoelastic properties of the gluten.

Meat and enzymes. Meat is formed from muscle through complex biochemical transformations after the supply of oxygen has stopped. Proteases (cathepsins) play an important role in these transformations. From a consumer's point of view, a meat that is juicy, easy to chew, and of a pleasing texture, color, and taste are the major concerns. In Western cultures, this was traditionally achieved by storage and marination. Other cultures wrap meat in papaya leaves or dip it into pineapple juice to tenderize it and enhance the aroma. Spraying meat with papain, a sulfhydryl protease isolated from papaya leaves, has the same effect. Alternatively, inactivated papain, in which the sulfhydryl group is oxidized with H_2O_2, can be injected into the bloodstream of animals shortly before they are killed. Once the animal has died, the sulfoxy groups of the papain are reactivated by the reduction of oxygen in blood, resulting in rapid partial hydrolysis of the meat.

Meal processing and enzymes

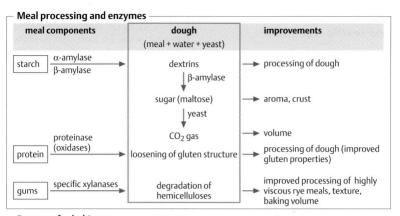

meal components	dough (meal + water + yeast)	improvements

starch — α-amylase / β-amylase → dextrins ↓ β-amylase → sugar (maltose) ↓ yeast → CO₂ gas → loosening of gluten structure

protein — proteinase (oxidases) →

gums — specific xylanases → degradation of hemicelluloses

- dextrins → processing of dough
- sugar (maltose) → aroma, crust
- CO₂ gas → volume
- loosening of gluten structure → processing of dough (improved gluten properties)
- degradation of hemicelluloses → improved processing of highly viscous rye meals, texture, baking volume

Enzymes for baking

enzyme	effect on meal	method of determination	effect on baked product
α-amylase from malt, Bacillus, Aspergillus glucoamylase from *Aspergillus oryzae*	opening up of damaged starch grains and sticky starch to maltose and glucose	gas formation*, viscosimetry (amylograph or falling number), baking experiments	increased volume, better crust and taste
neutral proteases from molds	degradation of gluten	viscoelastic properties in farinograph, extensogram etc.	enhanced viscosity of dough, gas is better retained
xylanases from *Trichoderma viride*	hydrolysis of gums	viscosimetry in amylograph	improved fermenting tolerance
β-amylase from *B. stearothermophilus*, Aspergilllus	prevents interaction of starch and gluten	baking experiments, organoleptic test	prolonged freshness, better texture, antistaling

Saccharomyces cerevisiae cannot metabolize higher oligosaccharides, but only maltose and glucose

Extensability of dough

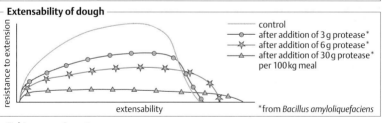

resistance to extension

- .... control
- ⊙ after addition of 3 g protease*
- ✳ after addition of 6 g protease*
- △ after addition of 30 g protease* per 100 kg meal

extensability

*from *Bacillus amyloliquefaciens*

Baking experiment

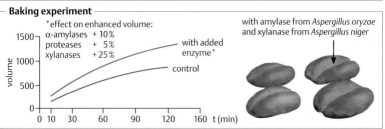

*effect on enhanced volume:
α-amylases + 10 %
proteases + 5 %
xylanases + 25 %

with added enzyme*

control

volume — t (min)

with amylase from *Aspergillus oryzae* and xylanase from *Aspergillus niger*

95

Enzymes in leather and textile treatment

General. The preparation of leather from animal hides can be traced back to antiquity. Since rather harsh chemicals such as lime, alkali, and sulfur have been used throughout history, and feces and urine were unknowingly used as an enzyme source, tanning was considered an "unclean profession". Otto Roehm in Germany was the first to lay the foundation for a science-based technical approach to leather treatment, resulting in a highly improved image of this craft. Proteases for leather treatment today are used at a level of several 100 t per year (world).

Leather. Animal skin consists of 60–65 % water, ca. 30 % proteins (> 90 % collagen, with keratin, elastin, and others as minor components), and 2–10 % fat. The epidermis (upper skin) is just ca. 1 % of the skin; it is further divided into the outer *stratum corneum* (horny layer), a granular layer, and a mucosal layer. The dermis/*coreum* (leather skin) makes up 85 % of the skin's thickness; it is composed of the papillary layer, with collagen fibers, and the reticular layer, made up of connective tissue. The final layer is the hypodermis, which contributes about 15 % to the skin's thickness and contains collagen, muscle, and fat tissue, blood vessels, and nerves. Leather is made from the leather skin layer from which hairs, fat tissue, nonfibrous protein, and water have been removed and which has been stabilized by various process steps. In a humid environment, non-stabilized hides decay quickly and lose their malleability upon drying.

Leather treatment and enzymes. Immediately after their removal from the carcass, hides are conserved by removing water (e. g., by salting) to prevent microbial attack. In a so-called water shop, several steps ensue: during soaking, blood, dirt, salts, fats, and nonfibrous protein are removed by the addition of water, surfactants, and reducing agents. In this step, the leather skin reabsorbs water.

Proteolytic enzymes support this softening process significantly: they help to remove pigments, fat, and sweat glands – a prerequisite to arrive at high-value "aniline leather" with a clear scar pattern. The enzymes used in this process must exhibit high proteolytic activity without attacking collagen. Trypsin (pancreatic extracts) and proteases from *Bacillus subtilis* or *Aspergillus sojae* are well suited for this purpose. In the following liming process, epidermis and remaining hairs are removed, and the skin is opened up for the later tanning procedure. Lime and sodium sulfite are especially suited for this step. In the traditional craft, they were applied by hand on the inside of the hide. Today, strongly alkaline reagents composed of calcium hydroxide and sodium sulfite are used, and alkali-stable proteases are added (e. g., from *Bacillus alkalophilus*). During the ensuing bating process, alkali is removed by the addition of ammonium salts or organic acids. Pancreatic enzymes, as well as neutral and alkaline bacterial or mold proteases, may assist in this step to remove remaining noncollagen proteins and to loosen the collagen for dying. Mold proteases are also used for loosening chromium-treated leathers (wet blues). Attempts to combine soaking, liming, and tanning into one processing step based on intense enzyme treatment have not yet been accepted in the industry, although the number of steps, their duration, and the consumption of water were reduced by ca. 50 % in each step.

Desizing of textiles. Weaving textile fibers exposes the fabric's warp and weft threads to substantial mechanical stress. To prevent breakage of the threads during this step, they are usually reinforced with sizing. For cost reasons, starch is usually used as the sizing agent. It adheres to the threads very well and can be removed easily after weaving, before the ensuing steps of dying, bleaching, and fortifying. Bacillus α-amylase is usually used for this purpose. Stone-washed textiles are also treated with this enzyme. In addition, cellulases may be used for this process.

Leather manufacturing

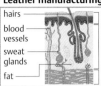

hairs — upper skin (epidermis)

blood vessels

sweat glands — leather skin (coreum)

fat — hypodermis connective tissue

salt curing (preservation)

raw hide stock depot

soaking, liming (hair loosening, opening up corium)

[1]

 fleshing: hides in this state called "white" or "limed" stock

 splitting skins: corium is cut horizontally by guiding skins along band knife splitter

 grain layer

corium

flesh layer (split)

splitting skins

deliming, bating, pickling, chrome-tanning

[2]

 wringing

 shaving knife

shaving

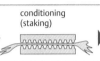 at this stage of manufacturing, leather is called "wet-blue"

neutralizing, retanning, dyeing, fat liquo-ring

[3]

 wringing

setting

drying

conditioning (staking)

stretching

at this stage of manufacturing, leather is called "crust"

Enzymes used in leather manufacturing

	processing step	pH	target proteins	enzymes
[1]	soaking (degradation of proteins, opening up of corium, hair loosening)	7–9	non-fibrillary proteins, fats	proteases from molds, bacteria; trypsin (pancreas extracts)
[1]	dehairing	10	keratin	keratinases, proteases
[1]	liming	12.5	non-collagen proteins	proteases from bacteria, molds; elastase
[2]	deliming, bating	8–9	residual proteins, fat cells	proteases from bacteria, molds; trypsin (pancreas extracts)
[2]	pickling (enhancement of elasticity, improved dyeing)	5–6	collagen	proteases from bacteria, molds; trypsin (pancreas extracts)
[3]	wet blue	6	chromium-tanned leather	mold proteases

Desizing of textiles

 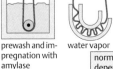

prewash and im-pregnation with amylase

water vapor

starch degradation

wash bath

normal or thermophilic α-amylase from Bacillus strains, T = 60–110 °C, depending on enzyme

Procedures for obtaining novel technical enzymes

General. The isolation of new enzymes has accelerated, due to improved screening procedures, expression in host organisms which can be handled easily, and enhanced productivity using mutagenesis and selection or genetic engineering techniques. In spite of this progress, enzymes are still only rarely competitive with organic synthesis. One reason is that, for some important synthetic unit operations such as C–C coupling, only a few enzymes are available that do not require expensive cofactors; aldolases are one positive example. On the other hand, the time required to develop and optimize an enzyme process is often too long to suit the short time requirements encountered in the establishment of GMP-attested protocols as required in the pharmaceutical and agricultural industries. Current ideas for improving this situation include 1) rational protein design, 2) directed evolution, and 3) novel screening procedures that are better suited to enhance the high natural diversity of biocatalysts.

Novel habitats. Organisms are adapted to very diverse habitats (ecological niches) and usually have evolved enzymes that are perfectly suited to this environment. Microorganisms are especially versatile at adapting to habitats exhibiting a wide range of pH values, temperatures, and osmotic and other conditions. A systematic screening in unusual habitats such as highly acid or alkaline environments, deserts, deep-sea sediments, hot springs on the surface of the earth or on the bottoms of the oceans have allowed the identification of a large number of unusual microorganisms and the isolation of novel enzymes from them. For example, the thermostable DNAse I used in most PCR reactions (Taq polymerase) was isolated from the thermophilic prokaryote *Thermus aquaticus*, which was first isolated from the 90 °C waters of Old Faithful, a geyser in California's Yellowstone Park. Deep-sea submarine expeditions to hydrothermal vents, 1500 m deep, on the bottom of the Mediterranean Sea have led to the isolation of hitherto unknown microorganisms that belong to the group archaea. Due to the extreme environmental conditions in their environment, their DNA polymerases show even better fidelity in DNA replication than that of *Thermus*. Some of these novel enzymes have become available commercially (e. g., Pfu- and Tma-polymerase).

Novel screening methods. Besides traditional screening methods, the modification of enzymes by protein design or directed evolution has become a powerful tool for optimizing an enzyme for its intended technical application. Directed evolution in particular has allowed rapid adaptation of enzymes to the desired temperature, pH, organic solvent or substrate. The flood of sequence information resulting from genome sequencing, analyzed by the tools of bioinformatics, has also opened up the potential to identify new enzymes. In the ever-increasing number of completely sequenced genomes, many thousands of enzymes could be identified due to their fingerprints (consensus sequences, as already known from enzymes with similar functions). They can be isolated rather easily using PCR techniques or hybridization probes for a suitable combination of fingerprints. The search for their natural substrate, however, can often be quite troublesome. By analyzing sequences in DNA samples isolated from soil which code for conserved regions of the 16S- or 23S-ribosomal RNA, it has been concluded that < 1 % of all microbial species have yet been cultivated and classified. To circumvent this limitation, methods have been successfully established to isolate new enzymes from noncultivated microbes by isolating DNA from soil, cutting it with restriction enzymes, and expressing it in suitable host organisms, using appropriate high-throughput assays to screen for the desired enzyme activity. Although these methods are not yet fully mature, they have already resulted in the detection and isolation of novel enzymes with unusual properties.

Diversity (all known organisms)

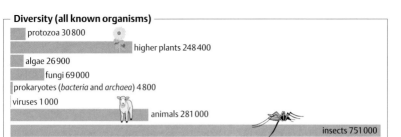

protozoa 30 800

higher plants 248 400

algae 26 900

fungi 69 000

prokaryotes (*bacteria* and *archaea*) 4 800

viruses 1 000

animals 281 000

insects 751 000

Properties of various DNA polymerases

origin	half life time at 95 °C (min)	3'→5'- exo-nuclease activity	5'→3'- exo-nuclease activity	type of terminus
Thermus aquaticus	40	not present	present	3'A
Thermotoga maritima	20	not present	present	3'A
Pyrococcus furiosus	>120	present	not present	n.i.
Thermococcus litoralis	1 380	present	not present	>95% blunt end

Diversity of lipases (Conolly structures and consensus sequences)

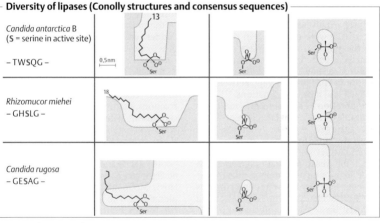

Candida antarctica B
(S = serine in active site)

– TWSQG –

Rhizomucor miehei
– GHSLG –

Candida rugosa
– GESAG –

Strategies towards new enzymes

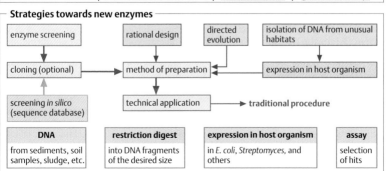

| enzyme screening | rational design | directed evolution | isolation of DNA from unusual habitats |

cloning (optional) → method of preparation ← expression in host organism

screening *in silico* (sequence database)

technical application → traditional procedure

DNA	restriction digest	expression in host organism	assay
from sediments, soil samples, sludge, etc.	into DNA fragments of the desired size	in *E. coli*, *Streptomyces,* and others	selection of hits

99

Bakers' yeast and fodder yeasts

General. The preparation of dough from flour is documented back to early history: Egyptian clay tablets tell about the baking of "beer breads", where moist barley exposed to a yeast fermentation was used. Until the 19th century, bakers in Europe also used brewers' yeast for baking. It was obtained either by the "Dutch process" (ca. 1750), based on filtration of a turbid mash, or by the "Vienna process" (ca. 1800), where yeast was skimmed off the fermentation vat. Around 1870 Louis Pasteur in France discovered that the production of bakers' yeast (*Saccharomyces cerevisiae*) in high yield with little concomitant formation of ethanol was possible with strong aeration. Since then, bakers' yeast has been produced in stirred vats under aeration. Molasses is widely used as a carbon source. The first processes for preparing yeasts as animal feeds (fodder yeasts) were developed in the 1930s for the purpose of creating a self-sufficient economy. In the post-war period, however, a new driving force appeared: closure of the "protein gap" between developed and developing nations.

Fermentation raw materials. Due to its favorable price and high content of sugar, nitrogen sources, vitamins, and minerals, molasses (from cane or beet sugar) has become an important raw material for yeast production. It remains after extraction of chopped sugar cane or sugar beets by hot water and contains 40–50 % saccharose, which is hydrolyzed by yeast invertase into glucose and fructose-sugars that can be metabolized by yeast.

Technology. To obtain a high yield coefficient (Y_S) for the transformation of molasses into yeast cell mass, two factors are essential: strong aeration during fermentation to suppress ethanol formation, and strict control of the glucose concentration in the medium. If glucose concentrations exceed 100 mg L^{-1}, respiration is decreased even at high oxygen concentrations, and ethanol is formed (catabolite repression, "aerobic fermentation", "Crabtree effect"). As a result, modern fermentation procedures use optimized aeration and stirrer systems and a computer-based feed control for the molasses. In an optimized process, 1 kg H_{27} of yeast per kg molasses is formed. H_{27} is a relative value used in the yeast industry to calculate yields, indicating pressed yeast with 27 % dry weight. Since molasses contains ca. 50 % sugar and since the yield of yeast related to sugar substrate is ca. 54 g dry weight/ 100 g sugar, the yield of yeast is ca. 1 kg H_{27} kg^{-1} molasses. Traditionally, workup procedures led to a pressed yeast having limited storage stability, necessitating local production sites for distribution. Today, the cell mass is dried in a vortex dryer to yield dried bakers' yeast that is stable for several months and is reactivated by water and sugar within a few minutes, an important convenience feature for bakers who begin working early in the morning.

Fodder yeasts compete in price with other protein-rich residual materials such as fish meal and especially soy meal. As a result, the price of the carbon source is critical for the competitiveness of a fodder yeast manufacturing process. Oil fractions, natural gas, ethanol, methanol, celluloses, hemicelluloses, starch, and whey were studied as carbon sources. Of these, several have gained economic success. Thus, the Symba process, based on potato starch and the yeasts *Endomycopsis fibuliger* and *Candida utilis* and the Rank-Hovis McDougall process have been implemented. The latter, which is based on the fungus *Fusarium graminearum*, transforms glucose to the low-calorie food *Quorn*. Another carbon source is spent sulfite liquor that remain from the extraction of wood chips during pulp manufacture and which contain ca. 4 % polyoses. In the Finnish *Pekilo* process, the fungus *Paecilomyces variotii* is used for this purpose. In the Canadian *Waterloo* process, the fungus *Chaetomium cellulolyticum* is used to produce cell mass from agricultural and forest waste materials such as straw, bagasses, manure, or sawmill chips.

Baker's yeast

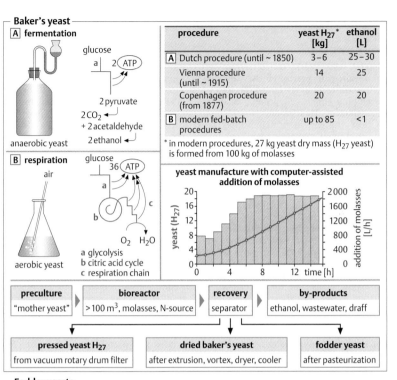

A fermentation

glucose

a → 2 ATP

2 pyruvate

$2 CO_2$ + 2 acetaldehyde

2 ethanol

anaerobic yeast

B respiration

glucose

a → 36 ATP

O_2 H_2O

a glycolysis
b citric acid cycle
c respiration chain

aerobic yeast

procedure	yeast H_{27}* [kg]	ethanol [L]
A Dutch procedure (until ~ 1850)	3–6	25–30
Vienna procedure (until ~ 1915)	14	25
Copenhagen procedure (from 1877)	20	20
B modern fed-batch procedures	up to 85	<1

* in modern procedures, 27 kg yeast dry mass (H_{27} yeast) is formed from 100 kg of molasses

yeast manufacture with computer-assisted addition of molasses

preculture	bioreactor	recovery	by-products
"mother yeast"	>100 m³, molasses, N-source	separator	ethanol, wastewater, draff

pressed yeast H_{27}	dried baker's yeast	fodder yeast
from vacuum rotary drum filter	after extrusion, vortex, dryer, cooler	after pasteurization

Fodder yeasts

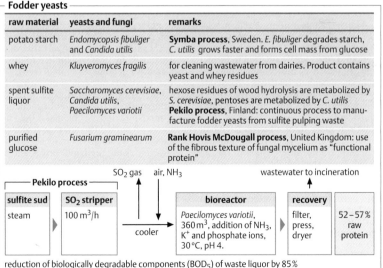

raw material	yeasts and fungi	remarks
potato starch	*Endomycopsis fibuliger* and *Candida utilis*	**Symba process**, Sweden. *E. fibuliger* degrades starch, *C. utilis* grows faster and forms cell mass from glucose
whey	*Kluyveromyces fragilis*	for cleaning wastewater from dairies. Product contains yeast and whey residues
spent sulfite liquor	*Saccharomyces cerevisiae*, *Candida utilis*, *Paecilomyces variotii*	hexose residues of wood hydrolysis are metabolized by *S. cerevisiae*, pentoses are metabolized by *C. utilis* **Pekilo process**, Finland: continuous process to manufacture fodder yeasts from sulfite pulping waste
purified glucose	*Fusarium graminearum*	**Rank Hovis McDougall process**, United Kingdom: use of the fibrous texture of fungal mycelium as "functional protein"

Pekilo process

SO_2 gas   air, NH_3 wastewater to incineration

sulfite sud	SO_2 stripper	bioreactor	recovery	
steam	100 m³/h	*Paecilomyces variotii*, 360 m³, addition of NH_3, K^+ and phosphate ions, 30 °C, pH 4.	filter, press, dryer	52–57% raw protein

cooler

reduction of biologically degradable components (BOD_5) of waste liquor by 85 %

Single cell protein, single cell oil

General. After World War II, technological advances in petrochemistry and fermentation technology led to the idea of developing single cell protein (SCP) by fermentation. This product would help to close the gap in protein supply between affluent and poor populaces in a world where the human population showed exponential growth. The potential of single cell oil (SCO) had been explored during the war in Germany, in answer to the Allies' sea blockade which had completely halted plant-oil imports from Asia and the Americas.

Single cell protein: raw materials and microorganisms. Yeasts such as *Candida tropicalis* and *C. bombicola* can grow on alkanes, which are terminally oxidized within the peroxisomes to mono- and dicarboxylic acids and then metabolized further. Based on these observations, the use of high-boiling petroleum fractions as a C-source for yeast protein formation was intensely explored between 1965 and 1975. As a result of the two oil crises, petroleum was gradually replaced by methanol. This C-source can be assimilated by a wide range of methylotrophic bacteria and yeasts such as *Hansenula polymorpha, Pichia pastoris, Candida boidinii, Methylophilus methylotrophus*, and *Methylomonas clara*. The metabolism of methanol starts with oxidation to formaldehyde, which is incorporated into the pentose phosphate pathway. Formaldehyde can also be further oxidized to CO_2 via formic acid, providing reduction equivalents.

Yeast cell mass from alkanes. High-boiling alkanes are only sparingly soluble in water, but emulsifiers should not be added since they might end up in the yeast product and may also lead to foam problems during fermentation. As a result, optimization of aeration and of the mechanical stirrer system is of key importance in process development. Since the aeration level in the process is high (10 vvm, due to the requirement of 16 mol O_2 per mol hexadecane), formation of foam must be carefully controlled. The 40 m^3 bioreactors in Grangemouth,

UK (BP) and a 100 m^3 bioreactor plant in Sardinia used Intermig stirrers and mechanical foam separators. With fed-batch cultures fermented for 5 d, 0.9 kg of moist *Candida tropicalis* kg^{-1} alkane was produced. Cell mass was recovered using separators. Since the high RNA content (4%) of the product was critical from a dietary point of view, it was reduced by careful autolysis. Using vortex dryers, the yeast was stabilized for transport.

Methanol-based processes. Methanol has a low boiling point (64.5 °C) and is toxic even for methylotrophic microorganisms in concentrations > 100 mg L^{-1}. Using a narrow, tall airlift fermenter (ICI Billingham, UK: 8×60 m) and up to 600 computer-controlled nozzles, the distribution of methanol was controlled, and its solubility was enhanced due to the high hydrostatic pressure. The considerable aeration rate required was achieved by a loop system. Using the bacterium *Methylophilus methylotrophus*, a continuous 2-d process resulted in 0.5 kg biomass kg^{-1} methanol. Cell mass was separated using separators, the RNA content was reduced by careful autolysis, and a vortex dryer was used to render the protein product stable for transport.

Acceptance and economics. The use of alkane yeasts as a food was quickly criticized, in view of a potential enrichment of carcinogenic polyaromatic compounds from petroleum. Though in-depth toxicological studies did not support this view, registration in 1974 was limited to use as an additive in domestic animal feeds. The price increases of petroleum during the two oil crises finally led to termination of the projects. The market introduction of methanol-derived biomass did not meet with such difficulties, as it was declared a feed additive from its beginning. However, rising methanol prices and an EU-wide subvention of milk powder as a feed protein made this process also uneconomic.

Single cell oil can be obtained from glucose-using yeasts such as *Rhodotorula glutinis* or fungi such as *Mortierella isabellina*. Oil yields > 60% of dry cell mass have been reported. The composition of the triglycerides is similar to that of plant lipids and thus does not have an economic advantage on an open world market.

Single cell protein (SCP)

carbon-source	microorganisms	problems
high-boiling alkanes	*Candida lipolytica, Candida tropicalis*	isolation difficult (formation of emulsions), oil residues, taste, consumer acceptance
methane	*Methylococcus capsulatus*	high O_2 consumption, strongly exothermic process (cooling), danger of explosions
methanol	*Hansenula polymorpha, Pichia pastoris, Candida boidinii, Methylophilus methylotrophus*	high content of RNA

Composition (%) of soy meal and milk powder as compared to single cell protein

SCP-product	raw protein	amino acids	fats	nucleic acids	salts, water
TOPRINA (BP)	60	54	9	5	10
PRUTEEN (ICI)	83	65	7,4	15	10
soy meal	45	40	2	-	18
whole milk powder	25	-	27	-	10

SCP yields (pilot and production plants)

microorganism	carbon source, metabolism	Y_s*	productivity [kg/m³·h]
Candida lipolytica (TOPRINA)	alkanes	0.95	2
Methylophilus methylotrophus (PRUTEEN)	methanol, RMP	0.53	8 – 10
Candida utilis	ethanol	0.8	4.5
Saccharomyces cerevisiae (bakers' yeast)	molasses, FDP	0.85	5.2
Paecilomyces variotii (PEKILO)	sulfite suds, FDP	0.6	2.7

*g cell dry weight/g carbon source
RMP: ribulose monophosphate pathway, FDP: fructose diphosphate pathway

PRUTEEN plant of ICI

Manufacture of PRUTEEN

raw materials	airlift reactor	flocculation and flotation	isolation	
methanol from 600 nozzles, ammonia, salts	3 000 m³ working-volume; 60 m high, 600 t	30 min at 64 °C, RNA content is reduced from 80 to 2 mg/g	separation, drying	83 % raw protein, <1 % nucleic acids

Aerobic wastewater treatment

General. Aerobic wastewater treatment was introduced about 100 years ago, using trickling filters and aeration basins. In combination with the construction of sewage systems, which were started in ca. 1850, these measures resulted in a dramatic increase in human life expectancy, due to a decrease in epidemics. Today, in most industrial countries almost all wastewater is biologically treated before entering surface waters. (Germany 2000: ca. 10 000 municipal sewage plants), but worldwide there remains much to do in many coastal regions, where considerable amounts of unpurified wastewater are still entering the sea, and in those developing countries where industrialization and rapid population growth coincide.

Composition of wastewater. Domestic and industrial wastewater have different compositions. Although the latter varies depending on the type of industry producing it, domestic wastewater is surprisingly constant in composition, after temporary deviations have been averaged. It contains an organic load of 60 g BOD_5 per inhabitant per day ("inhabitant equivalent"). BOD_5 (biochemical oxygen demand after 5 days) is measured as oxygen consumption using a standardized procedure. Other important parameters are the COD (chemical oxygen demand) and the TOC (total organic carbon), which are determined by chemical oxidation of a wastewater sample and thus represent the total organic load as well as the oxidizable inorganic components of a wastewater sample. Many wastewaters are of mixed domestic and industrial origin.

Microbiological aspects. In the sludge of an aerobic wastewater treatment plant, a large variety of microorganisms, algae, and protozoa form a symbiotic microcosm (biocenosis). If only domestic wastewater is treated, these populations are quite constant over long periods of time. The addition of industrial wastewater may dramatically alter the composition of these populations. Recently, sludge starter cultures have become available which are specialized for the oxidation of certain industrial waste components and thus aid the purification of industrial wastewater.

Trickling filter process. In this process, mechanically pretreated wastewater is percolated through a tower filled with materials having large surface areas (e.g., lava stones). As a result, sludge consortia start to form on the surface of these materials, oxidizing wastewater components in a continuous manner. The oxidative capacity of the system is limited by the diffusion of oxygen. Excess sludge is removed by flushing under pressure and further treated by anaerobic sludge digestion.

Aeration basins (aeration tanks). In this process, wastewater is retained in a stirred aerated basin. This technology is often preferred over trickling filters because its oxidation efficiency is ca. 5 times higher, due to the active introduction of air into the process. Performance can be further increased by injecting oxygen instead of air. The sludge formed in this process is removed in a subsequent sedimentation basin and transferred to anaerobic sludge digestion. Disadvantages of this procedure are its open construction, resulting in limitations for process modifications and in odor nuisances for the neighborhood. In some countries such as Germany, tertiary treatment of sewage has become standard. It is used to remove phosphate, a eutrophication factor, by precipitation or biological processes, and nitrate, a risk factor in drinking water, by a biological process using consortia of nitrifying and nitrate-reducing (denitrifying) bacteria.

Tower biology. If large volumes of industrial wastewater must be purified, a highly engineered process involving closed fermenter towers can be used. In technologies developed by Hoechst and Bayer in Germany, 30-m high towers with loop pumps are used. Aeration is enhanced by thorough optimization of the sewage feed and air impellers. The oxidative capacity of such towers is ca. 50-fold higher than that of aeration basins, and odors are retained inside the closed system. Domestic wastewater may be added to enhance overall biodegradation.

Composition of waste water

origin	inhabitant equivalent*	remarks
domestic wastewater	1	per inhabitant
breweries	150–350	per 1000 L beer
dairy (without cheese production)	25–70	per 1000 L milk
starch factory	500–900	per 1000 t corn wool
wool	200–4500	per t wool
paper factory	200–900	per t paper
sugar factory	1000–2000	per t sugar

*one inhabitant equivalent corresponds to 60 g BOD_5

Tower biology*

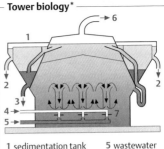

1 sedimentation tank   5 wastewater
2 purified wastewater   6 exhaust air
3 excess sludge   7 injector
4 compressed air   *e.g. Bayer AG

Biological sewage plant

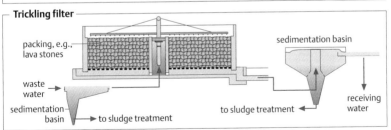

mechanical sewage treatment

biological sewage treatment

screen   grit chamber   sedimentation basin   aeration basin   sedimentation basin ← air

sludge

biogas

digested sludge (sapropel)

chemical precipitation

receiving water

incineration, agricultural use, landfill site

Trickling filter

packing, e.g., lava stones

sedimentation basin

waste water

sedimentation basin → to sludge treatment

to sludge treatment ←

receiving water

Typical construction and performance data

	trickling filter	aeration basin	tower biology
height or depth (m)	2–4	3–6	30
diameter (m)	up to 30	up to 30	30
working volume (m³)	~ 10	~ 100	15000
residence time (h)	~ 4	6–10	14
reduction in BOD_5 (t/d)	0,5	2	100

Anaerobic wastewater and sludge treatment

General. Sludges formed during aerobic wastewater treatment are usually subjected to anaerobic fouling before they are incinerated, deposited in landfills, or used in agriculture. Anaerobic sludge digestion is probably the highest-volume process in biotechnology. In Germany alone, ca. 5 000 sludge treatment units with a total volume of 1 million m^3 produce 100 million m^3 biogas per year. Wastewater can be treated anaerobically as well, using anaerobic fluid bed reactors. In some developing countries such as China and India, the removal of waste by anaerobic digestion in small decentralized facilities has been implemented to produce biogas (methane) as a local energy source.

Microbiological aspects. Methane is formed from sludge by the action of three types of bacterial consortia: 1) various obligatory or facultative anaerobic bacteria (Clostridia, Enterobacteria, Streptococci) degrade starch, fats, and proteins to organic acids, H_2, and CO_2; 2) acetogenic bacteria transform higher fatty acids into acetic acid, H_2, and CO_2; and 3) methanogenic bacteria form methane and CO_2 from acetic acid; the latter are obligate anaerobic strains and frequently belong to the Archaebacteria.

Analytical aspects. The essential parameters in sludge digestion are the reduction of organic carbon (TOC, DOC, or COD), and biogas formation. Total organic carbon (TOC) is determined by infrared spectroscopy after persulfate oxidation of a sample, dissolved organic carbon (DOC) as TOC after filtration, and chemical oxygen demand (COD) by oxidation with dichromate, followed by titration. All three methods thus represent the amount of CO_2 formed or O_2 consumed that is required for the complete chemical oxidation of a sample (here, sludge), or its dissolved fraction (DOC), to CO_2 and H_2O. Biogas is the end product of anaerobic sludge digestion. It is composed of ca. ⅔ CH_4 and ca. ⅓ CO_2, with traces of H_2, N_2, H_2S, and other gases. Its composition is usually determined by gas chromatography.

Technical aspects. Compared with the aerobic treatment of wastewater, anaerobic digestion of sludge is considerably slower (residence time ca. 20 d). The development of the microbial consortia required for sludge digestion occurs only in a very narrow range of pH and temperature, necessitating thorough control of these parameters. The advantage of this process is, however, that the sludge is nearly completely (> 90 %) transformed into CH_4 and CO_2 with the formation of little biomass and with little odor. The digested sludge remaining after fouling is incinerated, deposited in landfills, or used as an agricultural fertilizer. The energy content of the biogas produced largely exceeds the energy requirements of the whole sewage-treatment plant.

Anaerobic fluid bed reactor. Wastewater with a high load of biodegradable organic residues can be advantageously treated by anaerobic sludge treatment, i.e., without prior aerobic oxidation. In such processes, tower reactors are used where the microbial consortia aggregate into sludge particles, either on their own or after addition of particles with a high surface area. These particles have a high density and thus tend to sediment, resulting in the enrichment of bacteria in the lower part of the tower. Since the wastewater is added from the bottom, and the biogas formed leads to further mixing of the column, the process operates with high efficiency. In the upper part of the column, gas is separated from the particles by a gas separator. With some wastewaters, e.g., those from paper mills or the sugar and starch industries, a TOC reduction > 95 % and excellent biogas production can be achieved within less than a day even when loads are high.

Biogas in developing countries. In China alone, more than 7.6 million households run biogas digesters, which can generate 200 million $m^3 y^{-1}$ of biogas and provide enough fertilizer to reduce the consumption of firewood significantly. Biogas plants are based on simple technology. Sludges are mostly derived from agricultural biomass such as manure, harvest residues, or domestic waste.

Degradation reactions

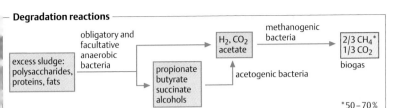

excess sludge: polysaccharides, proteins, fats → obligatory and facultative anaerobic bacteria → H₂, CO₂ acetate / propionate butyrate succinate alcohols → acetogenic bacteria → methanogenic bacteria → 2/3 CH₄* 1/3 CO₂ biogas

*50–70%

Digestor

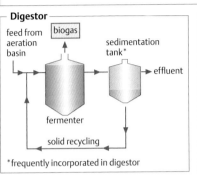

feed from aeration basin → biogas — fermenter — sedimentation tank* → effluent

solid recycling

*frequently incorporated in digestor

Anaerobic fluidized bed reactor

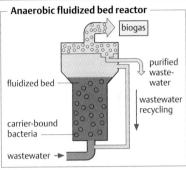

biogas

fluidized bed

carrier-bound bacteria

wastewater →

purified waste-water

wastewater recycling

Typical construction and performance data

	conventional digestor	contact fouling	fluidized bed reactor
height, volume	up to 30 m, 16 000 m³	pilot plant stage	up to 20 m 2 000 m³
load (kg COD/m³·d)	1–8	1–5	5–30
residence time of liquid (d)	10–30	0.5–25	0.2–1.5
residence time of microorganisms (h)	10–30	>20	>100
reduction in COD (%)	30–70	60–90	80–90

Biogas plants and agriculture

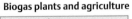

biomass	sedimen-tation tank	fouling chamber	60–70% conversion to biogas*
excrements, liquid manure, harvest residues, waste materials		C/N ~ 30:1; 20–60 d, 30–35°C	

*2–5 m³ biogas/d from 1000 kg domestic animal, ~ equivalent to 1–3 L diesel oil

opening — digested sludge as fertilizer — gas reservoir — fouling chamber

opening — biogas — digested sludge

Biological treatment of exhaust air

General. With ever-intensifying legislation concerning pollution control and waste gas emissions, new ways to remove volatile compounds, such as those causing odors, from exhaust air have been explored. Microorganisms are capable to degrade such components, after they have been absorbed in an aqueous phase. They are used in biowashers to oxidize water-soluble compounds and in biofilters to remove sparingly water-soluble organic waste gas components.

Exhaust air and gas. Biofilters have been used for many decades to remove smells from some parts of the sewage plants, in particular from sludge thickeners or dewatering units of anaerobic digesters. Exhaust air from many other industrial and agricultural plants such as foundries, food factories, chicken and pig farms, and slaughterhouses is also being successfully treated. Typical organic odor components are the lower fatty acids, amines, and mercaptans (manure, animal rendering plants); phenols and low-molecular-weight amines; aldehydes and ketones (foundries); aromatic compounds (varnish firms); and furfural (food factories). Biofilters are sometimes also used in the cleanup of soils with a low load of organics.

Biofilters are usually units of rather simple construction. Compost, barks, peat, or wood with a high surface to volume are being used as carriers. Inorganic materials such as cinders or lava are added to reduce the packing density. If humid exhaust gas is passed through such filters, consortia of microorganisms start to grow and to oxidize odor components in the organic exhaust gas. After several years of use, the organic support materials of the biofilter may become biologically oxidized as well, resulting in reduced performance and increased pressure drop. With a new packing, the previous level of performance can be recovered. Depending on their physical form, single- and multi-level biofilters can be distinguished.

Bioscrubbers have a more complex construction than biofilters: the organic odor components are first absorbed in an aqueous phase, often enriched with nutrients, to be later metabolized in a an aeration basin, or a bioreactor by adapted microorganisms. Due to the two-step setup, more sophisticated monitoring and control of plant performance is required, which also leads to higher performance than in simple biofilters, since toxic metabolites or end products are continuously removed. If exhaust air containing H_2S is purified, rapid acidification takes place through the enrichment of *Thiobacilli* in the microbial consortium that oxidize H_2S to H_2SO_4. Bioscrubbers that are equipped with a pH control allow the generated acid to be neutralized and thus have a long operational lifetime. Their energy consumption remains quite low due to the limited amount of water consumed. For easily biodegradable solvents such as the lower alcohols, the residence time of the exhaust gas in the washer can be limited to 1–2 minutes. If a mixture of easily and poorly degradable components has to be treated, 2-stage scrubbers or a combination of a trickling filter with a washer are often used. For example, the exhaust gas from a lacquer plant can be treated in a 2-stage process where, in the first stage, the easily degradable alcohol and esters, and in the second stage the less water-soluble and biodegradable aromatic components such as toluene or xylene are metabolized. The recycling of exhaust air from buildings used in intense animal production not only serves to remove CO_2 and to replenish O_2, but also to regulate temperature, remove germs, and eliminate strong-smelling exhaust gas components, in particular, ammonia. Removal of ammonia takes place in a biowasher in which nitrifying bacteria oxidize ammonia to nitrate. The nitric acid is absorbed in water and neutralized, and the heat generated in this process is recovered with a heat exchanger. Typically, the ammonia concentration of exhaust air can be reduced to 2–4 ppm, and the emission of ammonia into the neighborhood can be reduced to 0.2 kg per animal per year, as compared to 5.3–5.6 kg from a farm not using biowashers.

Two-stage bioscrubber

purified gas ↑

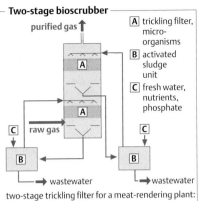

A trickling filter, micro-organisms

B activated sludge unit

C fresh water, nutrients, phosphate

raw gas →

→ wastewater

→ wastewater

two-stage trickling filter for a meat-rendering plant: odor reduction ~ 99 %

Bioscrubber for farm emissions

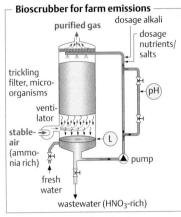

purified gas dosage alkali

dosage nutrients/salts

trickling filter, micro-organisms

venti-lator

stable-air (ammonia rich)

fresh water

pump

wastewater (HNO₃-rich)

Biofilter

multiple level biofilter

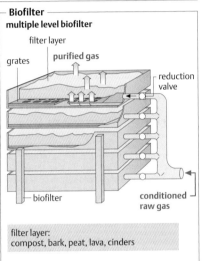

filter layer

grates

purified gas

reduction valve

biofilter

conditioned raw gas

filter layer:
compost, bark, peat, lava, cinders

Typical components of exhaust air

source	main com- ponentes	cleaning procedure
animal farm	lower fatty acids, ammonia	biofilters
industrial emissions	phenol	biofilters with Pseudo-monads
foundries	phenol, form-aldehyde, amines, ketones	biowashers
animal-rendering plants	lower fatty acids	biowashers
chemical plants	toluene, ammonia, aldehydes	biofilters or biowashers
digestors	H₂S	biowashers

single level biofilter

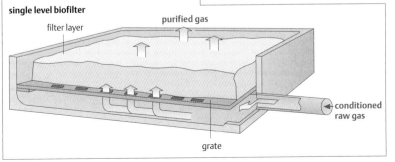

filter layer

purified gas

conditioned raw gas

grate

Biological soil treatment

General. Consortia of microorganisms play a crucial role in establishing an ecological balance in the environment through the degradation of biomass and the mineralization of organic matter. Although this capacity has been used for a century in wastewater treatment, the microbial decontamination of soil (bioremediation) has been investigated only in the past ca. 20 years. It competes with chemical and thermal processes. For biological soil treatment, natural or recombinant microorganisms can be used. Processes comprise the injection of microbial cultures *in situ* and the cleaning of soils after excavation.

Contamination and soil structure. Anthropogenic contaminants are mainly grouped into: 1) mineral hydrocarbons (MHC); 2) benzene, toluene, xylene, and ethyl benzene (BTXE); 3) polyaromatic hydrocarbons (PAH); 4) chlorinated hydrocarbons (CHC); and, 5) trinitrotoluene (TNT), in military areas. MHC and BTXE are usually easily biodegradable. Higher condensed PAH and CHC, in contrast, are not easily biodegraded. In any assessment of the biodegradability of such compounds, the soil composition must be considered. For example, sandy soils which can be easily penetrated are easier to purify than clay soils. So far, TNT can only be immobilized by applying a sequence of anaerobic and aerobic steps.

Soil treatment *in situ*. In this type of technology, nutrients and microbial cultures of high biodegradation activity, which have been isolated after enrichment in the presence of the contaminants, are added in solution through injection wells. Aeration and transformation are ensured by ground water. If the soil is mainly composed of clay and loam, oxygenation is not possible. In such cases, nitrate is used as an electron acceptor. The ecobalance of such processes, however, is doubtful in view of groundwater pollution with nitrate.

Soil treatment *ex situ*. Soils that are contaminated with readily biodegradable chemicals are usually treated after excavation in prepared long bed reactors. Suitable microorganisms are selected in precultures for high biodegradation efficiency and later used to inoculate the soil. Nutrients are added. Upon good aeration and mixing of the piles, which are ca. 2 m high, contaminants can be reduced by > 90 % within 2 weeks. Treatment costs amount to 75–150 € m^{-3} of soil.

Humification of TNT. Xenobiotics such as TNT, trichloro- or tetrachloroethylene, being substituted with a considerable number of electronegative substituents, are quite difficult to metabolize under aerobic conditions. They are rather well metabolized, however, by anaerobic bacteria. As a result, a procedure was elaborated for degradation of the TNT-contaminated soils of military camps which is based on a combination of both methods. In the first step, a reactor filled with 25 t of TNT-containing soil was kept anaerobic by adding saccharose as an electron donor. After 18 d, TNT was reduced to triaminotoluene to such an extent that, by aeration, it is covalently and irreversibly bound to soil components such as humic acids.

Recombinant microorganisms. Metabolic steps from different microorganisms can be combined by genetic engineering techniques to provide recombinant microorganisms that are better suited for the degradation of recalcitrant chemicals than the wild-type strains. This method has been explored with considerable success for the degradation of chlorinated aromatic and aliphatic compounds such as chlorobenzoic acids, chlorobenzofurans, and aliphatic CHCs. Pseudomonas strains are often used, since they contribute to the biodegradation of aliphatic and aromatic hydrocarbons in the environment to a significant extent. They often carry part of the genetic information for these steps in the form of plasmids. An example is the toluene-degrading plasmid TOL first found in *Pseudomonas putida*. The risks and advantages of releasing genetically modified organisms (GEMs) into the environment has repeatedly been investigated, but the practical potential of this technology is still under debate.

Biological soil remediation *in situ*

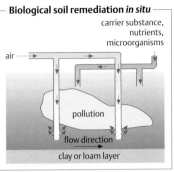

air

carrier substance,
nutrients,
microorganisms

pollution

flow direction

clay or loam layer

Anaerobic/aerobic degradation of TNT

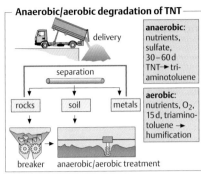

delivery

separation

rocks | soil | metals

breaker | anaerobic/aerobic treatment

anaerobic:
nutrients,
sulfate,
30–60 d
TNT→ tri-
aminotoluene

aerobic:
nutrients, O_2,
15 d, triamino-
toluene →
humification

Biological soil remediation *ex situ*

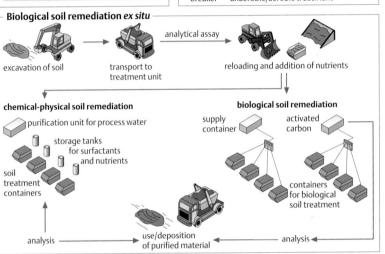

excavation of soil

transport to
treatment unit

analytical assay

reloading and addition of nutrients

chemical-physical soil remediation

purification unit for process water

storage tanks
for surfactants
and nutrients

soil
treatment
containers

analysis

use/deposition
of purified material

biological soil remediation

supply
container

activated
carbon

containers
for biological
soil treatment

analysis

Extension of degradation pathways

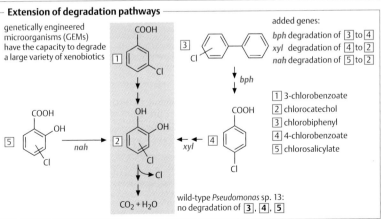

genetically engineered
microorganisms (GEMs)
have the capacity to degrade
a large variety of xenobiotics

COOH

[1]

Cl

[3]

Cl

bph

added genes:

bph degradation of [3] to [4]
xyl degradation of [4] to [2]
nah degradation of [5] to [2]

COOH
OH

[5]

Cl

nah

OH
OH

[2]

Cl

Cl

$CO_2 + H_2O$

COOH

[4]

Cl

xyl

[1] 3-chlorobenzoate
[2] chlorocatechol
[3] chlorobiphenyl
[4] 4-chlorobenzoate
[5] chlorosalicylate

wild-type *Pseudomonas* sp. 13:
no degradation of [3], [4], [5]

Microbial leaching, biofilms, and biocorrosion

General. Leaching of metals from low-grade ores through inoculation with Thiobacilli is carried out mainly in the USA, Mexico, and Australia: ca. 25 % of the global production of copper, ca. 10 % of uranium, and ca. 3 % of cobalt and nickel production are carried out by bioleaching.

Microbiology and physiology. Metal leaching is done with Thiobacilli, Gram-negative rods that are obligate chemolithotrophic organisms and that assimilate CO_2. They generate energy by oxidizing reduced sulfur compounds, such as sulfides, to sulfuric acid. Besides *T. thiooxidans*, *T. ferrooxidans* can oxidize not only reduced sulfur compounds, but also soluble Fe^{2+} salts. To synthesize 1 g of cell dry weight, *T. ferrooxidans* oxidizes 156 g-equivalents of Fe^{2+}. Both bacteria are well adapted to growth under acidic conditions and tolerate pH values down to 2.0. Sulfidic and oxidic ores such as pyrite (FeS_2), calchocite (CuS_2), covellite (CuS), sphalerite (ZnS), lead-, molybdenum-, and antimony sulfides (PbS, MoS_2, Sb_2S_3), cobalt and nickel sulfides (CoS, NiS) and also oxides such as pitchblende (UO_2) can be solublized by Thiobacilli. During the *direct bacterial leaching process*, Thiobacilli oxidize sulfide minerals directly according to the equation:

metal sulfide + $2 O_2 \rightarrow$ metal sulfate

via several intermediary steps. In contrast, the action of Thiobacilli in *indirect bacterial leaching* is catalytic, in that it assists the geochemical oxidation of sulfide minerals to Me^{2+} and sulfide oxidation according to the equation:

metal sulfide + $Fe_2(SO_4)_3 \rightarrow$
metal sulfate + $2 FeSO_4 + S°$

thus promoting oxidation at highly acidic pH. At pH 2-3, bacterial oxidation of Fe^{2+} is about 10^5–10^6-fold higher than chemical Fe^{2+} oxidation. Due to the complex compositions of ores, both types of leaching in field applications may be used.

Technology. For high efficiency, the following parameters must be met or optimized: the chemical composition and mesh size of the mineral, the mineral nutrient, very low pH, positive redox potential, temperatures around 30 °C, and a good supply of O_2. The technical process may be carried out *in situ* at mines, in ore piles, or in tanks. Alternatively, abandoned mine tunnels may be flooded for *in situ* leaching. The technology is most advanced for tank leaching, which competes best with pyrometallurgic processes if highly disperse concentrates of the metal predominate in the mineral and if environmental aspects come into play.

Biofilms and biocorrosion. Biofilms form when bacteria adhere to surfaces in aqueous environments and begin to secrete biopolymers that can anchor them to materials such as metals or tissues. Usually, a biofilm consists of many species of bacteria, as well as fungi, algae, protozoa, debris, and corrosion products. In the microbial corrosion of metallic iron, (Fe^0) is subject to "anaerobic oxidation" to FeS, catalyzed by anaerobic sulfate reducers such as *Desulfovibrio vulgaris* according to the equation:

$$4 Fe + SO_4^{2-} + 2 H_2O \rightarrow FeS + 3 Fe(OH)_2$$

Iron is oxidized under anaerobic conditions according to the equation:

$$4 Fe + 8 H^+ \rightarrow 4 Fe^{2+} + 4 H_2$$

and the H_2 layer produced during this process protects the metal from further oxidation. In the presence of sulfate, however, *Desulfovibrio* reduces sulfate according to the equation:

$$4 H_2 + SO_4^{2-} \rightarrow H_2S + 2 H_2O + 2 OH^-$$

leading to further corrosion of Fe^0 by the precipitation of iron sulfate and iron hydroxide:

$$4 Fe^{2+} + H_2S + 2 OH^- + 4 H_2O \rightarrow FeS + 3 Fe(OH)_2 + 6 H^+$$

The damage to iron pipes due to this microbial process is in the order of billions of €.

Microbial leaching

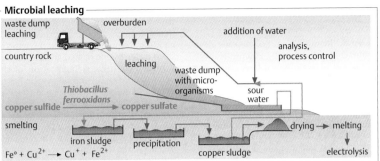

waste dump leaching
overburden
addition of water
analysis, process control
country rock
leaching
waste dump with micro-organisms
sour water

Thiobacillus ferrooxidans
copper sulfide → copper sulfate

smelting
iron sludge
precipitation
copper sludge
drying → melting
electrolysis

$Fe^0 + Cu^{2+} \longrightarrow Cu^+ + Fe^{2+}$

Biochemical reaction steps

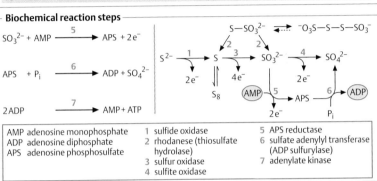

$$SO_3^{2-} + AMP \xrightarrow{5} APS + 2e^-$$

$$APS + P_i \xrightarrow{6} ADP + SO_4^{2-}$$

$$2\,ADP \xrightarrow{7} AMP + ATP$$

$$S-SO_3^{2-} \rightleftharpoons {}^-O_3S-S-S-SO_3^-$$

$$S^{2-} \xrightarrow{1} S \xrightarrow{3} SO_3^{2-} \xrightarrow{4} SO_4^{2-}$$

$2e^-$ $\quad$ $4e^-$ $\quad$ $2e^-$

S_8

$AMP \xrightarrow{5} APS \xrightarrow{6} ADP$

$2e^-$ $\quad$ P_i

AMP adenosine monophosphate	1 sulfide oxidase	5 APS reductase
ADP adenosine diphosphate	2 rhodanese (thiosulfate hydrolase)	6 sulfate adenylyl transferase (ADP sulfurylase)
APS adenosine phosphosulfate	3 sulfur oxidase	7 adenylate kinase
	4 sulfite oxidase	

Kinetics of uranium leaching

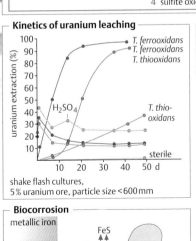

uranium extraction (%)

T. ferrooxidans
T. ferrooxidans
T. thiooxidans

H_2SO_4

T. thio-oxidans

sterile

10 20 30 40 50 d

shake flash cultures,
5% uranium ore, particle size <600 mm

Variations of ore leaching

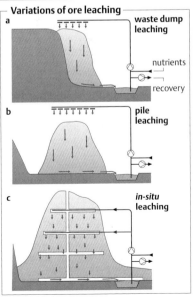

a waste dump leaching
nutrients
recovery

b pile leaching

c *in-situ* leaching

Biocorrosion

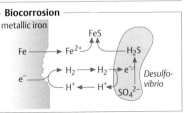

metallic iron

FeS

$Fe \longrightarrow Fe^{2+}$ $\quad$ H_2S

$H_2 \longrightarrow H_2$ $\quad$ e^- $\quad$ *Desulfovibrio*

$e^- $

$H^+ \longrightarrow H^+$ $\quad$ SO_4^{2-}

113

Insulin

General. Insulin is a polypeptide hormone that regulates the glucose level in the blood of vertebrates. It is also a key drug for treating diabetes mellitus (hyperglycemia). Until 1985, insulin was prepared by extraction from pancreatic glands of slaughtered animals. Since then, recombinant human insulin produced in *Escherichia coli* and *Saccharomyces cerevisiae* has become the dominant production technology. The global market volume is ca. $8 \, t \, y^{-1}$, with a value of ca. 1 billion US$ per year.

Diabetes mellitus is characterized by defects in the synthesis and release of insulin. In the more common case of type II diabetes (adult-onset diabetes), insufficient production of insulin by the pancreas can often be stimulated by medications. In type I diabetes, however, no insulin is formed, due to a genetic defect, virus infection, or autoimmune disease. As a result, the glucose level in the blood must be controlled by a steady supply of insulin via transcutaneous or intramuscular injection. About 140 million people suffer from diabetes, of which 60 million are type I patients (in Germany ca. 800 000). More than 2 % of the populations of industrialized nations suffer from type II diabetes (in Germany ca. 2.4 million).

Biosynthesis. Insulin is synthesized in the β-cells of the pancreas as preproinsulin, then is processed to proinsulin, and stored in the Golgi system. When triggered by a complex mechanism involving hypoglycemia, it is hydrolyzed by membrane-bound proteases into 3 polypeptide chains (A, B, and C). The A and B chains (21 and 30 amino acids, respectively) combine via 3 cystine bridges to form active insulin, and the C chain (31 amino) acids is liberated and catabolized.

Production. Insulin therapy was introduced in 1928. For *conventional production*, cattle or pig pancreas is extracted with 1-butanol. The insulin is precipitated as its Zn salt, which is easy to crystallize, and is further purified by gel chromatography ("single-peak insulin"). The amount of insulin that can be prepared from the pancreas of a pig covers the needs of a diabetic patient for 3 days, from a cow for 10 days, resulting in bottlenecks in its industrial production. In addition, human, porcine, and bovine insulin differ in one or two amino acids. As a result, continuous medication of diabetes patients with animal insulin has occasionally resulted in adverse allergic reactions. Although the *chemical synthesis* of human insulin succeeded as early as 1964, it proved economically unviable. A temporary solution to the allergies arising from the continuous use of animal insulin was found in 1975 by Novo-Nordisk Industri in Denmark, when porcine insulin was transformed into human insulin by *enzyme catalysis* using carboxypeptidase Y, exchanging the C-terminal ala^{30} with thr^{30}. Since 1985, however the production of *recombinant human insulin by fermentation* has become the method of choice. The DNA encoding preproinsulin was prepared by chemical synthesis, allowing for optimization of the codon usage with respect to the host organism *E. coli* K12. In an early procedure, the A and B chains were, for safety reasons, expressed and purified separately, and then chemically transformed to the active insulin molecule in an oxidative step. Today, recombinant proinsulin is produced as a fusion protein with tryptophan synthase, and processed by several steps into active insulin. Optimized production strains of *E. coli* synthesize up to 40 % of their cell mass as proinsulin fusion protein. Thus, a 40 m^3 bioreactor provides ca. 100 g of pure recombinant human insulin (about 1 % of the annual world demand). A different process starts from the expression of a shortened "mini-proinsulin" using recombinant strains of bakers' yeast.

New types of insulin. Human insulin with an inverted sequence of $lys^{28} \rightarrow pro^{29}$, obtained by protein engineering, is more rapidly available upon injection and thus facilitates the planning of food intake. Among further variants are $pro^{28}asp$ (Insulin Aspart, fast action, registered), $asn^3lys \, lys^{29}glu$ (Insulin Glulisin, fast action, phase III), $thr^{30}arg^{31}arg$ (B chain) $asn^{21}gly$ (A chain) (Insulin Glargin, prolonged activity), and insulin acylated at lys^{39} (NN304).

Primary structure of human insulin

A chain

gly - ile - val - glu - gln - cys - cys - thr - ser - ile - cys - ser - leu - tyr - gln - leu - glu - asn - tyr - cys - asn
 1   2   3   4   5   6   7   8   9   10 11 12 13 14 15 16 17 18 19 20 21

B chain

phe - val - asn - gln - his - leu - cys - gly - ser - his - leu - val - glu - ala - leu - tyr - leu - val - cys - gly - glu
 1   2   3   4   5   6   7   8   9   10 11 12 13 14 15 16 17 18 19 20 21

thr ← lys ← pro ← thr ← tyr ← phe ← phe ← gly ← arg ←
30   29   28   27   26   25   24   23   22

C-terminal thr can be cleaved with **carboxypeptidase Y**, the octapeptide with **trypsin**

Tertiary structure

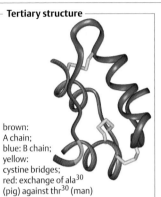

brown: A chain;
blue: B chain;
yellow: cystine bridges;
red: exchange of ala^{30} (pig) against thr^{30} (man)

porcine insulin, x-ray analysis at 0,18 nm resolution (9INS)

Biosynthesis and expression plasmid

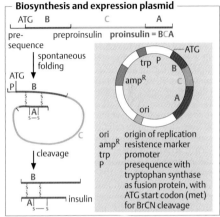

ATG   B   C   A

pre-sequence   preproinsulin   **proinsulin = BCA**

↓ spontaneous folding

↓ cleavage

insulin

ori	origin of replication
amp^R	resistence marker
trp	promoter
P	presequence with tryptophan synthase as fusion protein, with ATG start codon (met) for BrCN cleavage

Insulins of different origin (only C-terminal end of B chain)

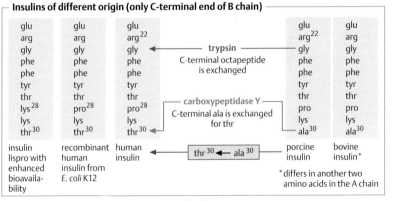

glu	glu	glu		glu	glu
arg	arg	arg^{22}		arg^{22}	arg
gly	gly	gly	← trypsin —	gly	gly
phe	phe	phe	C-terminal octapeptide is exchanged	phe	phe
phe	phe	phe		phe	phe
tyr	tyr	tyr		tyr	tyr
thr	thr	thr	— carboxypeptidase Y —	thr	thr
lys^{28}	pro^{28}	pro^{28}	C-terminal ala is exchanged for thr	pro	pro
lys	lys	lys		lys	lys
thr^{30}	thr^{30}	thr^{30} ←		ala^{30}	ala^{30}

thr^{30} ← ala^{30}

insulin lispro with enhanced bioavailability

recombinant human insulin from E. coli K12

human insulin

porcine insulin

bovine insulin*

*differs in another two amino acids in the A chain

Manufacture of human insulin with E. coli K12

bioreactor	recovery	processing	ca. 2,5 g
40 m^3, 30 hrs at 37 °C, C- and N-source	isolation of inclusion bodies, removal of cell fragments	1. CNBr cleavage of fusion protein at met site 2. oxidative sulfitolysis 3. reduction 4. cystine bridges formed (O$_2$, pH 10.6) 5. trypsin/carboxypeptidase B: removal of C-peptide	human insulin/ m^3 in 30 hrs

Growth hormone and other hormones

General. Next to insulin, growth hormone (GH, somatotropin) is the most important hormone produced by genetic engineering techniques. It is synthesized in the anterior pituitary gland and modulates a wide range of metabolic functions. Two mechanisms are of particular importance: if food intake is high, GH inhibits fat synthesis, thus channeling energy into protein biosynthesis, e.g., in the mammary gland or muscle tissue. As a result, GH has an anabolic effect, leading to enhanced protein and reduced fat formation, and GH enhances milk production in lactating animals. A second effect of GH is mediated by the insulin-like growth factor IGF-1, which is formed in the liver and induces cell division in most tissues, thus mediating the growth-promoting activity of GH.

Human growth hormone (hGH) is a polypeptide composed of 191 amino acids which contains 2 disulfide bridges. Therapeutic parenteral application is widespread in children with growth retardation in which the delay in longitudinal growth is explained by an insufficient endogenous GH secretion (0.1 % freqency). In most cases, in which growth retardation is due to GH receptor defects, however, supply of exogenous GH is ineffective. Overproduction of GH may lead to excessive growth in children and to acromegaly (an excessive growth of fingers, toes, ears, and the nose) in adults.

Animal somatotropin. Bovine growth hormone (bGH) differs from hGH by 67 amino acids and is also produced by genetic engineering techniques. Since 1990, it has been registered in the USA and some other countries (but not in the EU or Japan) for increased milk production in cows. This effect is due to enhanced energy supply to the udder. In pigs, porcine GH (pGH) is used to improve the fattening performance due to enhanced protein and reduced fat synthesis. It is predominantly used for fattening those pig breeds that produce high amounts of fat at the expense of protein. Although hGH is efficient in lower mammals, both bGH and pGH do not show anabolic effects in humans. Thus, the risk of hormonally active residues in the food chain from the consumption of GH-treated milk or pork is nil, especially since these proteins are taken up via the gastrointestinal tract, in which proteins are degraded. Even if parenterally administered, GH is not stored in tissues. In aquafarming, transgenic salmon have been raised by cloning salmon GH (sGH) behind the highly active promoter of an antifreeze protein. They grow 3–10-fold larger than untreated controls. It is claimed that, due to the closed aquaculture and manipulation of the chromosomal DNA leading to infertility, the commercial production of these fish does not pose any ecological risk.

Fermentation and recovery. Before the advent of genetic engineering techniques, GH was obtained by extraction from pituitary glands and thus was in very limited supply and poor quality. Since 1984, it has been produced by genetic engineering, mostly through recombinant *E. coli* strains. The cloning of hGH proved quite difficult in the beginning, since the hormone and its mRNA occur in only minute quantities in the pituitary glands, and cloning experiments must make do with one single restriction site in the cDNA of hGH. The problem was solved by total synthesis of the 5′-part of the cDNA and its combination with the 3′-fragment to a functional open reading frame (ORF) that could be expressed in *E. coli*. In present industrial processes, natural hGH is formed, to circumvent any immunological risks during continuous application. Purification of the recombinant hormone is carried out via a sequence of chromatographic steps.

Other recombinant hormones. Numerous other hormones have been cloned and are presently being investigated in more or less advanced tests for registration. Some promise new therapies, such as the use of recombinant parathyroid hormone for the treatment of osteoporosis. Hormones for animal reproduction, e.g., the gonadotropins (follicle-stimulating hormone FSH and luteining hormone LH) are, however, more economically replaced by naturally occurring analogs (e.g., horse serum gonadotropin PMSG from the blood of pregnant mares).

Somatotropin

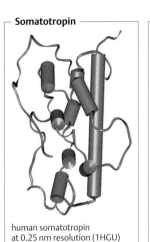

human somatotropin at 0.25 nm resolution (1HGU)

Cloning of human somatotropin

a preparation of the cDNA fragment for hGH

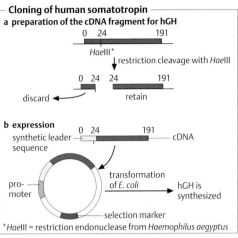

HaeIII* ↓ restriction cleavage with HaeIII

discard ◄ retain

b expression

synthetic leader sequence — cDNA

pro- moter

transformation of E. coli → hGH is synthesized

selection marker

*HaeIII = restriction endonuclease from *Haemophilus aegyptus*

Fermentation and recovery

inoculation culture	bioreactor	recovery	hGH
E. coli K12 with plasmid containing hGH gene	*E. coli* nutrient medium, 37 °C	cell lysate, chromatography, enzyme treatment resulting in charge difference, precipitation, gel and ion chromatography	(example Norditropin)

Somatotropin and milk production

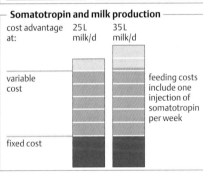

cost advantage at: 25 L milk/d 35 L milk/d

variable cost

fixed cost

feeding costs include one injection of somatotropin per week

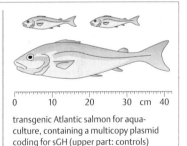

transgenic Atlantic salmon for aquaculture, containing a multicopy plasmid coding for sGH (upper part: controls)

Other recombinant hormones (selection)

	application	company	status
glucagon	hypoglycemia	Novo Nordisk	registered
follicle-stimulating hormone (FSH)	infertility	Serono, Organon, and others	registration pending
inhibin	contraceptive	Biotech Australia	
calcitonin	diseases of skeletal muscle	Suntory, Chugai, and others	
parathyroid hormone (PTH)	osteoporosis	Allelix	clinical tests phase II
leptin	appetite suppressant	Amgen	clinical tests phase II
thyroid-stimulating hormone	thyroid cancer	Genzyme	
atrial natriuretic peptide	kidney failure	Genentech/Scios	clinical tests phase III

Hemoglobin, serum albumen, and lactoferrin

General. Blood is composed of cells suspended in serum (blood plasma). In multicellular organisms, it is the most important medium for the transport of metabolites, pH buffering, regulation of body temperature and water balance, and defense against pathogens. In humans, ca. 20% of all genes code for blood proteins. Hemoglobin in the erythrocytes effects the transport of O_2 to some 10^{13} cells which make up our body. Sparingly water-soluble compounds are often transported in blood after binding to serum albumen. Blood proteins can be prepared by fractionating blood serum, but many blood proteins have already been prepared using genetic engineering techniques. Examples are α_1-antitrypsin, an elastase inhibitor contained in blood which protects lung tissue from proteolytic degradation, and lactoferrin, an antibacterial protein in milk. The propagation of blood cells involved in defense against pathogens, as well as the formation of antibodies from B cells, is regulated by cytokines. Hormones modulate numerous cell functions with high selectivity, as growth factors modulate the growth of selected types of cells. The viscosity of blood is regulated by a complex protein cascade, which prevents the formation of aggregated blood platelets through the formation of anticoagulants, but is also capable of forming a clot via fibrin formation if a blood vessel has been injured. Malfunctions in this delicate balance may lead to numerous diseases. With the advent of genetic engineering, it became possible to produce proteins involved in these complex processes in substantial quantities, both for medical research and for therapeutic use. **Hemoglobin** is the main protein of the erythrocyte. It is a nonglycosylated tetramer $\alpha_2\beta_2$ of M_R 64 kDa, composed of a pair of 2 identical ab subunits carrying 4 heme groups. Via allosteric regulation, binding of one O_2 increases the affinity of the other heme groups for oxygen. Hemoglobin is used in medical therapy after drastic blood loss for the transfusion of whole blood or erythrocyte concentrates obtained from blood donors – a treatment not without risks, since immunological side reactions occur and the donor blood may be contaminated with viruses. As a result, human hemoglobin has been cloned and expressed in *E. coli, S. cerevisiae*, or transgenic pigs. Although it has been obtained in high purity, the isolated protein is nephrotoxic and unstable outside the erythrocyte cell: it easily decomposes into $\alpha\beta$ dimers, which are rapidly degraded by proteolysis. Protein engineering and microencapsulation are currently being tested for removing these disadvantages.

Serum albumen (hSA). is a nonglycosylated protein of M_R 69 kDa, formed as preproalbumen in the liver; it constitutes ca. 60% of all serum proteins and thus has a significant influence on the osmotic pressure of blood. It binds and transports compounds of limited water solubility, e. g., lipids. In medicine, it is mainly used as a plasma expander in the treatment of shock after major loss of blood. It can be obtained by fractionation of donor blood, by a series of precipitation and chromatography steps. Viruses and other pathogens are removed by controlled heating of the sterile product at 60 °C for several hours. However, again and again, blood transfusions have resulted in severe infections. As a result, preparation of the functional protein was successfully attempted by fermentation of recombinant host strains such as *Bacillus subtilis, E. coli*, and *Saccharomyces cerevisiae*, and also in transgenic plants and in the milk of transgenic goats. Recombinant hSA has not yet been registered for human therapy.

Lactoferrin (M_R 77 kDa) is a protein with antibacterial and anti-inflammatory functions, whose activity is probably due to the high affinity with which it binds Fe^{3+} ions. The milk of lactating animals contains ca. 100 mg L^{-1} lactoferrin. In one series of experiments, it has been cloned behind the bovine α_{S1}-casein promoter, and a transgenic herd of cows producing up to 30 g L^{-1} lactoferrin was successfully established. The expression of this protein in transgenic tobacco plants has also been reported.

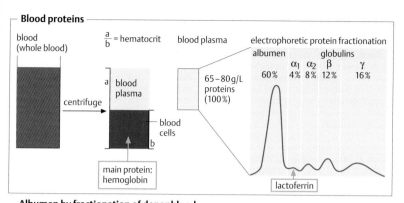

blood
(whole blood)

$\frac{a}{b}$ = hematocrit

centrifuge

a — blood plasma
blood plasma
blood cells — b

main protein: hemoglobin

blood plasma

65–80 g/L proteins (100%)

electrophoretic protein fractionation

albumen	globulins			
	α_1	α_2	β	γ
60%	4%	8%	12%	16%

lactoferrin

— **Albumen by fractionation of donor blood** —

human serum albumen (1 HSA) at 0.24 nm resolution;
5 bound molecules of myristic acid (red)

donor blood: human plasma

↓

filtration, desalting
desalted blood plasma

↓

DEAE Sepharose
pH 4.4: albumen fraction

→ **DEAE Sepharose**
pH 5.2: immuno-globulins

→ **DEAE Sepharose**
pH 4.0: glycopro-teins, ceruloplas-min

↓

CM Sephadex
ion gradient, pH 4.2 – 8.1: various proteins, pH 5.5: albumen

↓

ultrafiltration, Sephacryl
pure albumen

— **Lactoferrin** —

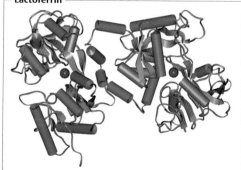

human lactoferrin (1BJ5) at 0.24 nm resolution.
Two chelated Fe ions in red.

transgenic cow
with human lactoferrin cDNA, cloned to regulation elements of the αS1-casein promoter of beef

↓

extraction
addition of 0.4 M NaCl to milk, extraction with S-Sepharose

↓

quality control
peptide mapping, MALDI-TOF mass spectrometry, 2D gel electro-phoresis, glycosylation pattern

Blood clotting agents

General. Once blood vessels have been damaged from within or outside, a blood clot forms quickly, preventing further bleeding. This process, called hemostasis, is regulated by a complex cascade of interactive steps, involving zymogen activation, proteolysis, and inhibition of proteolysis, which prevent blood clotting in the healthy organism. In an initial step to blood clotting, soluble fibrinogen is hydrolyzed by a protease, resulting in the formation of a soft clot (via an altered charge distribution) which is transformed into a hard clot through the action of transglutaminase (factor XIIIa), which forms additional amide bonds. The serine protease that hydrolyzes fibrinogen is called thrombin. It is liberated from a precursor protein, prothrombin, by the action of factor X_a, another protease, whose activity is modulated by the factor VIII complex. Most bleeders suffer from genetic aberrations that influence the synthesis or function of factor VIII or X.

Hemophilia. Descriptions of hemophilia were found on ancient Egyptian clay tablets. Today, 3 major diseases are distinguished: hemophilia A, hemophilia B, and von Willebrand disease. Hemophilia A occurs with a frequency of 1:5000 only in males. It is the consequence of a defect in the biosynthesis of the factor VIII complex: if < 1% of the normal amount of this complex is formed, spontaneous bleeding may occur, and life expectancy is low. An unusual inversion on intron F8A, preventing biosynthesis of the gene product in the liver, often seems to be the cause of this defect. The factor VIII gene is localized on the X chromosome. Factor VIII is a glycoprotein (M_R ca. 300 kDa), composed of 2332 amino acids forming a single peptide chain. The sugar content of the glycoprotein is ca. 35%, with 25 putative glycosylation sites. A structural model has recently been obtained by electron crystallography. Biosynthesis of factor VIII proceeds by splicing a DNA segment ca. 186 kbp long and containing 26 exons. During posttranslational glycosylation, the B domain is heavily glycosylated; it is later removed during activation by thrombin. von Willebrand disease is due to an erroneous biosynthesis of the von Willebrand factor (vWF) on the inner wall of blood vessels. The vWF gene is located on chromosome 12. As a result, the disease occurs with equal frequency in men and women (1:1000). The vWF protein is even larger than factor VIII, and as much glycosylated. Up to 100 monomers associate into large multimers, and each monomer binds one molecule of factor VIII, yielding the factor VIII complex (VIII:vWF). It accelerates over 100-fold the aggregation of blood platelets through the activation of the factor X/factor IX_a system. Hemophilia B occurs in a frequency of 1:25.000 and mostly among men. It is caused by a erroneous synthesis of factor IX, a glycoprotein of M_R ca. 55 kDa. Factor IX, like the factor VIII complex, participates in the activation of factor X to the active factor X_a. Its gene is localized on the X chromosome (Xq27) along a 34 kbp stretch.

Cloning. The cloning of factor VIII was achieved nearly simultaneously by Genentech and Genetics Institute, the low occurrence of an mRNA transcript (only 10^{-5} of the total mRNA of liver) constituting the major challenge. A complete cDNA transcript was eventually obtained from a lymphoma cell line through genome walking, cloned into a vector containing elements of SV40 and adenovirus promoters, and functionally expressed both in CHO- and BHK-cell lines.

Manufacture. Since ca. 1964, factors VIII, IX, and vWF have been isolated in pure form from donor blood by cryoprecipitation and fractional immunochromatography. After freeze drying, it is used for therapy. Since blood from several thousand donors per annum is required to supply a single hemophiliac A with factor VIII, the risk of viral infection is very high; estimates run as high as > 60%. Against this background, the successful manufacture of recombinant factors VIII and IX since 1992 constitutes a major breakthrough, and has led to annual sales of > 1 billion US$. Due to the high degree of glycosylation, animal cell cultures of CHO or BHK cells are used as hosts. The yields are very low, amounting to some mg of product per L of cell culture.

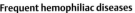

	hemophilia A	von Willebrand disease	hemophilia B
inheritance	1:5 000, males only	1:1 000, men and women	1:25 000, mostly men
clinical symptoms	bleeding at joints and muscles, brain hemorrhage	nose bleeds, strong menstrual bleeding, prolonged bleeding of wounds	spontaneous joint bleeding in childhood
locus	Xq 28	12p12	Xq 27

normal synthesis of factor VIII complex

X chromosome gene for factor VIII

chromosome 12 gene for vWF

liver-cell → VIII

inner wall of blood vessel → vWF

VIII + vWF ×100

complete factor VIII complex

incomplete synthesis (hemophilia A)

X chromosome gene for factor VIII

chromosome 12 gene for vWF

liver-cell → blocked

inner wall of blood vessel → vWF

only vWF

incomplete synthesis, von Willebrand disease

X chromosome gene for factor VIII

chromosome 12 gene for vWF

liver-cell → VIII

inner wall of blood vessel → blocked

in the absence of vWF, quick degradation

The factor VIII complex is composed of factor VIII and the von Willebrand factor (vWF).

Factor VIII: gene structure and expression vector

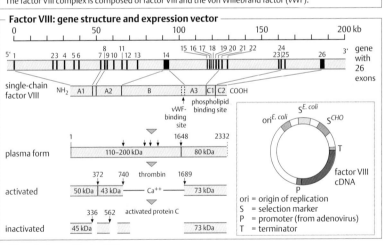

0   50   100   150   200 kb

5' 1   23 4 5 6 7 9 10 12 13 14 15 16 17 18 19 20 21 22 24 26 3' gene with 26 exons
                8 11   23 25

single-chain factor VIII NH_2 A1 A2 B A3 C1 C2 COOH

vWF-binding site

phospholipid binding site

plasma form 1 ... 1648 ... 2332
110–200 kDa | 80 kDa

activated 372 740 thrombin 1689
50 kDa | 43 kDa — Ca^{++} — 73 kDa

inactivated 336 562 activated protein C
45 kDa | 73 kDa

$ori^{E. coli}$ $S^{E. coli}$ S^{CHO} T factor VIII cDNA P

ori = origin of replication
S = selection marker
P = promoter (from adenovirus)
T = terminator

Manufacture of factor VIII

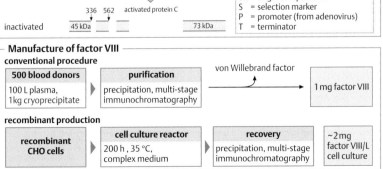

conventional procedure

500 blood donors	→	**purification**	→	von Willebrand factor →	**1 mg factor VIII**
100 L plasma, 1kg cryoprecipitate		precipitation, multi-stage immunochromatography			

recombinant production

recombinant CHO cells	→	**cell culture reactor**	→	**recovery**	**~2 mg factor VIII/L cell culture**
		200 h , 35 °C, complex medium		precipitation, multi-stage immunochromatography	

Anticoagulants and thrombolytic agents

General. The sudden formation of a thrombus (vein occlusion, heart attack, stroke) is among the most frequent causes of death in industrialized nations. In Germany alone, ca. 60 000 people die each year from the consequences of thrombosis. Anticoagulants prevent the formation of primary thrombi (e. g., after surgery). Thrombolytic agents dissolve thrombi by proteolysis. Important anticoagulants are heparin, coumarin derivatives, and thrombin inhibitors such as hirudin or human antithrombin-III (AT-III), produced by genetic engineering techniques. Thrombolytic agents are bacterial streptokinase and the recombinant preparations urokinase and tissue plasminogen activators (tPA, reteplase, TMK-tPA).

Heparin is a sulfated glucosaminoglucan of M_R 3–60 kDa. It can be obtained by extraction from porcine intestine or bovine lung. It is formed by mast cells and secreted into blood plasma, where it activates AT-III, which in turn inhibits the formation of fibrin by binding to thrombin.

Hirudin is a thrombin inhibitor, originally isolated from the saliva of leeches. It has been expressed in E. coli, Hansenula polymorpha, and other host organisms. As a result, it can be produced by fermentation (e. g., Lepirudin™). Like AT-III, it inhibits the formation of fibrin by binding to thrombin.

Tissue plasminogen activator, tPA. The degradation of fibrin, as observed during wound healing, is to a considerable extent catalyzed by the serine protease plasmin. However, this reaction proceeds only if plasmin has been formed from its zymogen plasminogen through the action of tPA, a serine protease. tPA thus is a thrombolytic agent. It has a molar mass of 72 kDa. It hydrolyzes, with high selectivity, the Arg^{561}-Val^{562} peptide bond of plasminogen. tPA forms five domains, whose functions were derived from their homologies to other proteins. The two "kringle domains" bind to the substrate fibrin, and the protease domain contains the functional center of the enzyme. Human tPA was first cloned in 1982 and has

been commercially available since 1986. Its 8 disulfide bonds and 3 sugar side chains, originally believed essential for substrate binding (ca. 5 % of the overall molar mass), rendered functional expression in E. coli impossible. Production of the recombinant enzyme is thus carried out in CHO cells, followed by a complex series of purification steps which include precipitation, ion-exchange chromatography, and immunoaffinity chromatography. More recently, however, it turned out that mutants of this enzyme can also be functionally expressed in E. coli. Thus, reteplase (Repilysin®), a mutant without the kringle-1 and the epithelial growth factor-domain, is manufactured in an E coli host and must be refolded from inclusion bodies; it exhibits a 3–4-fold prolonged residence time in blood serum and is not allergic. Another mutant, with 4 amino acid changes and moved glycosylation sites (TNK tPA), has improved fibrin selectivity and a longer half-life in serum, allowing for "single-bolus" dosage instead of infusion. It is also produced by a recombinant E. coli strain. tPA has also been expressed in the milk of transgenic animals, after the host animal had been transformed with a vector where the tPA cDNA had been cloned behind a lactalbumin promoter.

Other thrombolytic agents. Urokinase is a serine protease that is synthesized in the urogenital tract. It is formed as pro-urokinase in plasma and urine. Like tPA, urokinase hydrolyzes plasminogen to plasmin. Two variants of similar biological activity (M_R 54 and 30 kDa) can be purified separately, the lighter variant originates from the heavier one by autolysis. Urokinase can be prepared from urine, from human kidney cultures, or from recombinant E. coli. Streptokinase is a catalytically inactive protein of M_R 45 kDa that is formed by some hemolytic Streptococci. Once bound to plasminogen, it induces a conformation change in plasminogen, resulting in autolytic degradation to plasmin. Streptokinase is obtained from the supernatant of Streptococci cultures and purified by chromatographic procedures. While the production cost of this thrombolytic agent is favorable, streptokinase bears the risk of immunological reactions.

Thrombolysis (much simplified)

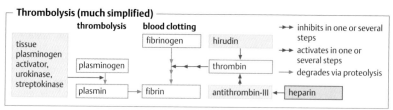

thrombolysis **blood clotting**

tissue plasminogen activator, urokinase, streptokinase

fibrinogen   hirudin

plasminogen

thrombin

plasmin → fibrin

antithrombin-III ← heparin

→→ inhibits in one or several steps
→→ activates in one or several steps
→ degrades via proteolysis

Anticoagulants (A) and thrombolytic agents (T)

	type	effect	enterprise (selection)*
heparin	A	sulfated polysaccharide binds to anti-thrombin III and inactivates thrombin	Celsus
hirudin	A	inhibits thrombin	Novartis, Schering*
antithrombin-III	A	inhibits thrombin	Genzyme
streptokinase	T	activates plasminogen	Kabi Upjohn*
urokinase	T	activates plasminogen	Grünenthal*
tPA	T	activates plasminogen	Genentech*, Boehringer Ingelheim*, Roche*

*registered   ▨ natural compound   ▨ recombinant protein

Tissue Plasminogen Activator, tPA

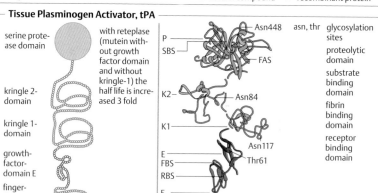

serine protease domain

with reteplase (mutein without growth factor domain and without kringle-1) the half life is increased 3 fold

kringle 2-domain

kringle 1-domain

growth-factor-domain E

finger-domain F

P
SBS
FAS
K2 — Asn84
K1
Asn117
E
FBS — Thr61
RBS
F — N-terminus

— Asn448

asn, thr glycosylation sites

proteolytic domain

substrate binding domain

fibrin binding domain

receptor binding domain

Manufacture of tissue plasminogen activator (examples)

cell reactor	**recombinant CHO cells** vector CHO NEOSPLA	**mammalian cell reactor** >100 m³ fixed bed reactor, complex medium, 37 °C, 20 d	**recovery** several immunochromatographic steps	several mg/L·d
trans-genic animal	**transgenic gout** vector LSAP + PA	**animal farming** 240 d milk production 3–4 L/d	**recovery** acid precipitation, butyl-Sepharose, immuno-chromatography	up to 30 g/L milk

quality control

complex quality control using RP-HPLC, SDS-PAGE, ELISA, peptide mapping, analysis of glycosylation pattern, and other methods

Enzyme inhibitors

General. Enzyme inhibitors have found a firm place in therapy. For example, the protease inhibitor aprotinin, obtained from bovine tissues, is used to treat pancreatitis or shock. In the future, recombinant α_1-antitrypsin might be used for treating emphysema. Although protease inhibitors of microbial origin, such as the leupeptins, pepstatin, antipain, chymostatin, and elastinal have not yet found a therapeutic use, the microbial metabolite acarbose, a glycosidase inhibitor, is a valuable antidiabetic. And tetrahydrolipstatin, a chemical derivative of the microbial metabolite lipstatin, has found broad use in the treatment of obesity, since it inhibits human pancreatic lipase.

Aprotinin is a polypeptide built from 48 amino acids (M_R 6511), which inhibits various proteases such as trypsin, chymotrypsin, and plasmin (pancreatic trypsin inhibitor, PTI). The inhibition constant K_i for trypsin is ca. 10^{-11} M. Aprotinin (Trasylol®) is used for treating pancreatitis and also for strong bleeding, shock, and organ transplantation. Still another application is in fermentations using mammalian cells, where aprotinin may prevent proteolysis of the recombinant proteins secreted into the medium. Aprotinin is isolated by extracting bovine pancreas or lungs, followed by chromatographic purification. It is not glycosylated and thus can also be functionally expressed in *Escherichia coli* host cells.

α_1-Antitrypsin (αAT). This large glycoprotein (M_R 54 kDa) is coded on chromosome 14 (14q32). It is synthesized in the liver and circulates in the blood serum at a concentration of ca. $2 \, g \, L^{-1}$, corresponding to > 90 % of the α_1-globulin fraction. It inhibits elastase, a protease secreted from the neutrophilic granulocytes of the immune system, preventing the proteolysis of lung tissue, which is largely composed of the protein elastin. In a genetic defect that occurs predominantly in Northern Europe, αAT is mutated at position 53 (lys^{53}→glu, "Z-type"). In this muta-

tion, the secretion of αAT by liver cells is greatly reduced, resulting in a serum level of just 15 % of the normal value. As a consequence, elastase starts to hydrolyze lung tissue, leading to life-threatening emphysema and to usually fatal adult respiratory distress syndrome. Smokers with Z-type αAT are especially endangered, since components of tobacco smoke oxidize met^{358}, which is essential for elastase inhibition. By intravenous application of the inhibitor (ca. 200 g/patient/year), the progress of disease can be retarded. αAT is predominantly obtained by plasma fractionation of donor blood. The inhibitor is also prepared in recombinant form. Since it requires a complex glycosylation pattern for its biological function, expression in *Saccharomyces cerevisiae* led to a better product than *E. coli*. The economically most attractive alternative seems to be expression of the recombinant glycoprotein as a fusion product with β-lactoglobulin, secreted into the milk of transgenic sheep.

Acarbose (Glucobay®), a pseudotetrasaccharide produced by the Actinomyces strain *Actinoplanes utahensis*, is a competitive inhibitor of invertase, maltase, α- and β-amylase, and various glucosidases. It reduces the glucose content in the gastrointestinal tract and thus is used as an oral antidiabetic and anti-adiposis drug, with annual sales on the order of 300 million US$. It is manufactured by microbial fermentation. Several genes involved in its biosynthesis have been cloned.

Lipstatin is a lipophilic ester with a midchain β-lactone ring and an N-formyl-L-leucine side chain, produced by *Streptomyces toxytricinii*. Through catalytic hydrogenation, it is transformed into tetrahydrolipstatin (Xenical®). Both compounds bind covalently to the serine residue in the active site of many lipases. After oral ingestion, it blocks pancreatic lipase, leading to an inhibition of triglyceride hydrolysis without reducing the uptake of free fatty acids. Tetrahydrolipstatin is used as a treatment for obesity and has a sales value of several 100 million US$.

α_1-Antitrypsin – a protease inhibitor

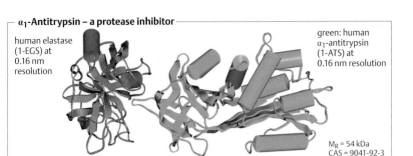

human elastase
(1-EGS) at
0.16 nm
resolution

green: human
α_1-antitrypsin
(1-ATS) at
0.16 nm resolution

M_R = 54 kDa
CAS = 9041-92-3

Acarbose – an α-glucosidase inhibitor

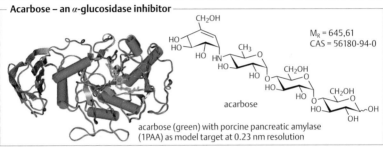

M_R = 645,61
CAS = 56180-94-0

acarbose

acarbose (green) with porcine pancreatic amylase
(1PAA) as model target at 0.23 nm resolution

Tetrahydrolipstatin (Xenical®) – a lipase inhibitor

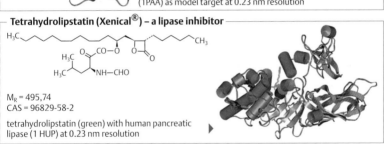

M_R = 495,74
CAS = 96829-58-2

tetrahydrolipstatin (green) with human pancreatic
lipase (1 HUP) at 0.23 nm resolution

Some processes and yields

α_1-antitrypsin	acarbose	tetrahydrolipstatin
transgenic founder sheep Tracy	**preculture** high-performance mutants of *Actinoplanes utahensis*	**preculture** high-performance mutants of *Streptomyces toxytricinii*
milk	**bioreactor** several m³ starch or maltose, 5–6 d at 28 °C	**bioreactor** several m³ starch or dextrins, soy meal, 124 h at 28 °C
recovery precipitation of casein, chromatography	**recovery** filtration, ion-exchange chromatography	**recovery** filtration, extraction with ethyl acetate, reverse-phase chromatography
10 mg pure αAT per L milk	several grams/L	several grams/L

The immune system

General. The immune system protects higher organisms from infections and provides immunity against many pathogens. It consists of specialized cells (cellular immune response) and messenger chemicals (humoral immune system) that communicate with them. Cytotoxic cells of the immune system destroy pathogens that have invaded the body, and also native cells of an organism that have been irreversibly damaged (apoptosis). They also participate in the immune defense against transplanted organs. To comply with changing environmental conditions, the immune system displays high plasticity, which is genetically determined. Misguidance may lead to a wide range of diseases such as insufficient immune response, allergies, autoimmune diseases, and malignant degeneration. The immune system is regulated by many messenger proteins (cytokines, growth factors). Many of them can be prepared as recombinant proteins and are being evaluated for therapeutic use.

Cell types. Hematopoetic stem cells are formed in the bone marrow. They differentiate there into myeloic and lymphatic stem cells. The former give rise to red blood cells, granulocytes, macrophages, and other cell types. The lymphatic stem cells, however, give rise to the lymphocytes that emigrate into the blood and lymph system. A healthy adult possesses ca. 10^{12} of these "naive" lymphocytes (meaning those which have not yet come into contact with antigens). Once a lymphocyte has been activated by an antigen (and some other signals), it forms, by clonal expansion, a large number of antigen-specific daughter cells. Lymphocytes differentiate further into B cells and T lymphocytes. Once B cells have matured in bone marrow, in lymph nodes or in the spleen, they form antibodies upon contact with an antigen (humoral immune response). In contrast, T cells mature in the thymus, where they differentiate upon contact with molecules of the major histocompatibility complex (MHC), a protein complex of the cell membrane which is exposed on the cell surface. The MHC-T-cell complex forms specific surface structures, which are distinguished by their function. T-cells are the main carriers of the cellular immune response. They secrete different cytokines. As an example, T helper cells may secrete various interleukins, thus activating, expanding and differentiationg B-cells. CD4 is a typical glycoprotein marker on the surface of T helper cells. In contrast, cytotoxic T lymphocytes carry the glycoprotein CD8 on their surface. They can lyse virus-infected cells and secrete, among other substances, the cytokines interferon-γ and lymphotoxin-α.

Immune response and cytokines. The immune response to infections differs, depending on whether viruses, bacteria, or parasites are the pathogen. Extracellular pathogens or their toxins are first tagged by antibodies, triggering a cascade that results in their endocytosis and degradation by macrophages. Intracellular pathogens, such as mycobacteria or viruses, are destroyed by a different mechanism (similar to the elimination of transformed cells): as soon as they have infected one of the omnipresent macrophages, it will exposes lysed fragments of the pathogen/cell on its surface, initiating a complex cascade that results in the destruction of infected cells by cytotoxic T lymphocytes. Autoimmune diseases follow a similar mechanism. For example, in type I diabetes, proteins of the β cells of the pancreas have been modified and are consequently misinterpreted as foreign proteins, resulting in their destruction by CD8 lymphocytes. The coordination of the immune response is largely effected by the cytokines and their receptors on the surfaces of cells of the immune system. Regulation of the immune response is highly complex. Cell-specific growth regulators and their receptors determine in a highly specific manner which cells of the immune system must be synthesized at a given time. The advent of genetic engineering has led to the possibility of producing cytokines and growth factors as recombinant proteins, initiating a novel area of medical research and, in some cases, new possibilities for medical therapy.

Blood formation and immune response: cell types

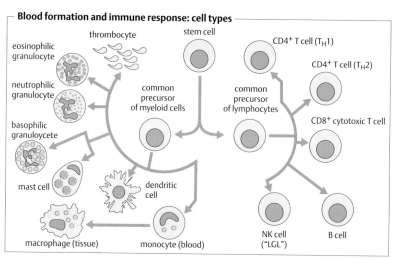

Cytokines and growth factors (selection)

general function	type	from cell type	tragets and effects
activation of lymphocytes	interleukin-2	T_H1, (CTL)	promotes growth of T-cells
	interferon-γ	T_H1, CTL	activates macrophages
	interleukin-4	T_H2	activates B cells, promotes growth of B- and T-cells
	interleukin-3	T_H1, T_H2, (CTL)	stimulates growth of hematopoetic precursor cells
local inflammation	interleukin-9	T-cells	increases mast cell activity
	interferon-α	leukocytes, fibroblasts	increases the expression of MHC class-I molecules
	TNF-α	macrophages, NK cells	causes local inflammation reactions
systemic and bone-marrow-specific effects	interleukin-1α, interleukin-1β	various cell types	causes fever, promotes growth of hematopoetic precursor cells
	interleukin-6	T_H2, macrophages	liberates acute-phase proteins
	erythropoietin	kidney	stimulates growth of erythroblasts
	granulocyte-macrophage colony-stimulating factor (GM-CSF)	T_H1, (T_H2), (CTL)	increases generation of granulocytes, macrophages, and dendritic cells

Therapeutic potential of cytokines and growth factors

infections	tumor treatment
shock	autoimmune diseases
defects in the immune system	allergic reactions
defects in cell growth	transplantation medicine

Stem cells

General. Stem cells have the capacity to divide continuously, if held in culture, and to develop into various kinds of specialized cells. Embryonic stem cells appear in the fertilized egg during an early stage of development, and adult stem cells occur in most tissues of adult animals or humans. Stem cells are an important tool in fundamental research, as they may teach us the molecular events that occur during development. They also may have great therapeutic potential in treating diseases related to tissues or organs (cell therapy).

Embryonic stem cells (ECS). All cells developing from a fertilized egg cell have in their early (morula) stage the capacity to differentiate into any kind of specialized cell (they are totipotent). Homozygous twins are the natural consequence of two totipotent cells separating from one morula. By ca. 4 d after fertilization, the morula has developed into a blastocyst, whose inner cells are multi- or pluripotent – they are still able to form a wide range of different cell types – and whose outer cells have already started to differentiate. Upon further cell division, the inner cells form a large reservoir of multipotent ECS that are able to differentiate into a wide range of specialized cells, e. g., bone marrow, nerve, or heart muscle cells. In the human embryo, this development is complete after ca. 8 weeks; most ECS by now have differentiated. As a consequence, human ECS may be isolated 1) from human blastocysts that have been generated by *in-vitro* fertilization (IVF) of infertile couples but have not been implanted; 2) from fetal tissue after miscarriages or abortions; 3) by transfer of the nucleus of any diploid human cell into an enucleated human egg cell and cultivation of this cell to the blastocyst stage (see also "cloned animals").

Adult stem cells. The bone marrow of children, and even of adults, contains multipotent stem cells. In low numbers, they reach the circulatory system and differentiate into various types of blood cells. More recently, stem cells have also been found in many other tissues of the adult human, e. g., neuronal stem cells in dissected brain samples. Based on animal experiments, it is now believed that adult stem cells, if transplanted into a different type of tissue, may adapt their differentiation to the host tissue, thus behaving like multipotent cells. A great advantage of adult stem cells is their immunocompatibility, provided the donor and acceptor are the same person. However, they are much harder to isolate than ESC. In addition, they possess the inborn or acquired genetic defects of the donor. It thus seems at present that ECS have a far wider application potential than adult stem cells.

Applications. ECS permit fundamental research on the molecular basis of cell differentiation to be carried out. In addition, they are a valuable tool for studying pathological situations, e. g., birth defects or tumors. They also could serve to develop a wide range of human cell lines most useful for testing drug efficacy and safety on a molecular level. Finally, their use opens the possibility of curing diseases through cell therapy. For example, the transplantation of pancreatic island cells, obtained from cultivation of stem cells in pancreatic tissue culture, could permanently cure children suffering from type I diabetes. It must be said, however, that many technical questions are still unresolved, e.g., reliable differentiation of ECS *in vitro* and their immunocompatibility with the host.

Ethical concerns. Whether human life has started already at the multicellular stage of a fertilized egg such as the morula or the blastocyst, which is subject to legal protection, is a controversial issue and subject to ethical debates. In the USA and most industrialized nations, a limited acceptance of embryonal stem cell research in therapeutic cloning has been reached, which is paralleled by emphasizing research on adult stem cell cloning and therapy.

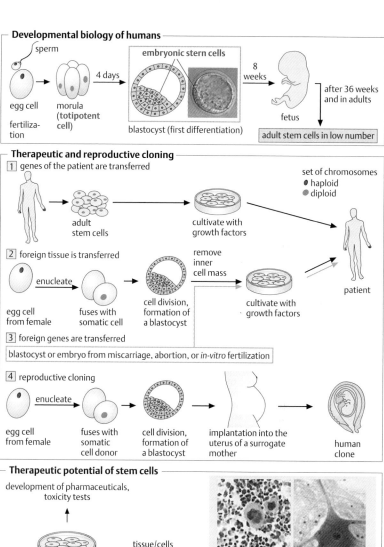

Developmental biology of humans

sperm

egg cell — morula (**totipotent cell**) — 4 days → embryonic stem cells — blastocyst (first differentiation) — 8 weeks → fetus — after 36 weeks and in adults → adult stem cells in low number

fertilization

Therapeutic and reproductive cloning

1 genes of the patient are transferred

set of chromosomes
- haploid
- diploid

adult stem cells → cultivate with growth factors → patient

2 foreign tissue is transferred

egg cell from female — enucleate → fuses with somatic cell → cell division, formation of a blastocyst → remove inner cell mass → cultivate with growth factors → patient

3 foreign genes are transferred

blastocyst or embryo from miscarriage, abortion, or *in-vitro* fertilization

4 reproductive cloning

egg cell from female — enucleate → fuses with somatic cell donor → cell division, formation of a blastocyst → implantation into the uterus of a surrogate mother → human clone

Therapeutic potential of stem cells

development of pharmaceuticals, toxicity tests

multipotent stem cells in culture → tissue/cells for therapy

fundamental research on developmental biology, gene regulation

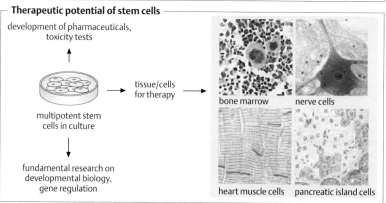

bone marrow

nerve cells

heart muscle cells

pancreatic island cells

Tissue Engineering

General. Tissue engineering aims at the functional regeneration of tissues through implantation of tissue cultured *in vitro*. Initial efforts on skin replacement for treating burns have been followed by tissue-engineered bone, blood vessels, liver, cartilage, cornea, muscle, and nerve cells. Biocompatible matrices and omnipotent stem cells play an important role in this research.

Traditional approaches. In *autografting*, tissue is transplanted in the same patient from one location to another. For example, leg veins are grafted into coronary bypasses (ca. 300 000/y in the US) or bone grafts from the hip are turned into a spine segment. Although this procedure does not pose immunological problems, it is costly and painful. In *allografting*, tissue or organs (heart, kidney, liver, bone, pancreas, etc.) from a foreign donor (e. g., a deceased person) are transplanted into a patient. With the advent of immuno-suppressives, this technology has become widely used (> 22 000 organ transplants in the US in 1998; 780 liver transplantations in Germany in 2000). *Artificial materials and devices* are widely used in, for example artificial valves, prosthetic hips and knees, or breast implants.

Scaffold-guided tissue regeneration. Extracellular matrices (ECM) are the form- and shape-determining parts of the human body, for example, the composite fibrous network of collagen of bone. Artificial scaffolds such as ceramics, collagen tubes, or synthetic materials (films, membranes, sponges, beads) have been used for seeding donor cells. In the presence of appropriate cell growth factors, the donor cells may grow on these scaffolds, yielding artificial tissue which can be implanted into a patient. Nourishing the interior cells that have grown into such scaffolds requires the formation of capillary vessels (angiogenesis) – a process which at present is not sufficiently mastered. CAD/CAM methods have been employed to fabricate arbitrarily complex scaffold shapes from CAD models (solid freeform fabrication, SFF).

3D cell cultures. Various types of progenitor and primary cells can be co-cultured, forming a more complex tissue equivalent. For example, by growing (expanding) epithelial cells in the presence of keratinocytes, an artificial epidermis was produced. A 3D skin equivalent was obtained by co-culturing dermal fibroblasts, embedded in a collagen matrix, with keratinocytes, forming a cornified epidermis.

Stem cells. Pluripotent human embryonic stem cells have a major potential in tissue engineering, because they can develop into a wide range of different cell types, depending on their environment and the cell growth factors present. Adult bone marrow is a traditional source of adult stem cells, with little ethical reservations. Stem cells can also be harvested from umbilical cord blood of newborn children.

Applications. Engineered tissues are mostly used to 1) prepare test systems based on human cell culture, in the pharmaceutical and cosmetic industries; thus, for testing the irritancy potential of substances or for drug development, 3D skin and cornea, with cells of human origin, are used as a replacement for animal tests, providing more reliable results, and for 2) transplantation. Artificial human skin has been commercially available for some time. As an example, Dermagraft® is an engineered human dermal tissue composed of a bioabsorbable scaffold and human fibroblasts. It is used to treat burns and to support wound healing in different types of ulcers (diabetic foot ulcers, venous ulcers and pressure ulcers) by minimizing infections and retaining fluids until a sufficient amount of the patient's own skin is available for autografting. Mesenchymal stem cells residing in the adult bone marrow have been induced by cell growth factors to differentiate into chondrocytes, leading to cartilage tissue. Traumatic cartilage defects are treated by autologous cartilage transplantation (ACT), a method in which chondrocytes are injected into the injured area under a newly sutured periost flap. Artificial nerve grafts or nerve guidance channels are being developed for nerve regeneration.

Targets in tissue engineering

stem cell-based tissue engineering		non stem cell-based tissue engineering	
blood vessels	liver	bladder	meniscus
bone	pancreas	cartilage (ear, nose and jonts)*	oral mucosa
cartilage	nervous tissue	heart valves	salivary gland
cornea	skeletal muscle	intestine	trachea
dentin	skin	kidney	ureter
heart muscle			urethra

*in clinical trials or clinical observational studies

Some tissue engineering companies

USA	Advanced Tissue Sciences	human tissues and organs, e.g., skin, cartilage, bones, liver
USA	Curis	wound healing, regenerative medicine
USA	MatTek	skin test systems
Germany	BioTissue Technologies	skin, cartilage
Netherlands	IsoTis	skin, cartilage, bones
France	SkinEthic Laboratories	skin test systems

Generation of a 3-dimensional skin equivalent

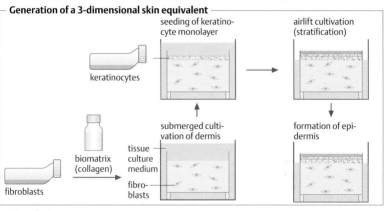

keratinocytes

seeding of keratino-cyte monolayer

airlift cultivation (stratification)

biomatrix (collagen)

submerged culti-vation of dermis

tissue culture medium

fibro-blasts

fibroblasts

formation of epi-dermis

Manufacture

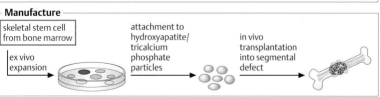

skeletal stem cell from bone marrow

ex vivo expansion

attachment to hydroxyapatite/tricalcium phosphate particles

in vivo transplantation into segmental defect

"Wound healing" using artificial skin

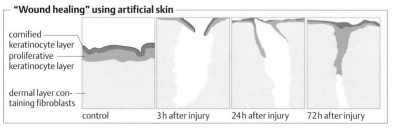

cornified keratinocyte layer
proliferative keratinocyte layer

dermal layer con-taining fibroblasts

control | 3 h after injury | 24 h after injury | 72 h after injury

Interferons

General. Interferons (IFN) are proteins synthesized in various cells of the immune system. They serve as messengers that initiate an immune response once they have bound to IFN receptors. Higher animals form several types: IFN-α, IFN-β, IFN-γ, and IFN-ῶ, which participate in the regulation of ca. 20–30 genes and exhibit a broad spectrum of immunomodulating, antiviral, and antiproliferative properties. IFN-α, IFN-β are stable at pH 5 and bind to the same receptor (type I-IFN), whereas IFN-γ is acid labile and binds to type II-IFN receptors.

Properties and applications. α-Interferons (IFN-α) are formed by leukocytes as a family of >20 nonallelic genes that exhibit high sequence homology. Their protein chains consist of 165–166 amino acids with a molar mass of ca. 16 kDa, which can increase up to 26 kDa by glycosylation. Up until now, the most important clinical applications of IFN-α are in the treatment of hepatitis B and C and of cancers such as bladder tumors, melanoma, leukemia, and lymphoma. The market value (world) is ca. 1.4 billion US$ y^{-1}. β-Interferon (IFN-β) is synthesized by fibroblasts. Its protein chain consists of 166 amino acids; but due to glycosylation, its molar mass is ca. 20 kDa. Its major clinical use is in the treatment of multiple sclerosis, with a market value of ca. 1 billion US$ y^{-1}. γ-Interferon (IFN-γ), sometimes termed "immuno interferon", is formed by activated T lymphocytes and in turn activates lymphocytes. Its peptide chain, made up of 143 amino acids, can be glycosylated to a various extents, resulting in molar masses between 15 and 25 kDa. IFN-γ1b is registered for the therapy of chronic granulomatosis and osteoporosis; its market value (world) is ca. 200 million US$. Further therapies based on the use of interferons are in various stages of clinical testing, e. g., the therapy of several cancers (IFN-α, -β and γ), of autoimmune diseases and viral infections (IFN-α, -β and -ῶ), and of rheumatoid arthritis, idiopathic pulmonary fibrosis and asthma (IFN-γ).

Cloning and expression. Although the therapeutic potential of the interferons was quickly recognized, their classical preparation by fractionation of donor blood prevented any large-scale clinical investigations. Only the advent of genetic engineering led from 1986 onward to the industrial manufacture of pure interferons as clinical preparations. The cloning of these proteins, which are formed only in very low quantities, was first successful for IFN-α in 1982, with the following methods: 1) isolation of leukocyte mRNA and reverse transcription into cDNA 2) expression in *E. coli* and hybridization with mRNA coding for IFN-α, 3) translation of the hybridizing mRNA by cell-free protein synthesis and testing of the antiviral properties of the resulting protein. Since the glycosylation pattern of the recombinant interferons does not seem to exert a major influence on their clinical effects, *E. coli* is often used as the host organism. However, interferons are also expressed in other host cells, such as *Saccharomyces cerevisia*, *Pichia pastoris*, and mammalian cells.

Manufacture and recovery. In ca. 1978, the industrial manufacture of IFN-α began based on human lymphoblastoma cell lines infected with Sendai virus (*Nawalma* cells). This procedure, however, led to the formation of at least 8 IFN-α isoforms. Today, interferons are produced using recombinant *E. coli* cells or, if glyosylation is important, as in the case of IFN-ῶ, by recombinant CHO cells. If *E. coli* cells are used, high yields of inclusion bodies can be obtained in high-cell-density fermentations. The cost-determining step, however, is in the down-stream processing, which includes refolding of inclusion bodies and chromatography. As an example, Roche prepares interferon-α2a (Roferon A®) using recombinant *E. coli* K12 cells. After harvesting, the cells are destroyed by deep freezing, rendering the cell pellets containing the recombinant inclusion bodies stable for storage. After stirring in buffer, the microbial debris is removed by centrifugation, and the protein mixture is purified in several chromatographic steps. Quality control of the products, in particular their correct folding, is precise and costly.

Registered interferons

	indications	manufacturer
IFN-α2a	hepatitis B and C	Roche
PEG-IFN-α2a	antiviral and anticancer	Roche
IFN-α2b	hair cell leukemia, myeloma, melanoma, hepatitits B and C	Biogen, Schering-Plough, Yamanoouchi, Enzon
IFN-β1a	multiple sclerosis	Serono
IFN-β1b	multiple sclerosis	Chiron, Schering AG
IFN-γ1a	rheumatoid arthritis	Rentschler
IFN-γ1b	chronic granulocytomatosis	Genentech, Boehringer-Ingelheim
consensus IFN	hepatitis C virus	Amgen
IFN-ω	hepatitis, viral defense	Boehringer-Ingelheim

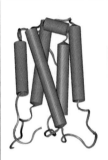 

human IFN-α (1ITF), NMR data

bovine IFN-β (1A41), 0.28 nm resolution

human IFN-γ, (1DC9), 0.3 nm resolution

Manufacture

example IFN-α2a (Roche)	example IFN-γ1b (Boehringer Ingelheim)	example IFN-γ
E. coli cells recombinant, various promoters and leader sequences	**CHO cells** SV40 promoter	**transgenic goat** vector: β-lactoglobulin promotor, human IFN-γ cDNA
▼	▼	
bioreactor high-cell density fermentation	**bioreactor** suspension culture	
▼	▼	
recovery isolation and refolding of inclusion bodies, conventional or immuno-chromatographic steps	**recovery** 4–5 chromatographic steps	
▼	▼	▼
up to 10 mg/L culture broth in 24 hrs	up to 1 mg/L culture broth	500 mg/L milk

▼

quality control
immuno assays, peptide mapping, SDS-PAGE, reverse HPLC, CD spectra, MALDI-TOF, biological assays

Interleukins

General. Interleukins (IL) are proteins formed by various cell types of the immune system. They are often called "the hormones of the immune system", since they modulate the activity of other cells in the immune system by binding to specific receptors. In humans, > 20 types of interleukins have been discovered (IL-1 – IL-23). IL-2 has already been registered for therapeutic use; others are in clinical trials.

Properties and applications. Interleukin-1 (IL-1) is synthesized in two forms (IL-1α and IL-1β) by phagocytic cells (macrophages, monocytes). It is formed as a pre-protein of molar mass 31 kDa, to be proteolytically processed to biologically active IL-1 of molar mass 17.5 kDa. It shows proinflammatory activity and stimulates the growth of lymphocytes, fibroblasts, hematopoetic cells, and thymocytes. IL-1 receptors have been identified on the surfaces of T-cells, fibroblasts (type I), and B-lymphocytes (type II). Upon phagocytosis and proteolysis of an antigen, macrophages are thought to secrete IL-1 and thus initiate the increased formation ("expansion") of immunodefensive cells against this type of antigen. During the subsequent amplification of the immune defense, interleukin-2 (Il-2, T-cell growth factor) plays an important role. This best-studied interleukin is formed by antigen-activated T-cells and stimulates the growth and differentiation of T- and B-lymphocytes. The immune cascade is further enhanced by the presentation IL-2 receptors on the surface of T-cells. Since NK-cells ("killer cells") and monocytes present IL-2 receptors constitutively, their activity is also potentiated by IL-2. The crystal structure of the hydrophilic glycoprotein composed of 133 amino acids (M_R 15.5 kDa) has been solved. IL-2 is a registered drug for the treatment of kidney carcinoma. Interleukin-3 (IL-3) is a glycoprotein made up of 133 amino acids. It is synthesized by activated T-lymphocytes and induces the differentiation of bone marrow stem cells to mature leukocytes. Since it is also involved in the differentiation of other cell types of the immune system, it is often called a "multipotent growth factor". Interleukin-4 (IL-4), a glycoprotein of M_R 20 kDa, not only stimulates the formation of B- and T-lymphocytes, like IL-1, but also the secretion of immunoglobulins IgG and IgE. In addition, it enhances the presentation of antigens by monocytes. Interleukin-6 (IL-6) has properties similar to IL-1 and IL-2, but, in contrast to them, induces the expression of several acute-phase proteins in hepatocytes; it is thought to play a role in several autoimmune diseases. Interleukin-10 (IL-10) plays an inhibiting role in the biosynthesis of other interleukins ("cytokine-synthesis inhibiting factor"). Interleukin-12 (IL-12) stimulates the synthesis of γ-interferon in T-lymphocytes and NK-cells and seems to play a pivotal role in the orchestration of immune defense. A major part of the clinical studies based on recombinant interleukins concerns cancer therapy. Further potential applications are seen in wound healing and the improvement of immune defense in AIDS patients and in immunosuppressed patients after bone marrow transplantations. The crystal structures of IL-19 and IL-22 have been solved. Their structures differs considerably from the structure of IL-2.

Manufacture of IL-2. Since the glycosylation pattern of this interleukin does not seem to be decisive for therapy, the protein is produced with transformed cells of *Escherichia coli*. Using medium- or high-cell density fermentations, high concentrations of inclusion bodies are obtained, which are first purified by gel permeation chromatography, then solubilized under reducing conditions. This is followed by refolding in the presence of oxidants. For further purification, precipitation reactions, HPLC, and gel filtration steps are used. The activity of IL-2 is usually determined in a complex bioassay based on the ingestion of ^{3}H-labeled thymidine into Il-2 dependant T-cells of the mouse.

Interleukins (as of ~ 2000)

	clinical investigations	manufacturer	status
IL-1	hematopoesis	Roche/Immunex	preclinical
IL-1R	asthma	Immunex	phase I
IL-1RA	inflammations	Amgen	phase II
IL-2PEG	HIV	Chiron	phase II
IL-2	kidney carcinoma	Chiron	registered
IL-3	stem cell transplantation	Sandoz	phase III
IL-4	lung carcinoma	Schering-Plough	phase II
IL-6	thrombocytopenia	Serono	phase I/II
IL-8RA	inflammations	Repligen	preclinical
IL-10	carcinoma, autoimmune diseases	Schering-Plough	clinical
IL-11	thrombocytopenia	Genetics Institute, Schering-Plough	phase III
IL-12	HIV, carcinoma	Genetics Institute, Wyeth-Ayerst	phase I/II
IL-15	mucositis, infections	Immunex	preclinical

Interactions in the immune system

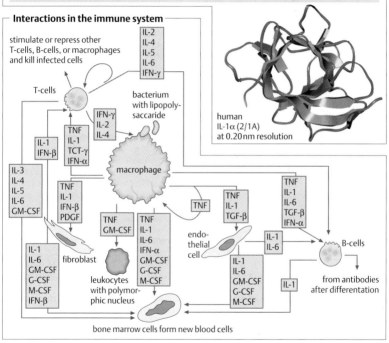

stimulate or repress other
T-cells, B-cells, or macrophages
and kill infected cells

IL-2
IL-4
IL-5
IL-6
IFN-γ

T-cells

bacterium
with lipopoly-
saccharide

human
IL-1α (2/1A)
at 0.20 nm resolution

IFN-γ
IL-2
IL-4

TNF
IL-1
TCT-γ
IFN-α

IL-1
IFN-β

IL-3
IL-4
IL-5
IL-6
GM-CSF

TNF
IL-1
IFN-β
PDGF

macrophage

TNF

TNF
IL-1
TGF-β

TNF
IL-1
IL-6
TGF-β
IFN-α

IL-1
IL-6

endo-
thelial
cell

IL-1
IL-6

B-cells

TNF
GM-CSF

TNF
IL-1
IL-6
IFN-α
GM-CSF
G-CSF
M-CSF

IL-1
IL-6
GM-CSF
G-CSF
M-CSF

IL-1

from antibodies
after differentation

IL-1
IL-6
GM-CSF
G-CSF
M-CSF
IFN-β

fibroblast

leukocytes
with polymor-
phic nucleus

bone marrow cells form new blood cells

Production and downstream processing of IL-2

E. coli	bioreactor	recovery	quality control	
recombinant, tet^R	up to 1000 L, medium- or high-cell density fermentation	refolding of inclusion bodies, chromato-graphic procedures	SDS-PAGE, peptide mapping, ELISA etc.	pure IL-2

135

Erythropoietin and other growth factors

General. During the development of animal tissue culture, various factors were found to stimulate the growth of a particular cell type (colony-stimulating factors, CSF). They are usually formed in minute amounts, act by binding to surface receptors of target cells, and belong to the cytokine group. Only after the advent of genetic engineering techniques did it become possible to prepare them in large amounts, to study their composition, and to explore their therapeutic potential in enhancing the growth of specific cell varieties, e.g., from skin, nerve, blood, or bones. During these early studies, erythropoietin (EPO), granulocyte CSF (G-CSF), and granulocyte-macrophage CSF (GM-CSF) were discovered to have a major medical use: since EPO induces the formation of erythrocytes, G-CSF of neutrophilic granulocytes, and GM-CSF of eosinophilic and neutrophilic granulocytes, these proteins can be successfully employed in various kinds of anemia, in particular for dialysis patients suffering from kidney failure. In Germany alone, the incidence of renal anemia (dialysis) patients receiving EPO is ca. 60 000, of neutropenia patients receiving GM-CSF is > 300 000 (neutropenia: neutrophils in blood are reduced, e.g., after chemotherapy or dialysis; as a result, the risk of infection increases). The world market for both products is about 6 billion US$ per year.

Erythropoietin is a factor that stimulates the growth of erythrocytes. It is synthesized in kidney endothelial cells and in the Kupffer cells of the liver and is regulated by the partial pressure of oxygen in the blood. In hemapoetic stem cells of the bone marrow, EPO induces the loss of nuclei and the concomitant formation of hemoglobin, resulting in the formation of erythrocytes. As a result, EPO is a valuable therapeutic agent for anemic conditions, in particular for secondary anemia, which is a major consequence of blood dialysis in patients dependant on an artificial kidney. If combined with other bone marrow growth factors such as GM-CSF or G-CSF, which stimulate the growth of granulocytes, eosinophilic granulocytes, and monocytes, EPO has become an indispensable restorative drug for dialysis patients. The gene coding for EPO was first cloned in 1984 from human bone marrow. The glycoprotein is based on a single peptide chain of just 165 amino acids; 4 large sugar chains in three asparagine and one threonine residue of this peptide contribute ca. 40 % to its molar mass of 34 kDa and are indispensable to its function. Due to the flexible sugar chains, all attempts to crystallize the functional protein have been futile, and structural concepts are based on NMR data. Due to the need to preserve the highly specific glycosylation pattern in any artificial host system, mammalian cell culture is presently the only way to manufacture EPO commercially, mostly in Chinese hamster ovary (CHO) cells.

Growth factors. Similar to EPO, other glycoproteins stimulate the growth and differentiation of other types of cells, e.g., in nerve, skin, bone, heart muscle, and blood. A wide range of growth factors have been cloned and prepared in sufficient quantities for clinical tests, and several have either been registered as drugs or are in an advanced stage of clinical testing, e.g., for the treatment of various neuropathies, gastric ulcers, and osteoporosis. As an example, the use of epithelial growth factor (eGF) is being explored in humans for treating cataracts, and in sheep for a "biological wool clipping", although this procedure does not seem to have an economic advantage.

Manufacture. EPO is produced in roller bottles or in bioreactors of 1000 L volume or more, using recombinant mammalian cells. CHO cells are the preferred cell line. The fermentation process takes up to 30 d. EPO is secreted into the nutrient medium and purified from the culture broth by a series of chromatographic steps. Because an authentic glycosylation pattern is of key importance for function, the pure recombinant product is thoroughly tested by appropriate methods. In many other growth factors (such as GM-CSF and G-CSF), the glycosylation pattern is of less or no importance for pharmacological function. To produce them, recombinant microorganisms such as *E. coli* can be used.

Growth factors

	protein type	company	applications	status
erythropoietin	glycoprotein, 34 kDa	Amgen, Roche, Boehringer Ingelheim	anemia after dialysis, chemotherapy	on the market
granulocyte CSF	glycoprotein, ~ 20 kDa	Amgen	infectious diseases	on the market
granulocyte-macrophage CSF	glycoprotein, ~ 18–30 kDa	Wyeth, Schering-Plough	infectious diseases, etc.	on the market
macrophage CSF	dimeric gyco-protein, ~ 90 kDa	Immunex	acute leukemia, cancer, fungal diseases	
keratinocyte GF		Amgen	mucositis	phase III
neuronal GF		Genentech, Amgen	peripheral neuropathies	phase II/III
epidermal GF		Johnson & Johnson	wound healing, cataracts	phase II
fibroblast GF		Scios Nova	skin ulcer	launched in J
bone morpho-genic protein		Genetics Institute	bone and cartilage diseases, bone marrow transplantations	phase I
thrombopoietin		Genentech	thrombocytopenia	phase III

CSF = colony-stimulating factor   GF = growth factor

Isolation of growth factors

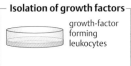

growth-factor forming leukocytes

overlaid with a culture of target cells

one type of target cell exhibits en-hanced growth

Wool clipping with eGF

Function of erythropoietin

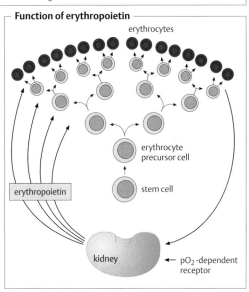

EPO and GM-/G-CSF: manufacture and purification

EPO	recombinant CHO cells	cell reactor	recovery
GM-CSF G-CSF	recombinant E. coli cells	bioreactor	recovery

quality control
immunoassays, peptide mapping, SDS-PAGE, reverse HPLC, MALDI-TOF, biological activity

Other therapeutic proteins

General. Among the hundreds of recombinant proteins that are presently under study for therapeutic applications, the cytokine tumor necrosis factor (TNF), DNase I, and glucocerebrosidase are discussed here.

Tumor necrosis factor. The discovery of TNF relates to early observations that some tumors are halted in their development after the patient has suffered a bacterial infection. It has since been shown that bacterial endotoxins (lipopolysaccharides) stimulate the formation of TNF in activated macrophages, monocytes, natural killer cells (NK cells), and also in liver and brain cells. TNF occurs in two variants, which exhibit similar biological activity although they show $<30\%$ sequence homology. TNFα (M_R 17.3 kDa, 157 amino acids) is formed by macrophages. Its crystal structure is dominated by an unusually large number of β sheets. Two types of specific TNFα receptors have been found, which occur in quite different cell types. TNFβ (171 amino acids, a glycoprotein) is formed by lymphocytes ("lymphotoxin"), but binds to the same receptors. Its function is less well understood. Originally, the interest in TNFα was focussed on the observation that it has a cytotoxic effect on isolated transformed cells which can be further stimulated by interferons. Clinical studies, however, did not corroborate these findings and, in contrary, showed highly toxic side effects. Indeed, many undesired side effects of the cytokines (inflammation, arthritis, elevated blood pressure, etc.) seem to originate from the formation of TNFα. TNF has also been involved in reactions such as septic shock (which is also initiated by lipopolysaccharides), in cachectic conditions (emaciation) following chronic infections, or in tumor development. It regulates the development of other cytokines, is involved in the development of autoimmune diseases such as rheumatoid arthritis, and plays a role in transplant rejection. This interesting ambivalent role of TNF and its easy production by recombinant hosts (e.g., *E. coli*) have led to significant interest in these proteins and its variants for clinical investigations.

DNase I (Pulmozyme®). Cystic fibrosis (CF) is a genetic disease affecting ca. 30 000 children and adults in the US alone. One in 28 Caucasians is an unknowing, symptom-less carrier of the defective gene. CF causes the body to produce an abnormally thick, sticky mucus, due to the faulty transport of sodium and chloride from within cells lining organs, such as the lungs and pancreas, to their outer surfaces. The thick CF mucus also obstructs the pancreas, preventing enzymes from reaching the intestines. Excessive formation of mucus results in dramatic hindrance of respiration. The viscosity of the sputum is further increased by extracellular DNA originating from leukocytes. A therapy used in cystic fibrosis is the inhalation of an aerosol containing recombinant human DNase I (260 amino acids). The recombinant enzyme is produced in CHO host cells, using a DHFR-containing insert and methotrexate-containing medium for high yields. Since DNase I is inhibited by the G-actin occurring in sputum, mutant enzymes have been engineered which are not inhibited and thus have 10-fold higher activity in sputum. Cystic fibrosis is a monogenetic disease and thus may lend itself eventually to gene therapy.

Glucocerebrosidase. Gaucher's disease is an inheritable storage disease. It results from a lack of formation of the enzyme glucocerebrosidase, resulting in the deposition of large amounts of cerebrosides in some cell types. Based on clinical symptoms, three types of Gaucher's disease are distinguished. Patients of the most frequent variety, Gaucher form 1, have pain in the bones and the digestive tract, but no nerve symptoms. They can be successfully treated by intravenous application of human glucocerebrosidase. The enzyme can be obtained from human placenta, but recently, production of the recombinant enzyme in CHO cells has become preferred. The use of recombinant plant cells for production has also been described.

Tumor necrosis factor (TNFα)

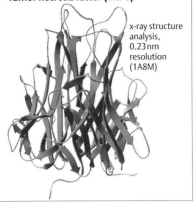

x-ray structure analysis, 0.23 nm resolution (1A8M)

DNase I

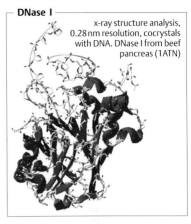

x-ray structure analysis, 0.28 nm resolution, cocrystals with DNA. DNase I from beef pancreas (1ATN)

Cystic fibrosis

reduced secretion of chloride through epithelial cells, mucus formation in the lung. Caucasians: 1 homozygous in 2000 live births, 1 in 28 heterozygous. Life expectancy for homozygous condition: < 30 y

therapy
recombinant human DNase I as an aerosol

cystic fibrosis membrane regulator (CFTR), a membrane protein made up of a single polypeptide chain of ~2 170 amino acids, coded on chromosome 7 (7q31), is mutated, most frequently (~ 70%) by a deletion of Phe[508] (AF508)

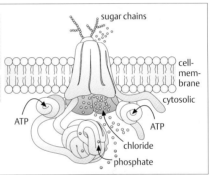

sugar chains

cell-membrane

cytosolic

ATP

ATP

chloride

phosphate

Morbus Gaucher

missing functional enzyme ß-glucocerebrosidase (coded on chromosome 1 (1q31), leads to blood anomalies, lung malfunction, bone and nerve defects.

therapy
recombinant human β-glucocerebrosidase

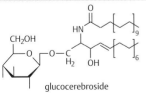

glucocerebroside

bone marrow spread with dyed macrophages (blue), whose form is typical for this microsomal storage disease

CHO cells	cell reactor	recovery	enzymatic transformation
with plasmid pGB20, coding for β-glucocerebrosidase (1 191 bp)	up to 2 000 L	ultrafiltration, chromatography	partial degradation of sugar chains

Vaccines

General. Temporary protection against viruses, bacteria, or toxins (e. g., after a snake bite) can be achieved by injection of antibodies specific to the antigen in question (passive immunization). Much better, and often life-long, protection can be obtained, however, if the immune system is stimulated to produce suitable antibodies by the injection of vaccines (active immunization). After vaccination, the immune system produces B lymphocytes, which secrete pathogen-specific antibodies; T lymphocytes, which destroy foreign antigenic material; and long-lived B- and T memory cells, which react promptly with an immune response if the antigen should show up again. Vaccines can be whole cells, cell components (e. g., cell wall lipopolysaccharides), or toxic proteins (toxins). Thus, inactivated antigenic materials or weakened (attenuated) viruses or microorganisms that are still immunogenic but have lost their pathogenic properties are used as vaccines. Attenuated viruses are obtained through a sequence of cell passages. For many decades, a wide number of vaccines have been produced for human protection, e. g., against measles, diphtheria, tetanus, whooping cough, tuberculosis, cholera, and polio. In some countries, farm animals are vaccinated e. g., against foot-and-mouth disease. In spite of this progress, a significant number of diseases have no vaccines available. Thus, a wide range of tropical diseases and AIDS have resisted vaccination. In addition, several infectious diseases which were believed extinct have reappeared, e.g., tuberculosis. The increasing resistance of some diseases to any treatment with antibiotics adds to this dangerous development. Fortunately, genetic engineering methods have opened a new way to prepare novel and highly pure vaccines.

Vaccine preparation. The conventional method consists in the formulation of inactivated or attenuated antigenic material for subcutaneous, intramuscular, or oral administration. Usually, strains of a pathogen that have lost their pathogenic properties but still stimulate an immune response are used. Alternatively, the pathogen is first cultured in the laboratory and then inactivated by heat or formaldehyde treatment, while keeping its immunogenic properties. For the preparation of vaccines against microbial pathogens or their toxins, the microorganisms are cultured in bioreactors. Until ca. 1970, viruses were preferentially produced in embryonated chicken eggs, and the virus coat protein, after purification from the egg albumen, was used as a vaccine. Today, another preferred method is to propagate the virus in animal cells using animal cell culture technology. Attenuated viruses are usually used in both processes, and after isolation from hen's eggs or cell culture, they are further weakened or inactivated by treatment with heat or formaldehyde. In view of the potential risks of these fermentation and recovery processes, all steps, including formulation, are carried out under high safety standards. Activity and stability of the product are usually tested in animals ("release tests").

Examples. Tetanus occurs due to infection of a wound by *Clostridium tetani*. During anaerobic growth, this pathogen secretes a neurotoxic protein that is transported by blood to the nerves, resulting in spastic paralysis. For the preparation of tetanus toxin, a hypertoxinogenic strain (Harvard strain) is cultivated in a bioreactor. After growth is complete, the microorganisms are autolyzed, which releases the toxin. After removal of cell debris by filtration, the toxin is inactivated for 4 weeks in a ca. 0.5 % formaldehyde solution, resulting in a "toxoid". This is purified by diafiltration and salt precipitation and absorbed on aluminum salts, thus increasing its immunogenic properties (adjuvant effect). Immunogenicity and tolerability of the lot are tested in animals. To obtain a measles vaccine, animal or human cells in culture are inoculated with a virus strain of low virulence (Edmonton strain). After lysis of the host cells, the virus is isolated by continuous-flow zonal ultracentrifutation and further purified, leading to a freeze-dried or liquid preparation of high stability to storage.

Basic techniques

passive immunization

application of antibodies

active immunization

1. protective vaccination
2. oral vaccination using
   - inactivated pathogens
   - attenuated pathogens
   - pathogen-specific antigens
   - pathogen DNAs
   1. for systemic infections
   2. for local infections

Gaps in vaccination

disease	cases per year (millions)	death toll per year (thousands)
diarrhea	>4000	>400
various worm diseases	>2000	>20
respiratory diseases	>350	>4000
malaria	>300	>1
schistosomiasis	>250	>10
tropical measles	>44	>1000
Chagas disease	>25	high
tuberculosis	>6	>2000
AIDS	>5	>150

Available vaccines (examples)

	vaccine	application	manufacture
1	BCG	tuberculosis	attenuated live vaccine from *Mycobacterium bovis*
1	rubella	measles	live vaccine using attenuated *Rubella virus*
2	poliomyelitis	polio	live vaccine using attenuated Poliomyelitis virus
1 2	cholera	cholera	inactivated strains of Vibrio cholerae, live alterated vaccines
2	typhus	typhus	attenuated strains of *Salmonella typhimurium*
1	Haemophilus	meningitis	purified capsular polysaccharide from *Haemophilus influenzae*
1	FMD	foot and mouth disease	aziridine-inactivated FMD virus

*attenuation: weakening through repeated passages in cell culture ("passaging")

Manufacture of a virus vaccine

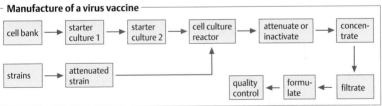

Fermentation and recovery

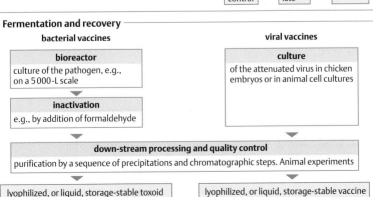

Recombinant vaccines

General. Genetic engineering procedures have opened up new possibilities for preparing vaccines. They allow for the manufacture of highly pure component vaccines but may also lead to completely new vaccination strategies. Examples are the incorporation of vaccine components into the coat of harmless viruses, the expression of vaccines in transgenic plants or in the milk of transgenic farm animals (both procedures would obviously lead to immunization via food intake), and vaccination by direct transfection with DNA. Up to now, however, only a single recombinant component vaccine, preventing hepatitis B infection, has reached the world markets.

Strategies. Using methods of genetic engineering, component vaccines can be obtained that are directed towards single-cell components of a pathogen, e.g., against its surface proteins. This procedure implies, however, that the immunogenic components of the pathogen are known. A successful example of this strategy is the production of a recombinant hepatitis B vaccine. In many countries of Asia, hepatitis B is an endemic disease, proceeding undiscovered in 90% of all cases and leading to chronic liver disease in about 5% of the patients. It is estimated that about one billion humans suffer from hepatitis B. To prepare a recombinant vaccine, the surface antigen HbsAg of the hepatitis B virus was isolated from the blood plasma of infected humans, sequenced, and cloned. It can be expressed in *E. coli*, or, preferentially, in *Saccharomyces cerevisiae*. Purification of the secreted protein is carried out by a sequence of chromatographic steps. A different strategy for recombinant vaccines is the preparation of genetically attenuated host strains. For example, *Vibrio cholerae*, the strain responsible for cholera, normally produces cholera toxin, a fusion protein having adenylate cyclase activity. After secretion within the small intestine, it leads to the formation of cAMP, resulting in a massive loss of liquids and electrolytes, with a clinical picture of violent diarrhea. Using genetic engineering techniques, a deletion mutant of *V. cholerae* was prepared that lacks adenylate cyclase activity but shares its immunogenic properties with the pathogenic strain and thus can be safely used for vaccination. As a third strategy, vector vaccines are being proposed: this would imply vaccination with viral DNA that has been modified so as to code for the desired immunogenic proteins but lack all pathogenic elements. As a vector, the cattle pox virus *Vaccinia* was chosen because it is highly infective but completely harmless to humans. Viral antigens were successfully engineered into the *Vaccinia* genome, resulting, after infection, in immunization against the G protein of the rabies virus, the hepatitis B virus surface antigen, the NP- and HA proteins of influenza virus, and other antigens. The concept, however, is considered not safe enough, in particular for infants or immunosuppressed patients.

Transgenic plants are being proposed for a vaccination program in developing countries (eatable vaccines, e.g., transgenic bananas). Since the vaccine would be taken up by the gastrointestinal tract (as occurs with oral vaccination), its efficacy depends on its stability during gastrointestinal passage and its transfer through the mucous membrane, where it must stimulate the immune system of the small intestine to produce antibodies. This concept also raises a number of regulatory questions, e.g., of consistent production and the behavior of the vaccine protein during ripening, decay, or processing of fruits or other food products.

DNA vaccines. After the injection of DNA coding for the surface structures of the malaria pathogen *Plasmodium falciparum* into the spleen of mice, the vaccinated animals produced antibodies against this parasite. Similar experiments with the genomic digest of *Mycobyceterium tuberculosis*, the tuberculosis pathogen, also led to a T cell response and allowed for the identification of gene products leading to an immune response. In both studies, the antigen-specific DNA was integrated into plasmids. This interesting new method is still in its infant stage.

Recombinant vaccines (selection)

		antigen	status
viruses	hepatitis B	surface antigens	registered
	Herpes simplex type 2	surface antigens	clinical studies
	rabies vaccine	surface antigens	not registered
	yellow fever virus	surface antigens	preclinical studies
	AIDS virus	surface antigens	clinical studies
bacteria	*Streptococcus pneumoniae*	polysaccharide conjugate	registered
	Clostridium tetani	tetanus toxin	not registered
	Mycobacterium tuberculosis	surface antigens	clinical studies
parasites	*Plasmodium falciparum*	(malaria)	clinical studies
	Trypanosoma sp.	(sleeping sickness)	clinical studies
	Schistosoma mansoni	(bilharziosis)	clinical studies

Vaccination by recombinant Vaccinia virus

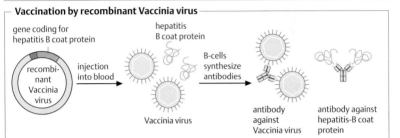

gene coding for hepatitis B coat protein

recombinant Vaccinia virus

injection into blood

hepatitis B coat protein

Vaccinia virus

B-cells synthesize antibodies

antibody against Vaccinia virus

antibody against hepatitis-B coat protein

Immunization with virus coat protein or DNA

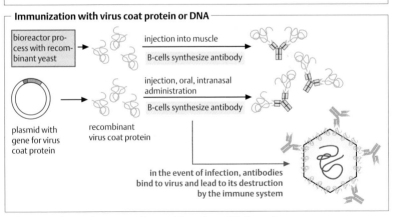

bioreactor process with recombinant yeast

injection into muscle

B-cells synthesize antibody

plasmid with gene for virus coat protein

recombinant virus coat protein

injection, oral, intranasal administration

B-cells synthesize antibody

in the event of infection, antibodies bind to virus and lead to its destruction by the immune system

Fermentation and recovery of recombinant hepatitis B vaccine

bioreactor	recovery	
recombinant *S. cerevisiae* expresses plasmid-coded rHBAg protein	by precipitation, diafiltration, chromatography	rHBAg vaccine

complex quality control (absence of pathogens, allergens, etc.)

143

Antibodies

General. Antibodies are specific defense proteins circulating in the blood and lymph of vertebrate organisms. They are formed upon contact of B-lymphocytes with immunogenic antigens and bind with high affinity to such antigens. Most foreign proteins, polysaccharides, and lipopolysaccharides may act as antigens, e. g., macromolecules that constitute the cell surface of viruses, microorganisms, and parasites. Toxic proteins (toxins) may also lead to antibody formation. Even low-molecular-weight compounds may give rise to antibody formation if they are presented on the surface of strong immunogenic structures ("haptens"). In autoimmune diseases, the organism's own proteins have become "foreign" and have developed antigenic properties. Antibodies have long been used as vaccines for the treatment of infections and toxins (e. g., snake bites) (passive immunization). They are of great value as reporter groups in immunoanalysis. They are industrially used for the purification of some recombinant proteins, e. g., factor VIII, by immunochromatography.

Structure. Antibodies belong to the immunoglobulins. In man, they are classified into 5 groups (IgG, IgM, IgA, IgE, and IgD), which play various roles in immunodefense. IgG, which predominates in serum, is a heterodimer composed of two identical light (L) and two heavy (H) chains, which are linked by cysteine bridges. Structures having constant (C_H, C_L) and variable sequences (V_H, V_L) domains can be distinguished in the heavy and light chains. The F_c-region of the antibody binds to a receptor, and the F_{ab}-region binds to the antigen. This region of the antibody is hypervariable: the 6 complementarity-determining regions (CDRs) consist of ca. 20 amino acids each; thus, each CDR allows for $20^{6 \times 20}$ permutations.

Biosynthesis. Antibodies are synthesized by B lymphocytes, which have nearly 1000 sets of gene segments available. The gene segments are combined by random recombination ("gene shuffling") to code for the variable region of the immunoglobulins. In addition, during the expansion of B-cell clones, mutations occur in the genes that are responsible for the variable regions. Thus, a relatively small genotype coding for antibodies is turned into a huge phenotypic diversity.

Preparation. Polyclonal antibodies are mixtures of different antibodies that are directed towards different epitopes of the same antigen. They are obtained by immunization of animals such as rabbits, sheep, goats, and horses. Through repeated immunization, in intervals of several weeks, and extraction of blood, similar lots of antibodies can be obtained repeatedly from the same animal (horse, cattle, sheep). Purification is done by precipitation and chromatographic procedures. In manufacturing highly purified antibodies, affinity chromatography based on protein A may be used. Protein A (M_R 42 kDa) is obtained from *Staphylococcus aureus*. It binds with high specificity and affinity to the F_c-region of IgG. Purified IgG solutions are portioned in a sterile manner and lyophilized in the absence of air. If stored under refrigeration, antibodies are stable for several years. Industrial production is done under GMP conditions.

Risks. For human therapy, antibodies are administered parenterally, since they are not stable to gastrointestinal passage. Antibodies obtained from animals are recognized by the human immune system as foreign and thus can give rise to an immune defense, especially after repeated injections. One solution to this problem is to shift among antibodies obtained from different animal species. Alternatively, antibodies can be obtained from blood donors. Although donated blood stored in blood banks is thoroughly scrutinized before use, a risk of viral contamination such as hepatitis or AIDS does exist.

Applications. Antibodies are mainly used for diagnosis and therapy of human diseases and as analytical tools in molecular and cell biology. More recently, they are also being applied in food and environmental analysis.

IgG antibody

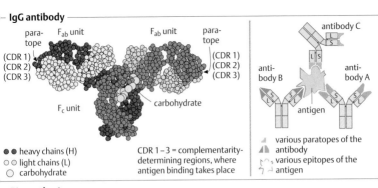

para-tope (CDR 1) (CDR 2) (CDR 3) — F$_{ab}$ unit — F$_{ab}$ unit — para-tope (CDR 1) (CDR 2) (CDR 3)

F$_c$ unit — carbohydrate

●● heavy chains (H)
○○ light chains (L)
○ carbohydrate

CDR 1 – 3 = complementarity-determining regions, where antigen binding takes place

antibody C
antibody B — antigen — antibody A

◢ various paratopes of the
◢◣ antibody
various epitopes of the
antigen

Biosynthesis

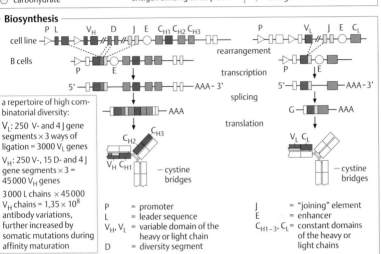

cell line — P L V$_H$ D J E C$_{H1}$ C$_{H2}$ C$_{H3}$ — P V$_L$ J E C$_L$

B cells — rearrangement

P E — transcription

5'—AAA–3' — splicing

—AAA — translation

C$_{H2}$ C$_{H3}$ V$_H$ C$_{H1}$ — cystine bridges

5'—AAA–3'

G—AAA

V$_L$ C$_L$ — cystine bridges

a repertoire of high combinatorial diversity:

V$_L$: 250 V- and 4 J gene segments × 3 ways of ligation = 3000 V$_L$ genes

V$_H$: 250 V-, 15 D- and 4 J gene segments × 3 = 45 000 V$_H$ genes

3 000 L chains × 45 000 V$_H$ chains = 1,35 × 10^8 antibody variations, further increased by somatic mutations during affinity maturation

P = promoter
L = leader sequence
V$_H$, V$_L$ = variable domain of the heavy or light chain
D = diversity segment

J = "joining" element
E = enhancer
C$_{H1-3}$, C$_L$ = constant domains of the heavy or light chains

Polyclonal antibodies

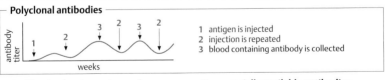

antibody titer

weeks

1 antigen is injected
2 injection is repeated
3 blood containing antibody is collected

Purification by chromatography

several L of blood
separate erythrocytes by centrifugation, obtain plasma

IgG fraction
precipitate with ethanol or ammonium sulfate

purificcation
chromatographic procedures, affinity chromatography using protein A

Commercially available antibodies

some commercially produced polyclonal antibodies for passive immunization

antibody	obtained from
tetanus antitoxin	horse serum
snake venom antisera	horse serum
measles virus immunoglobulin	human serum
immunoglobulin G	human serum

Monoclonal antibodies

General. In contrast to polyclonal antibodies, which are obtained from immunized animals and consist of a mixture of antibodies directed against the same antigen, monoclonal antibodies are homogeneous: they consist of a single type of antibody with defined selectivity and activity. They are manufactured using hybridoma cells.

Hybridoma technology. Antigen is injected into an experimental animal (usually mice, sometimes rats). The spleen lymphocytes are isolated and fused *in vitro* with murine lymphocyte tumor cells (myeloma cells), which can be held in culture and divide indefinitely. Some of the resulting hybridoma cells express antibodies against the antigen on their surfaces and thus can be isolated, using immunoassays and cell cloning procedures. The most productive clones are deep frozen, rendering them stable for many years. This methods allows the production, in a reproducible manner, of pure monoclonal antibodies against a nearly unlimited choice of antigens and haptens.

Manufacture of monoclonal antibodies. Hybridoma cells divide and can be propagated in cell culture permanently. Monoclonal antibodies synthesized by hybridoma cells are secreted into the culture medium at levels of $10-30$ mg L^{-1}. Larger quantities $(800->1000$ mg $L^{-1})$ are obtained through cultivation of cells in a bioreactor, using complex media. In addition to D-glucose and L-glutamine, fetal calf serum was originally used in the medium, since it contains vital cytokines and growth factors such as lactoferrin. More recently, complex synthetic media have been developed which mimic bovine serum factors (BSF). Hybridoma cells can be grown under aerobic conditions on solid surfaces, but also have been adapted to growth in suspension. They require the presence of oxygen and CO_2. On a laboratory scale, slowly rotating spinner cultures or roller flasks are mostly used. In industry, cell reactors are applied. They must be optimized for aeration and mixing, since mammalian cells are sensitive to shear forces. Originally, bioreactors with large internal surface areas, such as macroporous beads or hollow-fiber modules were used to protect the cells from mechanical disruption by aeration. Since then, conditions for suspension culture of free cells have been developed, using both stirred tanks and airlift reactors (up to 12 500 L). Fermentation is done in batch or continuous modes: in industry, fed-batch procedures are preferred and may yield several grams of monoclonal antibody per liter of culture. In a typical purification procedure, the medium is concentrated by ultrafiltration or diafiltration, followed by binding of the antibody to a protein A column. Further purification steps may include ion-exchange chromatography and the removal of aggregated antibodies and foreign proteins by gel chromatography.

Humanized antibodies. Monoclonal mouse antibodies (murine antibodies) bear the risk of provoking an immune response, if used in therapy or for *in-vivo* diagnostics in humans. In addition, the murine F_c-fragment may not bind properly to the human receptor. Since experimental immunization of humans is unethical and human myeloma cells are difficult to keep in culture, other procedures have been developed to achieve human antibodies. For example, 1) cell cultures of human lymphocytes have been shown to produce monoclonal antibodies, if an antigen and certain growth factors and cytokines are present (in-vitro immunization); 2) antibodies have been obtained from the spleen of immunodeficient mice in which human lymphocytes had been grown; and 3) humanized recombinant antibodies have been obtained, which carry murine CDR fragments within a human antibody core (grafting).

Applications. Monoclonal antibodies can be used as biopharmaceuticals for the treatment of many diseases, e.g. tumors. They are among the key pharmaceutical products prepared by mammalian cell culture. At present, > 10 monocolonal antibodies are registered therapeutics and > 100 have reached clinical testing.

Preparation

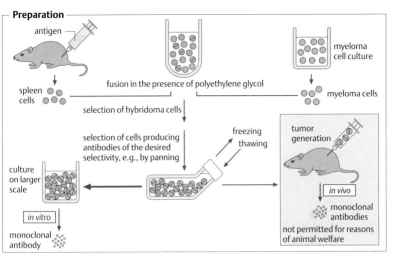

Fermentation and recovery

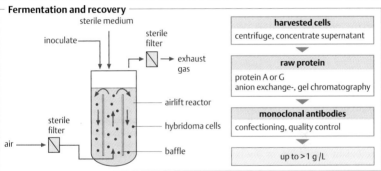

	harvested cells
	centrifuge, concentrate supernatant

	raw protein
	protein A or G
	anion exchange-, gel chromatography

	monoclonal antibodies
	confectioning, quality control

	up to >1 g /L

Some applications

application	brand name	target protein/function	origin
diagnostics	many	gonadotropin, growth hormone, carcinoembryonic antigen, prostatespecific antigen, Herpes, *Legionella* etc.	usually murine
transplant rejection	Zenapax™	kidney transplantation	humanized
oncology	CEA-Scan®	carcinoembryonic antigen	murine
	Rituxan™	B-cell lymphoma (Hodgkin's disease)	chimeric
autoimmune disease	Remicade™	Crohn's disease	chimeric
allergies, asthma	RhuMab E25,	anti-IgE	humanized
sepsis	Norasept II	anti-TNF	murine

murine: from mouse; chimeric: constant regions from human gene, variable regions from mouse gene; humanized: rodent CDR regions are grafted onto human framework regions

Recombinant and catalytic antibodies

General. Genetic engineering methods can be used to express native or modified antibodies in host organisms. If microorganisms are used, antibody fragments are usually formed, in particular, single-chain (scF$_v$) and F$_{ab}$ fragments. Complete recombinant antibodies have been obtained from eukaryotic cells, the baculovirus system, transgenic plants ("plantibodies"), and the milk of transgenic animals. Recombinant antibodies have great potential in both diagnostics and therapy. Bispecific and bifunctional antibodies have been studied for targeting drugs bound as antigens to the desired type of cell, e. g., for detoxification, immunosuppression, and cancer therapy. In proteome analysis, specific antibodies have been used to identify and quantify specific proteins after they were separated by 2D electrophoresis, or antibody arrays have been used directly. Catalytic antibodies are being investigated for use in biocatalysis.

Manufacture. The genetic starting material is cDNA, derived either from the mRNA of myeloma cells originating from an immunized laboratory animal or from the mRNA of "naive" B-lymphocytes. After cloning into the host organism, recombinant antibodies or antibody fragments are produced. For example, correctly folded scF$_v$ or F$_{ab}$ fragments can be expressed in the periplasmic space of *Escherichia coli* after cDNA, coding either for a fusion protein or for the separate V$_L$ and V$_H$ chains, has been integrated into a λ vector and competent cells have been transfected with this vector. However, these fragments lack two functional domains: the F$_C$ fragment, which attaches to a receptor, and the glycosylated C$_H$2 domain. Therapeutic antibodies are thus produced in animal cell culture, e. g., with CHO cells. The production of whole antibodies in transgenic plants is also an active area of research.

Combinatorial modification. A great advantage of the production of antibodies by recombinant techniques, compared to the hybridoma technique, resides in the potential to prepare very large antibody libraries. These are usually created by a technique called "phage display". In this method, the whole antibody repertoire of a B lymphocyte is isolated as cDNA, fused with the gene for a viral coat protein, and packaged into M13 expression vectors. After one infectious passage through an *E. coli* host, up to 10^{10} helper phages can be formed which, depending on the cDNA construct used, express on their surface an scF$_v$ or F$_{ab}$ fragment that is coded in their genome. Antibodies of high affinity can be easily isolated from this library by affinity chromatography, and the encoding gene remains within the isolated phage particle. Repeated cycles of mutation and selection based on chain shuffling, error-prone PCR mutagenesis, or mutant strains of *E. coli* have led in a short time to antibodies of remarkable affinity. Thus, by stepwise mutagenesis of the CDR regions of an antibody fragment (gp120) neutralizing the HIV-1 virus, its affinity could be increased 420 fold to 15 pM. The concept has already been successfully applied to the preparation of human antibodies of high affinity, starting out from a naive cDNA library from human B lymphocytes.

Catalytic antibodies have been prepared by immunizing animals with compounds related to the transition state of an enzyme-catalyzed reaction, and also by phage-display procedures. Monoclonal or recombinant antibodies have been obtained by this method which are catalytically active, even in reactions that do not seem to occur in nature (for example, Diels-Alder additions). X-ray analysis of these antibodies showed a functional similarity to the active site of the corresponding enzymes. For example, the catalytic antibody 17E8, which hydrolyzes formyl-norleucine phenyl ester, contains a "catalytic dyad" of ser–his instead of the catalytic triad of ser–his–asp that is found in serine hydrolases catalyzing the same reaction. However, the efficiencies of catalytic antibodies (determined as k_{cat}/K_M) have not yet reached the efficacy of the related enzymes. There is emerging evidence that catalytic antibodies may also occur in living organisms.

Recombinant antibody fragments

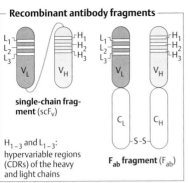

single-chain fragment (scF$_v$)

H$_{1-3}$ and L$_{1-3}$: hypervariable regions (CDRs) of the heavy and light chains

F$_{ab}$ fragment (F$_{ab}$)

Expression vector for *E. coli*

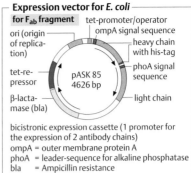

for F$_{ab}$ fragment

- tet-promoter/operator
- ompA signal sequence
- ori (origin of replication)
- heavy chain with his-tag
- tet-repressor
- phoA signal sequence
- β-lactamase (bla)
- light chain
- pASK 85 4626 bp

bicistronic expression cassette (1 promoter for the expression of 2 antibody chains)

ompA = outer membrane protein A
phoA = leader-sequence for alkaline phosphatase
bla = Ampicillin resistance

Antibody diversity

system	repertoire	clonal selection	increase of affinity
B-lymphocytes (immune system)	>10^8 genes	stimulation of specific B-cells through IgM on their surface	lymphocytes: somatic hypermutation
E. coli	cloned repertoire of lymphocytes plus synthetic genes with random CDR >10^8 genes	expression of antibody fragments on the surface of phages followed by affinity screening	gene shuffling, mutator strains, error-prone PCR

Clonal selection: phage-display technique

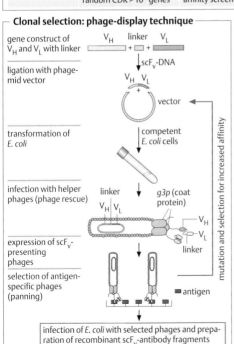

gene construct of V$_H$ and V$_L$ with linker

ligation with phagemid vector

transformation of *E. coli*

infection with helper phages (phage rescue)

expression of scF$_v$-presenting phages

selection of antigen-specific phages (panning)

mutation and selection for increased affinity

scF$_v$-DNA

vector

competent *E. coli* cells

g3p (coat protein)

■ antigen

infection of *E. coli* with selected phages and preparation of recombinant scF$_v$-antibody fragments

Antibody hybrids

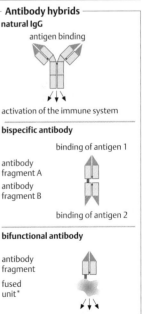

natural IgG

antigen binding

activation of the immune system

bispecific antibody

binding of antigen 1

antibody fragment A

antibody fragment B

binding of antigen 2

bifunctional antibody

antibody fragment

fused unit*

* enzyme, toxin, radioactive label, biotin binding domain, etc.

Immunoanalysis

General. Although enzymes may allow the rapid and quantitative determination of an analyte in a complex biological matrix such as serum, immunoanalysis has developed into an even better method, since it is more sensitive and much more versatile. The development of radio- and enzyme-linked immunoassays in the early 1970s led to a market which today is ca. 6 billion US$ (2002).

Methods. Polyclonal and monoclonal antibodies bind antigens and haptens with high affinity (k_d usually 10^{-6} to 10^{-8} M). The extent of binding, however, cannot be easily determined. Thus, it was therefore a great breakthrough when reporter mechanisms were designed that allowed to detect and quantify the competition between antibodies and antigen binding sites. Usually, homogenous immunoassays, in which no separation step is needed between the binding and the reporter reaction, are distinguished from heterogenous immunoassays, which include a separation step to remove excess reagent and interfering matrix compounds, but in return are usually more sensitive. Numerous test formats have been designed, allowing the specificity and sensitivity of the test to be adapted to the individual requirements, including the choice of the reporter group. When radioactive isotopes are used (radio immunoassay, RIA), the binding event detection signal is formed in a ratio of 1:1. Using enzyme immunoassays (ELISAs), additional amplification of the signal occurs through the enzyme reaction. Well implemented enzyme immunoassays allow for sensitivities down to picomolar or even femtomolar ranges (10^{-12}–10^{-15} M). Horseradish peroxidase or alkaline phosphatase are frequently used as reporter enzymes.

Readout. The results of immunoassays are read through calibration curves. Microtiter plates containing 96 or 384 wells are often used as reaction devices in combination with a microtiterplate reader (a highly parallel photo-, fluori-, or luminometer). In this format, calibration curves can be easily established together with a highly parallel immunoassay.

Analytes. In principle, all substances against which specific antibodies can be raised – either in free form or bound to special hapten carrier molecules – are detectable by immunoassays. Their molecular weight may range from some 10^2 (e. g., drugs, hormones) to about 10^6 Dalton (e. g., multimeric proteins).

Test strips. As with enzyme diagnostics, immunoassays can also be performed with test strips, many of which are available in drugstores. For example, pregnancy can be determined by the level of human chorion gonadotropin (HCG), the concentration of which in urine (and in blood) rises strongly after a fertilized egg has implanted in the uterus. Test strips are also used in the hospital for emergency testing ("point-of-care" situations). A typical example is the analysis of troponin T, a protein liberated from damaged heart muscle cells after a heart attack. A drop of blood is placed on the test strip, where it is soaked into the transport zone while erythrocytes are retained. By capillary force, the analyte is transported to the reaction zone, where it elutes specific antibody conjugates, forming a sandwich complex. In a subsequent specific reaction, this sandwich complex binds to a second gold-stained antibody, resulting in a red line. Controls assure that false-positive reactions are ruled out. The reaction is complete within about 15 min and can be qualitatively monitored by visual inspection or quantified with a CCD camera.

Other examples. Immunoassays have become key diagnostic aids for medical services. With the tremendous growth in the knowledge about the regulation of human metabolism, new diagnostic markers for the early immunodetection of diseases are discovered every year. In food analysis, immunoassays are used to detect the addition of disallowed proteins to food (e. g., of casein to sausages) in a rapid and quantitative manner. The presence of pathogenic microorganisms or toxins in food or water can often be rapidly analyzed using immunanalysis. Xenobiotics such as herbicides can also be easily detected at high sensitivity (nM) in food or water by these techniques.

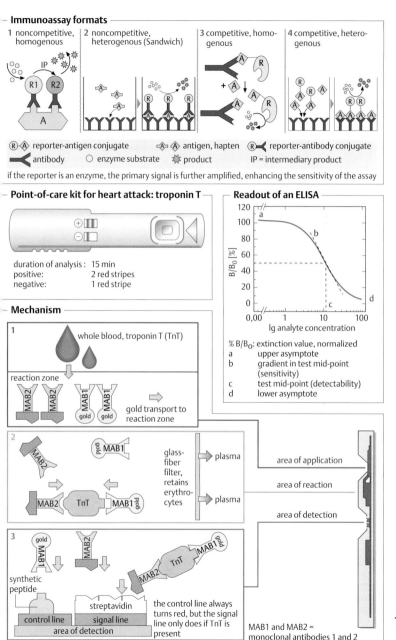

Immunoassay formats

1 noncompetitive, homogenous

IP

R1 R2

A

2 noncompetitive, heterogenous (Sandwich)

3 competitive, homo-genous

A R

+ A R

4 competitive, hetero-genous

R A R A

R R

Ⓡ Ⓐ reporter-antigen conjugate
Ⓐ antigen, hapten
Ⓡ reporter-antibody conjugate
antibody
○ enzyme substrate
✳ product
IP = intermediary product

if the reporter is an enzyme, the primary signal is further amplified, enhancing the sensitivity of the assay

Point-of-care kit for heart attack: troponin T

duration of analysis : 15 min
positive: 2 red stripes
negative: 1 red stripe

Readout of an ELISA

B/B_0 [%]

lg analyte concentration

% B/B_0: extinction value, normalized
a upper asymptote
b gradient in test mid-point (sensitivity)
c test mid-point (detectability)
d lower asymptote

Mechanism

1 whole blood, troponin T (TnT)

reaction zone

MAB2 MAB2 MAB1 MAB1
gold gold

gold transport to reaction zone

2

MAB2 gold MAB1

MAB2 TnT MAB1 gold

glass-fiber filter, retains erythro-cytes

plasma

plasma

area of application

area of reaction

area of detection

3

gold MAB1 MAB2

MAB2 TnT MAB1 gold

synthetic peptide

streptavidin

control line | signal line
area of detection

the control line always turns red, but the signal line only does if TnT is present

MAB1 and MAB2 = monoclonal antibodies 1 and 2

151

Biosensors

General. In a biosensor, biological recognition by an enzyme, an antibody, DNA, a microorganism, etc. is linked to a physical transducer such as an electrode, a fiber-optic device, or a piezo crystal. In spite of intense research and numerous concepts, only a few biosensors have been commercially successful. The market volume for biosensors is in the range of 2 billion US$, which is dominated by various types of glucose biosensors.

Electrochemical biosensors. Oxidases or hydrolases are used for electrochemical biosensors. By hydrolyzing their substrate, hydrolases cause a change in pH which can be monitored by ion-selective electrodes (pH electrodes) or field-effect transistors. Oxidases generate H_2O_2 and consume O_2. Both substances can be analyzed with an amperometric electrode, such as an oxygen electrode. By using so-called mediators, which transfer electrons to the flavine group of the oxidase, the oxidation potential can be significantly reduced. Thus, although the oxidation of H_2O_2 requires a potential of $+400$ mV vs Ag/AgCl, dimethylferrocene is oxidized at $+100$ mV, thus eliminating the risk of a false-positive signal if, e. g., L-ascorbic acid (normal potential $\varepsilon = +170$ mV) is present in the sample. The most commercially successful example of an enzyme electrode is the glucose biosensor, which is often used in hospitals as a standalone item for rapid glucose determinations, but is mainly used in pocket format for self-testing by diabetic patients. Glucose sensors can also be used for monitoring glucose consumption in bioreactors. Miniature implantable glucose sensors are available for diabetic patients but must be replaced within a few days due to their insufficient histocompatibility, rendering them unsatisfactory for controlling an automatic insulin pump. If a pure or mixed culture of aerobic microorganisms is immobilized on top of an oxygen electrode, respiration of the culture can be continuously monitored. This concept has been commercialized for the continuous measurement of wastewater BOD, resulting in data acquisition in minutes instead of days. Enzyme immunoassays can be based on electrochemical reporter reactions instead of color reactions. The intercalation of DNA with electrochemically active reagents such as daunomycin allow the electrochemical determination of hybridization events. Neither concept have yet been brought to market.

Optical biosensors. If an antibody interacts with an antigen, its mass increases, which can be observed as a change in the surface properties ("evanescent field") of an optical transducer. Instruments based on this principle have found good acceptance in the research market, since they allow a highly sensitive (nM) kinetic determination of antigen/antibody interactions. Other optical biosensors are based on fluorescence quenching by oxygen of polyaromatic chemicals, such as phenanthrene derivatives, providing a substrate specific signal in the presence of an oxidase system which consumes equimolar amounts of oxygen upon substrate oxidation.

Flow injection analysis (FIA). Although FIA in a strict sense is not a biosensor technology (the biological components and transducer are usually separated from each other), this method is extremely useful for enzyme, immuno, and DNA assays. It combines analysis with automated liquid handling and thus can be advantageously applied for repeated assays of one or a few analytes. The FIA concept has been successfully transferred to the microsystem and nanotechnology fields.

Natural biosensors. The chemoreceptors of bacteria and the sensory organs of higher animals are interesting examples of natural biosensors. These natural biosensors can reliably and quickly analyze highly complex mixtures of chemicals (e. g., rose fragrance, wine aroma). A technical simulation of natural biosensors resides in chemometric analysis based on neural networks. Although this technology is being investigated for use in biosensors, the results so far are a far cry from the performance of natural sensory systems.

Biosensors

	biological component	transducer
enzyme electrodes		
amperometric	mostly oxidases	O_2 electrode
potentiometric	mostly hydrolases	ion-selective electrode
enzyme FET (field effect transistor)	mostly hydrolases	field effect transistors
microbial sensors	microorganisms	O_2 electrode ion-selective electrode
piezosensors	antibodies	piezoelectric quartz crystal
optical sensors	antibodies, DNA	optical fiber with surface plasmon resonance or grating coupler

sample preparation is often done by liquid handling, e.g., flow injection analysis (FIA); signal processing is usually electronic, complex signals being analyzed by pattern recognition with a neuronal network

Oxygen electrode

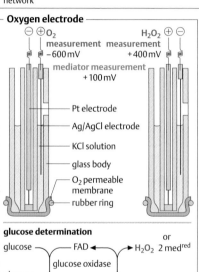

$\ominus \oplus O_2$ $H_2O_2 \oplus \ominus$
measurement   measurement
$-600\,mV$ $+400\,mV$
mediator measurement
$+100\,mV$

- Pt electrode
- Ag/AgCl electrode
- KCl solution
- glass body
- O_2 permeable membrane
- rubber ring

glucose determination

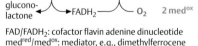

glucose → FAD ← → H_2O_2 2 med^{red} (or)

glucose oxidase

glucono-lactone ← → FADH$_2$ → O_2 2 med^{ox}

FAD/FADH$_2$: cofactor flavin adenine dinucleotide
med^{red}/med^{ox}: mediator, e.g., dimethylferrocene

Optical biosensor

using surface plasmon resonance

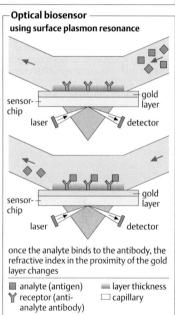

sensor-chip   gold layer
laser   detector

once the analyte binds to the antibody, the refractive index in the proximity of the gold layer changes

- ▪ analyte (antigen)
- Y receptor (anti-analyte antibody)
- ▬ layer thickness
- ▫ capillary

Flow injection system for determination of ethanol

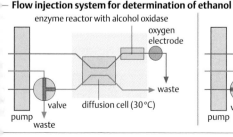

enzyme reactor with alcohol oxidase

oxygen electrode

waste

pump   valve   diffusion cell (30 °C)
waste

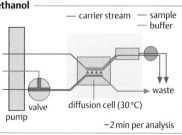

— carrier stream   — sample
              — buffer

waste

pump   valve   diffusion cell (30 °C)
~2 min per analysis

Animal breeding

General. Since the "Neolithic revolution" ca. 11 000 years ago, man has domesticated dogs, sheep, and goats, and ca. 8000 y ago and later, also cattle, pigs, and horses. In the beginning, taming and reproduction were the key issues of breeding. More modern targets of animal breeding are the generation of animal products (meat, milk, eggs, wool) with enhanced quantity and quality. For example, modern meat steers may gain > 300 kg y^{-1} of live weight, and dairy cows may yield $> 10 000$ L y^{-1} of milk – these values of performance traits were doubled within just 30 years by breeding programs. Mating of selected animals from prehistoric times has followed phenotypic criteria, whose values depend on genetic and environmental factors. The classical methods of population genetics and biometric analyses are more and more being complemented by modern methods of reproductive biology and gene diagnosis. Examples are 1) artificial insemination; 2) in-vitro fertilization (IVF) and embryo transfer; 3) the preparation of genetic maps comprising breeding markers; and 4) genotyping markers for performance traits or diseases. Transgenic and cloned animals have, up to now, been mainly used for research and industrial production.

Artificial insemination (AI). As early as 1729, the Italian doctor Lazzaro Spallanzani described artificial insemination for dog breeding, and even earlier, AI was known for horse breeding in Arabian countries. In 1942, the first insemination station for cattle breeding was established in Germany. AI is not costly and allows the selection of male animals with a high breeding value. From the ejaculate of one bull, 400 portions of sperm containing ca. 20 million sperms each can be obtained, which can be stored at $-196\,°C$. The selection of animals suitable for AI starts with young calves, which have to pass several screening steps based on weight gain, body form, and the milk performance of their mothers. The selected "test bulls" are mated with a number of cows, and the milk and meat performance of their progeny during a test period will decide whether the "bull in waiting" will become a "breeding sire", who, in the end, may replace as many as 1000 conventionally mated bulls for breeding. Cows are inseminated with a portion of thawed sperm by a veterinarian, an insemination technician, or the breeder. In most industrialized countries, $> 80\%$ of cows become pregnant through AI. In pig breeding, ca. 60% of sows are artificially inseminated.

In-vitro fertilization (IVF) and embryo transfer (ET) are mainly applied to increase the number of offspring from high-performing cows. For ET, dams are treated with suitable hormones, resulting in superovulation, which is followed by artificial insemination. This procedure may result in up to 8 embryos suitable for ET, yielding on average 4 live calves after the embryos have been transferred to foster mothers. Although the method is mature for practical application, it is rather complex and expensive and not widely used in agricultural practice. Another well studied method is in-vitro fertilization of egg cells outside the female genital tract, which requires methods for the cultivation and also – for many purposes – the conservation of the resulting embryos. With cows, the eggs are obtained without surgery through ultrasound-based follicle punctation. In other animals, however (sheep, pigs), surgery is required. Sex determination of the embryos can be done by PCR within $3-6$ h (sexing). For interested breeders, sex-sorted embryos are being offered for transfer into foster mothers.

Genetic maps. Detailed genetic maps were established for domestic animals during the past 10 many years, especially with concern for those genes that affect performance traits. Gene variants can be analyzed using PCR and RFLP methods. Examples of economic relevance are the ryanodine receptor gene of the pig, in which a mutant largely affects stress tolerance, and variants of the encoding genes that influence milk yield and quality in cows. Most performance traits however, are influenced by many genes and their analysis and application for genetic diagnosis need further efforts.

History of animal breeding

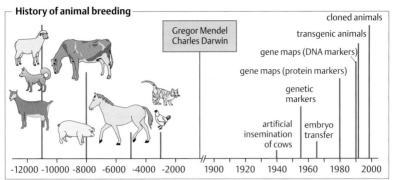

Gregor Mendel
Charles Darwin

cloned animals

transgenic animals

gene maps (DNA markers)

gene maps (protein markers)

genetic markers

artificial insemination of cows

embryo transfer

-12000 -10000 -8000 -6000 -4000 -2000 // 1900 1920 1940 1960 1980 2000

Biotechnological procedures in animal breeding

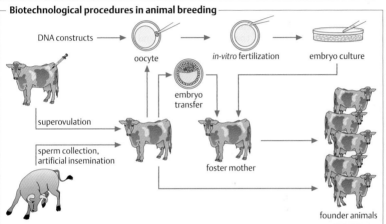

DNA constructs → oocyte → *in-vitro* fertilization → embryo culture

embryo transfer

superovulation

sperm collection, artificial insemination

foster mother

founder animals

Combination of biotechnological and conventional methods

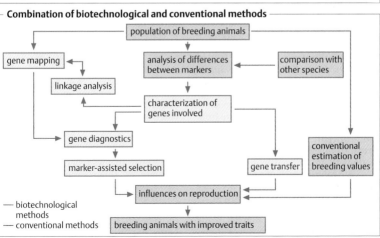

population of breeding animals

gene mapping

analysis of differences between markers

comparison with other species

linkage analysis

characterization of genes involved

gene diagnostics

conventional estimation of breeding values

marker-assisted selection

gene transfer

influences on reproduction

— biotechnological methods
— conventional methods

breeding animals with improved traits

155

Embryo transfer, cloned animals

General. This chapter includes methods for superovulation, the cultivation of embryos, and embryo transfer into foster mothers. Transgenic embryos and the production of cloned animals is also addressed.

Embryonic development in mammals. The egg of most mammals is released from the ovary during metaphase of the second meiosis (oocyte II). If fertilization by a sperm ensues, meiosis continues and the second polar body is released. The two haploid prenuclei, originating from the oocyte and the sperm, fuse to form the diploid nucleus of the zygote, and the first cell division ensues. During further divisions, up to the morula stage, the size of the nucleus continuously decreases. The morula stage sees the first cell differentiation, leading to the blastocyst, which implants, after the zona pellucida has been dissolved, into the mucousa of the uterus, resulting in an embryo.

Superovulation and embryo cultivation. With most mammalian species, the probability that excess oocytes are ovulated (superovulation) can be increased by hormone treatment of the mother. Often it is also possible to obtain egg cells, fertilize them in vitro (IVF), and develop them ex vivo up to the embryo stage, using suitable nutrient media. For economic reasons, most experiments have been done with cattle and sheep embryos. Both can be conserved for unlimited times at $-196\,°C$ (cryopreservation).

Embryo transfer (ET) and embryo splitting. ET is the transfer of foreign embryos into a foster mother of the same species. The embryos may originate from donor animals that were superovulated and artificially inseminated. This is different from embryo cloning, in which blastomers are isolated by microsurgery from one morula; each of a group of blastomeres is cultivated in vitro to the stage of a blastocyst. After transfer into a foster mother, these blastomers may develop into genetically identical animals. Using this procedure, usually 2, and sometimes 6–20 transferable embryos can be obtained from a single cow, of which ca. 50 % will develop into healthy calves. This method is widely practiced.

Transgenic embryos. During the oocyte or blastocyst stage of embryonic development, gene constructs can be introduced by microinjection into the prenuclei or into embryonic stem cells. By this method, new genetic material can be introduced into the recipient embryo. The transgenic embryos originating from this manipulation can be transferred into foster mothers, who eventually give birth to transgenic animals. Pioneering studies in this area were done with mice, since they are important laboratory animals for both fundamental and applied research. The procedure is also applied to farm animals and was a key issue for "gene farming".

Cloned animals. Asexual reproduction leading to identical clones is widely distributed in unicellular organisms, plants, and lower animals. In higher animals, genetically identical clones are quite rare, e. g., the frequency of homozygous twins in humans is only 0.3 %. For experimentally generating clones of higher animals, an egg (haploid chromosome set) is taken from a female donor, and its nucleus is removed with a micropipette. In somatic cells of the same animal species (e. g., obtained from a cell culture of udder epithelium), the G_o phase of the cell cycle (when no cell division takes place) is induced, and the cell, containing a nucleus with a diploid chromosome set, is fused to the enucleated egg cell. The resulting diploid cells are developed to the embryonic stage either in cell culture or in the oviduct of a sterile female, to be transferred into a foster mother. In 1997, for the first time an animal genetically identical to its mother was created based on differentiated somatic cells (Dolly the sheep). It was the only successfully developed lamb in an experiment using 277 enucleated eggs and a total of 27 embryos that were derived from them. In spite of these experimental drawbacks, the method was successfully transferred to mice, goats, pigs, cattle, and other species. It is mostly proposed for the generation of transgenic, monoclonal herds, e. g., to produce therapeutic proteins in their milk (gene farming).

Hormones controlling the female cycle of mammals

hormone	isolation	effect
prostaglandin PGF24	chemical synthesis	degradation of *corpus luteum*; release of sexual heat
follicle-stimulating hormone (FSH)	pituitary of pigs or recombinant protein	follicle stimulation leads to superovulation
pregnant mare serum gonadotropin (PMSG)*	from serum of pregnant mares	follicle stimulation leads to brunt and superovulation
gonadotropin-releasing hormone (GnRH)	chemical synthesis	induces ovulation

*synonym: eCG = equine choryogonadotropin

Embryonic development of a mammal and biotechnology

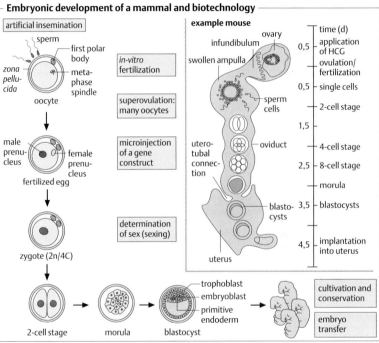

Cloned animals

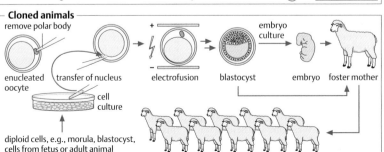

157

Gene maps

General. During the past 100 years, animal breeding has been carried out by selection and mating. Statistical methods were developed to analyze genetic and environmental influences. For the important domestic animals (horse, cattle, pig, sheep, goat, chicken, dog, cat), increasingly precise genetic maps were developed, which are based on the linkage of inherited traits or markers. Physical maps, in contrast, indicate the position of genes on the DNA of individual chromosomes. The genomes of domestic animals are usually on the order of 3 Gbp and of similar complexity as the human genome. So far, genome sequences have been completed for model species like mouse, Drosophila, and the worm *Caenorhabditis elegans*. Moreover, for chicken, cattle, and pig, genome-wide coupling maps do exist, which comprise for each species several thousands of microsatellite loci as markers and several hundred genes whose functions have been assigned. Using these data, the allelic variants of these genes can be typed, using PCR methods. Finally, the genetic variants can be associated with performance traits and used for breeding.

Genetic improvements. The assignment of desired traits (e.g., high milk production) to the genes resembles the task of assembling a puzzle made up of tens of thousands of pieces of high similarity but diverse origin (parents and progeny). To simplify genetic analysis, environmental factors involved in the trait development had first to be standardized. In breeding practice, the influence of genetic vs. environmental factors on the variation of trait values can be estimated by complex statistical methods (e.g., BLUP, best linear unbiased prediction). With methods such as artificial insemination and embryo transfer, groups of domestic animals can be obtained in which half the gene pool originates from one parent, allowing a more precise analysis of the genetic origin of traits in the progeny. However, none of these methods allows the effect of individual gene polymorphisms on complex traits to be analyzed.

Gene maps and genome sequencing. Since the genomes of most domestic animals comprise ca. 3 Gbp, and direct sequencing of DNA is limited to a length of ca. 600 bp per assay, complex procedures are necessary to localize individual genes. Of key importance is the identification of DNA marker sequences that are polymorphic in the parental DNA (i.e., they are composed of different alleles) and trace the inheritance of allelic markers in offspring generations. Microsatellites are the most important markers used for this purpose. They have variable numbers of tandem repeats (VNTR) and are frequently found in the genome of a mammal (about 50 000 – 100 000 microsatellite loci per genome). A second method for analyzing the positions of DNA markers is RFLP analysis (restriction fragment length polymorphism). To this end, parental and progeny DNA are digested with endonuclease, and the resulting patterns of fragments are compared by gel electrophoresis. This method leads to maps for polymorphisms that correlate, in favorable cases, with values of performance traits, rendering them useful for breeding experiments. Once the location on the genome of a polymorphism important for breeding has been identified, the locus can be amplified by PCR, and sequence analysis can be carried out using the PCR product. However, care must be taken that enough information is available about the intron – exon structure of the pertinent gene, preferentially through analysis of the transcribed mRNA or the cDNA derived from it. For economically important domestic animals such as chicken, cattle, and pig, gene maps are already available which have been obtained by observing the inheritance of traits, genes, and DNA markers. Linkage analysis provides further information about the relative positions of genes and markers on chromosomal DNA, since recombination events during meiotic crossing over are recorded by this method.

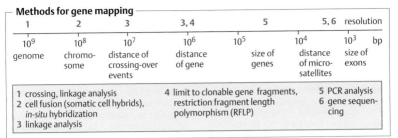

Methods for gene mapping

1	2	3	3, 4	5	5, 6	resolution
10^9	10^8	10^7	10^6	10^5	10^4	10^3 bp
genome	chromo-some	distance of crossing-over events	distance of gene	size of genes	distance of micro-satellites	size of exons

1 crossing, linkage analysis
2 cell fusion (somatic cell hybrids), *in-situ* hybridization
3 linkage analysis
4 limit to clonable gene fragments, restriction fragment length polymorphism (RFLP)
5 PCR analysis
6 gene sequen-cing

Gene map of chromosome 6 of the pig

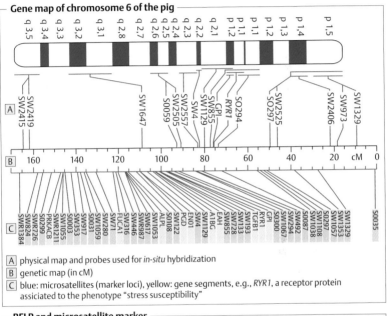

A physical map and probes used for *in-situ* hybridization
B genetic map (in cM)
C blue: microsatellites (marker loci), yellow: gene segments, e.g., *RYR1*, a receptor protein assiciated to the phenotype "stress susceptibility"

RFLP and microsatellite marker

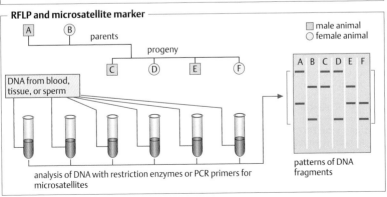

□ male animal
○ female animal

parents
progeny

DNA from blood, tissue, or sperm

analysis of DNA with restriction enzymes or PCR primers for microsatellites

patterns of DNA fragments

159

Transgenic animals

General. Transgenic animals, which express foreign genes (knock-in) or lack expression of indigenous genes (knockout), play an important role in fundamental research, as animal models for human diseases, and for animal breeding. Due to its physiological closeness to man and its simple breeding, the mouse (*Mus musculus*, adjective: murine) is the most important animal model.

Transgenic animals. To minimize side effects due to heterozygous genotypes, transgenic animals are preferentially derived from parental inbred lines. In mice, inbreeding between brothers and sisters for 7–10 generations results in largely homozygous populations. For the transfer of foreign genetic material, microinjection of gene constructs into the prenuclei of embryonic stem cells (ES) is mostly used. Although this method is fast, only a few embryos survive this treatment. Suitable vectors usually contain the gene construct in an intron–exon structure, a promoter, and a polyA signal sequence. Recombination of this sequence into the recipient animal genome often occurs as several copies at several positions simultaneously. Since this is undesired for knockout experiments, transfection of embryonic stem cells in vitro is preferred. Insertion vectors that contain DNA stretches homologous to the gene to be replaced, but lack sequences that would be important in coding for a functional gene, are used. Recombination is initiated by inducing double strand breakage or crossing over and results in inactivation of the functional gene. Selection markers or gene traps, where a reporter gene is cloned into an exon or regulatory sequences of the target gene, make it easier to control the experiment. Using such gene-targeting protocols, recipient cells can be selected which have only one gene copy integrated into a single well defined position in the DNA sequence. To silence genes, various kinds of antisense or interfering RNA constructs are also being intensely applied. Progeny containing the newly recombined or silenced genes in some of their sexual cells (germline chimeric animals) are used for further breeding of homozygous transgenic animals (founder animals).

Transgenic mice. Superovulation is induced in inbred females by injecting gonadotropin (HCG) intraperitoneally, then they are made pregnant by mating. From their oviduct, multi-cell stages, morula, or blastocysts can be obtained, depending on the time of preparation. These cells can be transformed by transfection or microinjection. In the transfection protocol, pluripotent embryonic stem cells (ES cells) are prepared from blastocysts and propagated in tissue culture. They can be transformed using suitable DNA vectors and reinjected into blastocysts, resulting in transgenic embryos. In the microinjection protocol, a suitable vector with foreign DNA is injected into the larger male prenucleus (which becomes visible in egg cells after fertilization). In both methods, the transformed blastocyst or egg cell is implanted into the uterus of a foster mother which was made pseudopregnant by mating with a vasectomized male mouse.

Applications. In 1982 for the first time, a gene construct for rat growth hormone was microinjected into the prenucleus of a fertilized mouse egg, which developed in a foster mother into a transgenic "super-mouse". Using this type of technique, particular genes can be analyzed for their involvement in genetic diseases. Thus, e. g., by knockout experiments, experimental evidence may accrue as to whether this gene is involved in the pathogenic state. In the "oncomouse", the activated v-Ha-ras oncogen was coupled to an embryonic promoter which, upon injury of the epidermis, led to the formation of skin tumors; oncomouses can thus be used for dermal testing of mutagens. Transgenic mice with mutations in the β-amyloid precursor protein (APP) serve as an animal model for Alzheimer's disease, the SCID (severe combined immunodeficiency) mouse with a genetic immunodeficiency as a model for immune diseases. By the end of 2002, over 800 transgenic mouse strains were commercially available. Since genome sequencing of various mouse strains has been completed, analysis of transgenic mice is a valuable tool for functional analysis of the human genome.

after microinjection of a gene construct containing the rat somatotropin gene into the prenucleus of a fertilized mouse egg (right side: control)

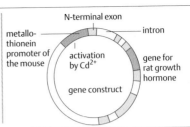

N-terminal exon

metallo-thionein promoter of the mouse

activation by Cd^{2+}

intron

gene for rat growth hormone

gene construct

Transgenic mice

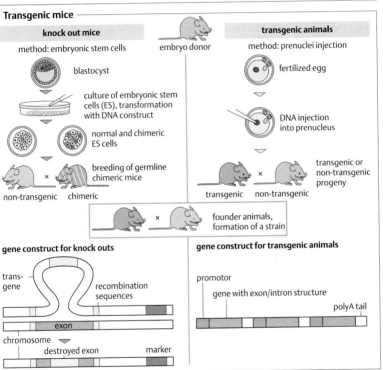

knock out mice

method: embryonic stem cells

embryo donor

blastocyst

culture of embryonic stem cells (ES), transformation with DNA construct

normal and chimeric ES cells

breeding of germline chimeric mice

non-transgenic × chimeric

transgenic animals

method: prenuclei injection

fertilized egg

DNA injection into prenucleus

transgenic or non-transgenic progeny

transgenic × non-transgenic

founder animals, formation of a strain

gene construct for knock outs

trans-gene

recombination sequences

exon

chromosome

destroyed exon   marker

gene construct for transgenic animals

promotor

gene with exon/intron structure

polyA tail

Mouse strains* for clinical research

changes/defects	applications
conventional breeding	nude or hairless mice for skin compatibility tests
expression of rat somatotropin	"supermouse"
deficient immune system	SCID mouse: immunodeficiency
defective p53 tumor-suppressor protein	oncomouse: carcinogenesis
angiotensinogen defective	high-pressure mouse
CTFR defect	cystic fibrosis mouse (for gene therapy experiments)

*as of late 2002, >800 types of transgenic mice had been reported. Many are commercially available

Gene farming and xenotransplantation

General. In the context of improved animal production, transgenic cattle (milk, meat), pig (meat), chicken (meat, eggs), and fish (meat) are being widely investigated. Horses (racing performance) and dogs are also being studied. Due to the demanding technology and public controversy, transgenic animals have been used in some areas only. For example, goats, sheep, and cattle have been engineered to produce pharmaceutically valuable proteins in their milk (gene pharming). Similar to related techniques in transgenic plants, yields are often surprisingly high and competitive with technical procedures using recombinant animals or animal or microbial cells in a bioreactor. Especially for heart transplantation, transgenic pigs are being considered for the provision of replacement organs for humans (xenotransplantation).

Breeding of transgenic animals originally focused on increasing growth by gene transfers resulting in increased endogenous production of growth hormone (somatotropin). Recently, other goals, such as enhanced resistance to disease and stress, as well as improvement of meat or egg quality, are being addressed. In mammals, gene transfer was done by microinjection into prenuclei of fertilized egg cells and embryo transfer. Chickens are transformed using recombinant retroviruses or by fertilization with recombinant sperm. Transgenic fish can be obtained by electroporation of egg cells with DNA. An examples for the application of these techniques is improvement of cold resistance of fish by cloning in an antifreeze protein.

Gene pharming. Biomedically important proteins can be secreted into the milk of transgenic animals. The desired gene is often cloned behind the β- or αS1-casein promoter, and a suitable gene construct is injected into the prenucleus of a fertilized egg cell, resulting in a transgenic embryo. Although the success rate of this technology is still modest (< 0.1 % of the treated egg cells develop into embryos), transgenic animals have already been developed that produce recombinant proteins such as α_1-antitrypsin, tPA, urokinase, IGF-1, IL-2, lactoferrin, or human serum albumen in yields up to 35 g L^{-1} milk. The products can be isolated from milk or used directly in milk. Since a high-performance dairy cow produces up to 10 000 L milk per year, a single transgenic dairy cow would produce the quantity of factor VIII required for the entire USA (120 g).

Xenotransplantation. Until now, > 200 000 people have obtained organ transplants. In the USA alone, ca. 45 000 people are waiting for a heart to be transplanted, although only ca. 2000 hearts per year are available. Against this background, transgenic animals are being studied as organ donors, the most suitable animal being the pig: its organs are similar to human organs in size, anatomy, and physiology. In xenotransplantation, the key issue is organ rejection by immunoreaction. In this context, we can distinguish 1) hyperacute immunorejection (seconds to minutes), based on rapid activation of the complement system in the recipient; 2) acute immunorejection (days), based on the reaction of T cells; and 3) chronic immunorejection (up to several years), whose precise mechanism is unknown. The prime goal when transplanting organs from other species is to prevent hyperacute immunorejection. To this end, transgenic pigs were obtained whose complement cascade is replaced by human factors. A key improvement was the cloning of hCD55 (decay accelerating factor, DAF). When hearts of transgenic pigs containing this factor in their complement system were transplanted into primates, the recipients survived for ca. 40 d, whereas controls died within a few minutes. Knock-out of α1,3-galactosyltransferase in pigs is another target, since α-linked terminal galactose residues occur only on porcine organs and tissues and, in humans, give rise to the formation of anti-α1,3Gal antibodies.

Improved performance in transgenic animals

	effect	target protein or coding gene
pig	malign hyperthermic syndrome Ca metabolism in muscle	ryanodin receptor gene (*RYR1*) on chromosome 6: cys → arg
cattle	κ-casein: milk quality	κ-casein gene
salmon	*antifreeze* protein (AFP) for enhanced temperature tolerance	expression of an AFP-encoding gene from the winter flounder

Manufacture of a pharmaceutical product with transgenic goats

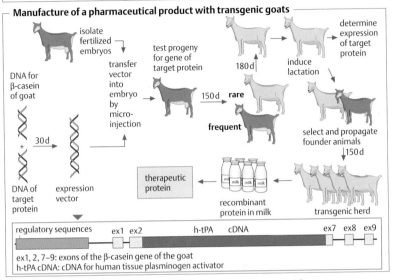

isolate fertilized embryos

DNA for β-casein of goat

30 d

DNA of target protein

expression vector

transfer vector into embryo by micro-injection

test progeny for gene of target protein

180 d

150 d **rare**

frequent

induce lactation

determine expression of target protein

select and propagate founder animals

150 d

therapeutic protein

recombinant protein in milk

transgenic herd

regulatory sequences ex1 ex2 h-tPA cDNA ex7 ex8 ex9

ex1, 2, 7–9: exons of the β-casein gene of the goat
h-tPA cDNA: cDNA for human tissue plasminogen activator

Espression of therapeutic proteins in the milk of transgenic animals

gene construct	recombinant protein	quantity expressed [mg/L]	transgenic animal
WAP	tPA (cDNA)	0 – 50	goat
WAP	protein C (cDNA)	1 000	pig
BLG	α_1-antitrypsin (genomic DNA)	35 000	sheep
BLG	human serum albumen	10 000	mouse
β-casein of goat	tPA (cDNA)	3 000	goat
αS1 casein of cow	urokinase (genomic DNA)	2 000	mouse
αS1 casein of cow	insulin growth factor (IGF-1) (cDNA)	10 000	rabbit

WAP = whey acidic protein   BLG = bovine β-lactoglobulin

Comparison of productivity

protein	market in USA (kg/y)	from human plasma (L)	number of dairy cows (milk production = 10 000 L/y)
factor VIII	0,120	1 200 000	1,2
protein C	100	20 000 000	100
fibrinogen	200	500 000	20
α_1-antitrypsin	800	4 000 000	80
serum albumen	100 000	2 000 000	5 000

Plant breeding

General. About 11 000 years ago, man began to cultivate plants. As the result of this long breeding process, today's cultivated plants yield much more biomass, fruits, and seeds than their wild ancestors. The nutrition of man and his domestic animals relies to a large extent on these cultivated plants. Even today, though, about ⅓ of the 6 billion humans on earth are inadequately nourished. Since the human population is expected to double within the next 50 years, an increase in plant productivity will be a key requirement for the future.

Plant breeding. The result of plant breeding is called a cultivar: a plant line with specifications that are typical for this variety and that are inherited upon propagation. A variety arises from crossing and selection. Depending on the type of propagation, pollen source, and the plant's genetic structure and composition, lines, populations, synthetics, clones, and hybrids can be distinguished. Genetically homogenous varieties are obtained with self-fertilizing plants such as wheat, rice, barley, and sugar cane. Other flowering plants such as maize, potato, soybean, and sugar beets, however, are outbreeding and thus highly heterozygous. If they can be propagated vegetatively, as is true for e. g., potatoes and sugar beets, synthetic or clonal varieties of a narrowed genotype can be obtained. Outbreeding plants can be bred into highly heterozygous but completely homogenous hybrid varieties through enforced self fertilization (inbreeding). In the breeding of some plants such as maize, the male inflorescences are removed for this purpose. If male and female organs are combined in one flower, however, this type of manipulation is difficult. A solution to this problem is the use of male-sterile parental lines, which are obtained by either of two methods: from varieties with cytoplasmic male sterility (CMS, coded by the mitochondrial genome) or by using self-incompatible (SI) lines, a widely distributed mechanism for the prevention of self-pollination. Heterozygous individuals, however, are often stronger than homozygous ones, presumably because their heteroallelic gene products are inactivated with less probability or else show a wider functional range. This property (heterosis) is often used for backcrossings, similar to the methods used for antibiotic-producing microorganisms. Similar protocols are used in the horticulture industry, which has developed more than 11 000 varieties and has a production value of 10 billion € in Germany alone.

Forestry. Already in the early history of man, the uncontrolled cutting of forests led to erosion and denudation of large areas of land. This process occurs also today in tropical rainforests. Only since ca. 1800 has a sustainable forest economy been established in Central and Northern Europe (e. g., pines must reach an age of ca. 100 y before being cut, oaks ca. 300 y). Wood (annual production ca. 7×10^{10} t) is a valuable renewable resource and will probably be used to a much greater extent in the future for the production of chemical base materials, including raw materials for fermentation. Today, it is predominantly used in the manufacture of lumber, chipboard, and for the paper and cellulose industries.

Modern biotechnological procedures. Male sterility, which is important for the breeding of homozygous hybrids, can also be achieved by genetic engineering techniques, e. g., through the expression of a highly active RNase from *Bacillus amyloliquefaciens* under a pollen-specific promoter, leading to pollen inactivation. This process can be modulated through expression of a restorer gene, which inhibits RNase activity. The preparation of callus, meristem, protoplasts, or haploid cultures, which can often be regenerated into intact dicot or monocot plants has revolutionized plant breeding, because it can dramatically accelerate the classical steps of selection. Transgenic plants can express resistance factors against viruses, fungi, bacteria, herbicides, or insecticides, but can also be engineered to express high-value products. The total sequencing of plant genomes is about to establish target-oriented strategies in plant breeding.

Methods of plant breeding

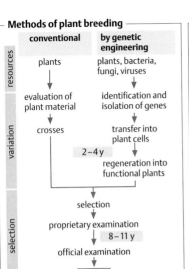

	conventional	by genetic engineering
resources	plants	plants, bacteria, fungi, viruses
	↓	↓
	evaluation of plant material	identification and isolation of genes
variation	↓	↓
	crosses	transfer into plant cells
	2–4 y	↓
		regeneration into functional plants
selection	selection	
	proprietary examination	
	8–11 y	
	official examination	
	cultivar	

Production of agricultural plants (2000)

product/plant		world production [10^6 tons]
sucrose	Saccharum officinarum	1 260
maize	Zea mays	594
rice	Oryza sativa	594
wheat	Triticum aestivum	584
flax	Linum usitatissimum	503
potato	Solanum tuberosum	321
beet sugar	Beta vulgaris	246
manioc	Manihot esculenta	175
soybean	Glycine max	161
sweet potato	Ipomoea batatas	142
barley	Hordeum vulgare	132
tomato	Lycopersicon esculentum	98
wine	Vitis vinifera	64
sorghum	Andropogonoideae	58

Technological advances

monocots	dicots
(inbreeding) e.g., wheat, rice, barley, sugar cane	(outbreeding) e.g., maize, potato, sugar beet, soybean
[1] [4] [5]	[1] [2] [3] [4] [5]
↓	↓
line variety (homozygous)	**hybrid variety** (highly heterozygous)

[1] genetic engineering

[2] cell cultures

[3] CMS, SI*

[4] hybrid breeding

[5] crosses and selection

1850   1900   1950   2000

* CMS cytoplasmic male sterility
  SI   self incompatibility

Size of plant genomes

	species	number of chromosomes	ploidy	genome 1 000 Mbp
monocots				
barley	Hordeum vulgare	7	2	4.8
rice	Oryza sativa	12	2	0.42
wheat	Triticum aestivum	7	6	16
dicots				
maize	Zea mays	10	2	2.5
Arabidopsis	Arabidopsis thaliana	5	2	0.1
rapeseed	Brassica napus	19	2	1.23
tomato	Lycopersicon exculentum	12	2	1
tobacco	Nicotiana tabacum	12	4	4.4
potato	Solanum tuberosum	12	4	1.8

Plant tissue surface culture

General. In the past 40 years it has become possible to propagate tissues and cells of plant organs (roots, leaves, etc.) as organ cultures. Through treatment with plant hormones, such cultures can often be regenerated into intact, fertile plants. This method is widely applied in fundamental research, but is also used to 1) produce large numbers of plants from a stock plant by micropropagation 2) breed plants with improved properties (in-vitro selection); 3) propagate ornamental and garden plants in a virus-free manner; 4) produce transgenic plants; and 5) conserve plant species menaced by extinction as regenerable cell cultures (germ plasms). Plant tissue culture can also be used for preparing secondary plant products.

Methods. In a procedure termed somatic embryogenesis, tissue from the desired organ of a plant that has been grown from a sterile seed (an explantate) is transferred under aseptic conditions to a growth medium. Since most cells and tissues are heterotrophic, growth requires a carbon source such as glucose or saccharose and nitrate as a nitrogen source. The medium also contains vitamins, trace elements, plant cytokinins such as kinetin or zeatin, and growth factors such as 3-indolyl acetic acid, 2,4-dichlorophenoxyacetic acid, or abscisic acid. For hard-to-cultivate cells, more complex media have been used. The cultures are usually grown in chambers under sterile conditions where light, humidity, and temperature are controlled (climate chambers). Depending on the cultivation conditions and starting material, callus or suspension cultures, or meristem or haploid cultures are obtained.

Callus cultures. A wounded tissue that grows in an uncontrolled manner is termed a callus. In plants, it originates from the planes of section of the explantates. Calluses can be cultured on the surface of agar plates and have been used as the starting material for the production of undifferentiated, omnipotent plant cells. Completely differentiated plants can be regenerated by treating callus cultures with plant hormones, provided that they have not become too old.

Suspension cultures. Similar to microorganisms or animal cells, plant cells can be propagated in sterile liquid nutrient media.

Meristem cultures. Meristems are the embryonic cells of plants, which can divide indefinitely. They can be isolated under sterile conditions as leafless vegetation cone from shoots, roots, or axillary shoots and will grow as callus or suspension cultures. A meristem culture is uniquely suited for obtaining and mass propagating specific pathogen-free plants. To this end, meristemic cells are subjected to a short heat treatment at 40 °C; high yields of pathogen- and virus-free cultures are obtained, probably through the formation of heat-shock proteins. From such cultures, pathogen- and virus-free plantlings can be regenerated. These are, however, not resistant and are subject to new infections. Meristem cultures of grapevines, strawberries, banana, sugarcane, or potatoes and of ornamental plants such as carnations, lilies, chrysanthemums, and orchids have revolutionized gardening practices. The world market for virus-free seed or plantlings is estimated as about 3 billion US$ per year.

Haploid cultures are cell cultures of plant sexual organs, in particular, the microspores. After propagation in surface cultures, they can be regenerated either into sterile haploid plants with only one set of chromosomes or, in the presence of the mitotic poison colchicine or after protoplast fusion, into homozygous diploid plants. Such plants are of great importance for the breeder since they pass on a set of constant traits to their progeny. Haploid cultures are being used in breeding potatoes, barley, rapeseed, tobacco, and some medicinal plants.

Somaclonal variation. Although clonal fidelity is high in conventional micropropagation, chromosomal instability of cultured plant cells can be used to recover novel genotypes useful for crop improvement from cell cultures.

Preparation and regeneration of a meristem culture

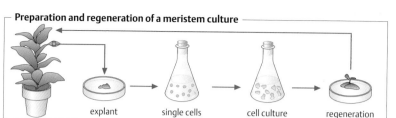

explant   single cells   cell culture   regeneration

Influence of phytohormones

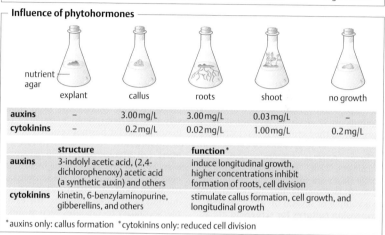

nutrient agar

explant   callus   roots   shoot   no growth

	explant	callus	roots	shoot	no growth
auxins	–	3.00 mg/L	3.00 mg/L	0.03 mg/L	–
cytokinins	–	0.2 mg/L	0.02 mg/L	1.00 mg/L	0.2 mg/L

	structure	function*
auxins	3-indolyl acetic acid, (2,4-dichlorophenoxy) acetic acid (a synthetic auxin) and others	induce longitudinal growth, higher concentrations inhibit formation of roots, cell division
cytokinins	kinetin, 6-benzylaminopurine, gibberellins, and others	stimulate callus formation, cell growth, and longitudinal growth

*auxins only: callus formation   *cytokinins only: reduced cell division

Virus-free plantlings by meristem culture

	elimination of			elimination of
tomatoes	tomato aspermic virus		chrysanthemum	chrysanthemum B virus, chrysanthemum stunt viroid
strawberries	nepovirus and others			

usually the meristem culture is heat-treated at 40 °C

Haploid culture

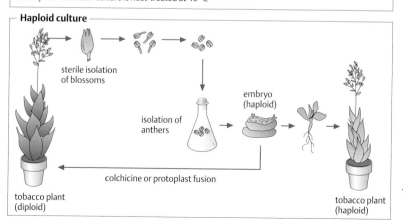

sterile isolation of blossoms

isolation of anthers

embryo (haploid)

colchicine or protoplast fusion

tobacco plant (diploid)

tobacco plant (haploid)

Plant cell suspension culture

General. Similar to microorganisms or animal cells, plant cells can be propagated under aeration in sterile liquid media to which plant hormones have been added (suspension culture). If cells from such cultures are plated on solid nutrient media, they may develop into embryoids, from which intact plants can be regenerated. Suspension cultures of plant cells are used for 1) rapid screening of variants with promising new properties; 2) preparation of protoplast cells and transgenic plants; and 3) production of secondary plant metabolites in a bioreactor.

Methods. Plant cells from stock or callus cultures are transferred to a liquid medium, where they grow in a heterotrophic mode, i. e., in the presence of carbon and nitrogen sources, minerals, and plant hormones. For suspension cultures, shake flasks are used; for the production of secondary metabolites, bioreactors up to a volume of several m³ are preferred.

Suspension cultures for screening are an excellent tool for quickly probing variants of cells for new properties, e. g., for enhanced salt tolerance, improved resistance to herbicides, or formation of secondary metabolites. Since such screening is much faster than classical selection of new variants, which is done by seeding, growth, and probing over several generations, the new method has become very popular in plant breeding, e. g., for the improvement of soybean, citrus, sugar cane, maize, wheat, and potato varieties, but especially for the preparation of novel stress- and pathogen-tolerant plant hybrids. Some disadvantages of this method are that mutants with undesired traits may also be selected. Furthermore, the regeneration of intact plants from suspension cells is not always easy and the properties of the regenerated plants do not always correlate with the properties observed in suspension culture. Suspension cultures are widely used for screening, e. g., in the selection step of transgenic plants, which express a desired genetic trait that was introduced by targeted gene transfer.

Protoplast cultures. The polysaccharide cell wall of plant cells can be removed in an isotonic solution by careful protocols using cellulases, hemicellulases, and pectinases. The resulting protoplasts can be fused with other protoplasts using chemical or electrical procedures (protoplast fusion), leading to somatic hybrids with combined genomes. If protoplasts are fused with enucleated cytoplasts, inheritable homozygous traits originating from the cytoplasmic organelles (plastids) can be transferred. This method is being successfully used for the transfer of cytoplasmic male sterility. It is highly valued in breeding, since it guarantees complete outbreeding. The fusion of protoplasts originating from different plant species has been intensely investigated (potato + tomato = pomato) but so far has not resulted in new plant varieties.

Plant cell bioreactors. The capacity of plant cells to propagate in the form of genetically omnipotent cell suspension cultures can sometimes be used for the production of valuable products. For example, secondary metabolites such as shikonin, berberin, and taxol are being industrially produced from plant suspension cultures using bioreactors up to several m³ in volume. Attempts to use suspended plant cells in single-step biotransformations, e. g., in regioselective glycosidation or hydroxylation reactions, were, however not economically successful, since these reactions can better be carried out in a simpler way using enzymes or recombinant microorganisms. In a typical case of production of a secondary metabolite, phytohormone-treated cells from callus cultures are transferred to suspension cultures, increasing the reactor scale step by step. Stirred reactors, airlift, bubble-column, and other reactors have been investigated. Although the process engineering challenges involved in this technology have usually been solved, the low stability of highly productive cell lines is often a problem, aggravated by the fact that the biosynthesis of most secondary plant products and its regulation is still poorly understood.

Protoplast fusion

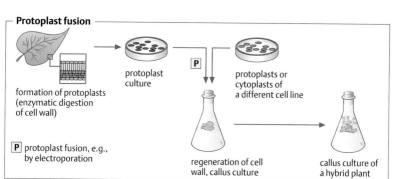

formation of protoplasts (enzymatic digestion of cell wall)

protoplast culture

P

protoplasts or cytoplasts of a different cell line

P protoplast fusion, e.g., by electroporation

regeneration of cell wall, callus culture

callus culture of a hybrid plant

Suspension culture for selection

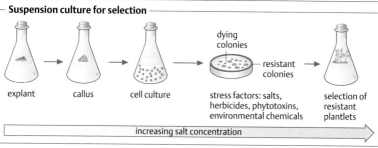

explant

callus

cell culture

dying colonies

resistant colonies

stress factors: salts, herbicides, phytotoxins, environmental chemicals

selection of resistant plantlets

increasing salt concentration

Bioreactor types for plant cell cultivation

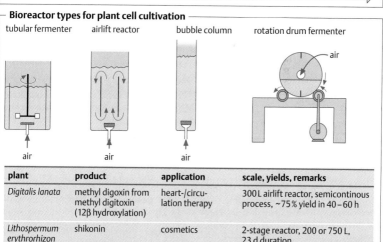

tubular fermenter

airlift reactor

bubble column

rotation drum fermenter

air

air

air

air

plant	product	application	scale, yields, remarks
Digitalis lanata	methyl digoxin from methyl digitoxin (12β hydroxylation)	heart-/circulation therapy	300 L airlift reactor, semicontinous process, ~75% yield in 40–60 h
Lithospermum erythrorhizon	shikonin	cosmetics	2-stage reactor, 200 or 750 L, 23 d duration
Berberis sp.	protoberberin	pharmaceutical	up to 1.7 g/L
Panax ginseng	ginseng pieces	health	30 L bioreactor
Coleus blumei	rosemarinic acid	pharmaceutical	
Taxus spp.	taxol	antitumor agent	up to 1 g/L after 14 d
Vanilla planifolia	vanillin	aroma compounds	16 mg/L after 45 d

Transgenic plants: methods

General. Various methods have been developed to transfer foreign DNA into a plant genome. In dicots such as tomato, tobacco, potato, pea, and bean, the Ti plasmid from *Agrobacterium tumefaciens* provides the method of choice. For dicots or monocots that can be regenerated from protoplasts, electroporation or protoplast transformation has been successfully used. Intact plant cells can be transformed by microinjection of DNA or by biolistic procedures. Successful transformation of plant cells is usually evaluated by PCR methods or by cotransformed reporter genes.

Ti plasmid. The soil bacterium *A. tumefaciens* can infect dicots, initiating cancerous cell propagation at the root neck, called root gall disease. In this process, a plasmid of ca. 200 kbp, the Ti (tumor-inducing) plasmid, is transferred into the plant cells, resulting in the integration of T-DNA, a DNA fragment 15–30 kbp long, into the plant chromosomal DNA. For the transformation of dicots, modified Ti plasmids have been developed which keep their infectivity but have lost their virulent properties. They contain, in addition to the gene to be transformed, the T-DNA, an origin of replication for *E. coli* or other laboratory host, and a reporter gene. Transformation is usually carried out by infecting susceptible plant cells in culture with recombinant strains of *A. tumefaciens*. Using organ-specific promoters and leader sequences enables T-DNA to be directed to any desired location in the target cell (leave, stem, root, or subcellular compartments such as chloroplasts or mitochondria). It has also been possible to induce plasmid DNA expression by external factors such as high temperature, drought, pathogen infection, light quality, or circadian cycles. A plasmid obtained from *A. rhizogenes*, the Ri plasmid, is sometimes used in a similar fashion. Plant viruses such as caulimo or gemini viruses are less well suited as vectors: they either have too low a capacity for foreign genes, too small an infection spectrum, or a replication mode that is too complex for practical use.

Transgenic monocots. Many monocots, such as wheat, barley, rice, and maize, are highly important agricultural plants, but until recently could not be transformed by *A. tumefaciens*. Recent work, however, has shown that the inducer syringeon can be used as a chemical signal for activation of the virulence region of the Ti plasmid in monocots. Using such protocols, cells of monocots can be transformed with the Ti plasmid system. Their regeneration into intact recombinant plants, however, poses another challenge, severely limiting all methods that depend on protoplasts, such as microinjection, electroporation, and liposome fusion. As a result, biolistics has become the method of choice. In this procedure, plant embryos are bombarded with highly accelerated tiny gold or tungsten particles on which the foreign DNA has been adsorbed.

Interference with gene expression. Two protocols are mainly used to obtain knockout plants. In one, integration of a foreign stretch of DNA into a target gene region through homologous recombination is used in a similar manner as in the preparation of knockout animals. Unlike in animals, however, the efficiency of this method is very low in plants. Much better results are obtained if a copy of the target gene, induced by a suitable promoter, is integrated in the reverse reading direction. Transformed plants transcribe this gene into an antisense RNA, which complexes with the mRNA of the correct gene and is destroyed by RNase as a nonfunctional RNA complex. Since this is a statistical process, the success rate is only in the range of 1–2%.

Plant genomes. The first plant genomes to be completely sequenced was *Arabidopsis thaliana*, with a genome size of ca. 100 Mbp, distributed among 5 chromosomes. Very recently, genome sequencing of two varieties of rice *(Oryza sativa)*, with ca. 420 Mbp on 12 chromosomes, was completed. Other plant genome projects now focus on the much larger genomes of maize, tobacco, cotton, and potato.

Transformation methods

	monocots	dicots
microinjection (protoplasts)	+	+
transfection (protoplasts, Ca^{2+})	0	+
electroporation (protoplasts)	0	+
biolistics	++	+
T-DNA	+	++
plant viruses	0	0

0 occasionally successful   + possible   ++ method of choice

Microinjection

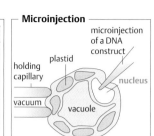

microinjection of a DNA construct

holding capillary
plastid
nucleus
vacuum
vacuole

Infection of a plant with *Agrobacterium tumefaciens*

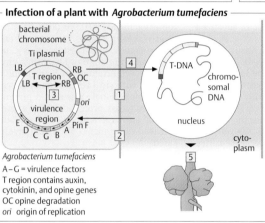

bacterial chromosome
Ti plasmid
LB   RB
T region   OC
LB   RB
virulence region
ori
E
D   C   G   B   A   Pin F

Agrobacterium tumefaciens
A–G = virulence factors
T region contains auxin, cytokinin, and opine genes
OC opine degradation
ori origin of replication

T-DNA
chromosomal DNA
nucleus
cytoplasm

1. *A. tumefaciens* attaches to an injured plant cell

2. plant cell sends chemotactic signals and activates virulence region

3. virulence region of Ti plasmid activates T region

4. T region is integrated into plant chromosome as single-stranded DNA

5. auxins and cytokinins initiate tumor growth. Opine becomes C-source for *A. tumefaciens*

Biolistics

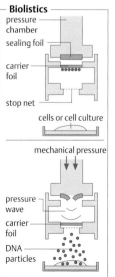

pressure chamber
sealing foil
carrier foil
stop net
cells or cell culture

mechanical pressure

pressure wave
carrier foil
DNA particles

Cloning vectors derived from *A. tumefaciens*

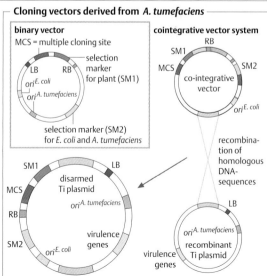

binary vector
MCS = multiple cloning site
selection marker for plant (SM1)
LB   RB
ori$^{E. coli}$
ori$^{A. tumefaciens}$
selection marker (SM2) for *E. coli* and *A. tumefaciens*

cointegrative vector system
RB
SM1
MCS   co-integrative vector   SM2
ori$^{E. coli}$

recombination of homologous DNA-sequences

SM1   LB
disarmed Ti plasmid
MCS   *ori*$^{A. tumefaciens}$
RB
SM2   *ori*$^{E. coli}$ virulence genes

LB
ori$^{A. tumefaciens}$
recombinant Ti plasmid
virulence genes

171

Transgenic plants: resistance

General. In the USA, > 30 transgenic plants have been registered for agricultural use and are being raised on > 35 million hectares of land. This includes transgenic cotton, potato, maize, rapeseed, soybean, and tomato. In most cases, foreign genes that convey tolerance to herbicides, insecticides, or viruses were incorporated. Plants with enhanced stress tolerance or ornamental plants with modified colors are in an advanced stage of development.

Herbicide-tolerant plants. About 10% of all harvests are lost by weeds. The ideal herbicide should be active at low concentrations, not inhibit the growth of the agricultural plant, be readily degraded, and not reach groundwater. Transgenic plants reduce the need for herbicides, since they are made genetically resistant by 1) containing the herbicide-sensitive protein in excess, 2) showing reduced binding of the herbicide, or 3) inactivating the herbicide through biodegradation. As an example, soybeans resistant to the broad-spectrum herbicide glyphosate (Roundup™) were produced by isolating glyphosate-resistant strains of *E. coli*, isolating the bacterial gene coding for 5-O-enolpyruvyl shikimic acid-3-phosphate synthase (ESPS synthase, the herbicide target), and expressing it in soybean under the control of a plant promoter. Resistance of tobacco, potato, rapeseed, and other plants to phosphinothricin (Basta™), an inhibitor of glutamine synthetase, was achieved by expressing a phosphinothricin acetyltransferase (PAT) from *Streptomyces hygroscopicus*.

Insect-resistant plants. *Bacillus thuringensis* synthesizes a protein of M_R 250 kDa (δ-endotoxin, BT toxin), which forms a highly toxic protein in insect intestines after proteolysis. In plants and mammals, this transformation does not occur. Thus, BT protein has been expressed in numerous plants as an insecticidal protein. Using optimized codons and strong constitutive promoters, e.g., for the 35S protein of cauliflower mosaic virus, the expression rate was increased ca. 1000 fold.

Cloned protease inhibitors were also successfully used as insect control agents.

Fungus-resistant plants. Fungal infections lead to important damages to crops. An important historical example is the potato blight (caused by *Phytophthora infestans*), which in the 19th century led to famines throughout Europe, especially in Ireland. By overexpressing glucanases or chitinases of plant origin, directed against the fungal cell wall, the fungal resistance of tobacco was increased. Good success was also obtained with ribosome-inactivated proteins (RIP).

Virus-resistant plants. Viruses can also lead to significant harvest losses, e. g., potato virus 4 in potato or *Rhizomania* virus in sugar beet. Attempts are being made to interfere with virus replication by expressing nonfunctional virus capsid proteins in plants (cross protection). The expression of antiviral antibodies or of hammerhead ribozymes is also being investigated.

Plants with enhanced stress tolerance. Many forms of physiological stress (strong light, UV radiation, heat, drought) are accompanied by the formation of oxygen radicals, in particular of the oxygen radical anion. Transformed plants, expressing the gene product superoxide dismutase under control of the 35S promoter of cauliflower mosaic virus, not only were more resistant to physiological stress, but also wilted more slowly.

Altered blossom colors, aging. Form and color are important characteristics of ornamental plants, as storage stability and aroma are of fruits. Incorporating foreign genes or silencing indigenous genes can affect these properties. For example, genes involved in the secondary metabolism of other plants have been expressed to modulate chromophore biosynthesis, mostly within flavonoid or anthocyan glycoside metabolism. In the "blue" rose, a P450 monooxygenase from carnations, which hydroxylates dihydroquercetin yielding a blue pigment, is expressed. To silence genes, the antisense technique has been successfully applied: a commercial example is the Flavr-Savr™ tomato in which a pectinase gene has been inactivated.

Resistent plants (examples)

plant	overexpressed/ foreign/modified or inactivated gene product	desired property	trans- formation method	company organization
cotton	acetolactate synthase	resistance to sulfonylurea		DuPont
maize	BT-δ-endotoxin	resistance to European corn borer		DeKalb, Monsanto
	glutamine synthase, transacetylase, EPSP synthase	phosphinothricine and glyphosate resistance		Bayer Crop Science, Zeneca Seeds, Monsanto
soy-bean	glutamine synthase, transacetylase, EPSP synthase	phosphinothricine and glyphosate resistance		Bayer Crop Science, Zeneca Seeds
potato		resistance to Colorado beetle		Monsanto
	polyphenol oxidase	prevention of browning	antisense construct	Bayer Crop Science, Zeneca Seeds, Monsanto
papaya		papaya ringspot virus resistance		Cornell-University/ University of Hawaii
tomato	polygalacturonase	retarded maturation	antisense construct	Calgene, Monsanto "Flavr Savr"
		thicker skin, retarded maturation	antisense construct	Zeneca Seeds
	monellin	enhanced sweetness	Ti plasmid	
petunia	dihydrokaempferol reductase from maize	ruffled petals	sense construct	Max-Planck-Institute
rose	dihydroquercetin-5'-hydroxylase	blue pigment	sense construct	Suntory, Calgene

Rhizomania of the sugar beet

resistance through ex-pression of the virus coat protein

plant infec-ted with the Rhizomania virus

Transgenic plants 2001

35.7 million ha in the USA, 11.8 in Argentia, 3.2 in Canada, 1.5 in China

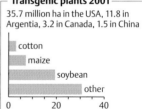

cotton
maize
soybean
other

0   20   40
million ha

Phosphinothricin-resistant soybean

by expressing phosphinothricine acetyltransferase from *Streptomyces hygroscopicus*, the plant becomes resistant to the herbicide phosphino-thricine (Basta™).
Right panel: a non-resistant control.

Transgenic plants: products

General. Modification of plants by genetic engineering includes not only the modification of plant chemicals (renewable resources) but also the synthesis of valuable chemicals in transgenic plants. Examples are 1) modification of the amino acid, starch, lignin, or oil composition; 2) expression of antigens, antigen fragments (plantibodies), vaccines, human serum albumen, or biopolymers.

Modification of plant chemicals. Several routes have been explored for supplementing plant proteins with essential amino acids required in human food (usually L-lysine and L-methionine): 1) expression of proteins from other plants, which have a more adequate composition; 2) site-directed mutagenesis of indigenous storage proteins, replacing nonessential with essential amino acids; and 3) cloning genes for deregulated key enzymes in branched metabolic pathways, e. g., aspartokinase from *Escherichia coli* and dehydrodipicolinic acid synthase from Corynebacterium for enhanced synthesis of L-lysine. ADP glucose pyrophosphorylase, a key enzyme of starch biosynthesis, is an allosteric enzyme in most plants. To increase the percentage and modulate the composition of starch, a nonregulated enzyme from *E. coli* was expressed in tomatoes, resulting in fruits with a 20 % higher starch content. The ratio of linear amylose to branched amylopectin is essential to the properties of starch in processed food and in technical applications. Expression of a gene that is responsible for α-1,6 bond formation, *glgB* from *E. coli*, in potatoes and under control of the granule-bound starch synthase promoter, led to the formation of starch with a 25 % higher percentage of amylopectin. The fatty acid composition of oil plants was also significantly modified. Thus, lauric acid (C12), an important renewable raw material for the manufacture of surfactants soluble in cold water, occurs only in tropical oils (palm kernel oil, coconut oil) as a triglyceride. By cloning chain-length-specific fatty acid ACP thioesterases from laurel (*Umbellularia californica*) into suitable rapeseed lines, complemented by other measures, transgenic rapeseed varieties were obtained whose seed contains 50 mole % of trilauroyl glycerol. The value of the world's total annual timber harvest is in excess of 400 billion US$. In view of faster growing trees and reduced pollution during the paper making process, the lignin content of woods was reduced by modulating genes required for lignin biosynthesis, e. g., in the cinnamic acid pathway. Fast growing hardwood trees like poplars and eucalyptus trees are the focus of such investigations.

Expression of valuable chemicals. Most of these experiments were carried out with transgenic tobacco plants or in *Arabidopsis thaliana*, since transformation of these organisms is relatively easy. The expression of human serum albumen was achieved with good yield, as was the expression of complete IgG antibodies, which, in view of developing an immunological means of preventing tooth decay, were directed against the adhesin of *Streptococcus mutans*. Concentrations of up to 1 % of total protein were obtained. The economic expression of antigens in plants has been discussed for carrying out inexpensive vaccination procedures in developing countries, using food intake as the vaccination step. In model experiments, the surface antigen of hepatitis B virus was expressed in tobacco at levels of ca. 0.01 % of the total soluble protein, and feeding mice with meal derived from such tobacco plants led to an immune response. Eating potatoes containing expressed heat-labile enterotoxin B from *E. coli*, a protein leading to diarrhea, also led to an immune response in human volunteers. When a *Ralstonia eutropha* operon coding for the synthesis of polyhydroxybutyric acid and which consists of three genes is expressed in the chloroplasts of *Arabidopsis thaliana* or rapeseed, this valuable polymer can be produced by photosynthesis. The economic processing of plants containing this polymer must, however, be further optimized.

Tomato puree

from tomatoes with
enhanced sugar content

Transgenic rapeseed with altered fatty acid composition

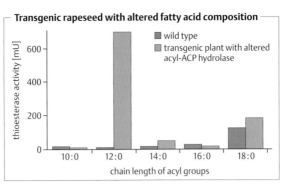

- wild type
- transgenic plant with altered acyl-ACP hydrolase

y-axis: thioesterase activity [mU] (0, 200, 400, 600)

x-axis: chain length of acyl groups (10:0, 12:0, 14:0, 16:0, 18:0)

Polyhydroxybutyric acid in the plastids of transgenic rapeseed

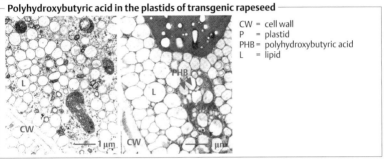

CW = cell wall
P   = plastid
PHB = polyhydroxybutyric acid
L   = lipid

Plants as bioreactors

plant	foreign/modified/inactivated gene	observed property	transformation method
modification of indigenous plant genes			
rapeseed, soybean	derepressed aspartokinase (*E. coli*) and dihydrodipicolonic acid synthase (*Corynebacterium*)	higher lysine content	Ti plasmid
potato, tomato	starch synthase	altered starch composition	antisense construct
rapeseed	acyl-ACP hydrolase	altered fatty acid profile	Ti plasmid
Arabidopsis	γ-tocopherol methylase	synthesis of vitamin E	Ti plasmid
pine tree	shikimic acid synthase	altered lignin content	antisense construct
expression of genes foreign to plants			**yields [g/kg]**
potato	coat protein of hepatitis B virus	immune response in mice	
	human serum albumen	formation of human serum albumen	
tobacco	IgG fragments against adhesin of *Streptococcus mutans*	formation of antibody fragments	10
Arabidopsis rapeseed	3 genes (*phb* operon) from *Ralstonia eutropha*	formation of polyhydroxy butyric or valeric acid	140

transgenic plant		harvest, extraction		recovery		up to 140 g/kg "phytopolymer"
operon for PHV	▶		▶			

Viruses

General. A virus is an infectious particle without indigenous metabolism. Its genetic program is written in either DNA or RNA, whose replication depends on the assistance of a living host cell. A virus propagates by causing its host to form a protein coat (capsid), which assembles with the viral nucleic acid (virus particle, nucleocapsid). Viruses can infect most living organisms; they are mostly host-specific or even tissue- or cell-specific. Viruses are classified by their host range, their morphology, their nucleic acid (DNA/RNA), and their capsids. In biotechnology, viruses are used for the development of coat-specific or component vaccines and for obtaining genetic vector and promoter elements which are, e. g., used in animal cell culture and studied for use in gene therapy.

Viruses for animal experiments. The first cloning experiments with animal cells were done in 1979, using a vector derived from simian virus 40 (SV40). This virus can infect various mammals, propagating in lytic or lysogenic cycles (lysis vs. retarded lysis of host cells). Its genome of ca. 5.2 kb contains early genes for DNA replication and late genes for capsid synthesis. Expression vectors based on SV40 contain its origin of replication (ori), usually also a promoter, and a transcription termination sequence (polyA) derived from the viral DNA. For the transfection of mouse cells, DNA constructs based on bovine papilloma virus (BPV) are preferred. In infected cells, they change into multicopy plasmids which, during cell division, are passed on to the daughter cells. Attenuated viruses derived from retro, adeno, and herpes viruses are being investigated as gene shuttles for gene therapy. Retroviruses, e. g., the HIV virus, contain an RNA genome. They infect only dividing cells and code for a reverse transcriptase which, in the host cell, transcribes the RNA into cDNA. HIV-cDNA is then integrated into the host genome where it directs, via strong promoters, the formation of viral nucleic acid and capsid proteins. Over 200 experiments with retroviral vectors having replication defects have already been carried out for gene therapy. A disadvantage of using retroviral vectors lies in their small capacity to package foreign DNA (inserts), whereas vectors derived from adenoviruses can accommodate up to 28 kb of inserted DNA. For ca. 170 gene therapy experiments, adenovirus-derived vectors were used. In contrast to retroviruses, adenoviruses can infect nondividing cells, but their DNA does not integrate into the host chromosomal DNA. Since a gene therapeutic experiment using an adenoviral vector led to the death of an 18-y-old volunteer through overreaction of his immune system, this method is now rarely used. For gene therapy targeted to neuronal cells, e. g., in experiments related to Alzheimer's or Parkinson's disease, *Herpes simplex*-derived vectors are often used. Their large genome of 152 kb allows them to accommodate rather large inserts of foreign DNA.

Viruses for plant experiments. Most plant viruses have an RNA genome. Only two groups of DNA viruses are known that infect higher plants, caulimo virus and gemini virus. Caulimo viruses have a quite narrow host range: they infect only crucifers such as beets and some cabbage varieties. Their small genome reduces their potential for accommodating foreign inserts. Gemini viruses infect important agricultural plants such as maize and wheat and thus bear significant risks for application. Moreover, their genome undergoes various rearrangements and deletions during the infection cycle, rendering the correct expression of foreign DNA inserts difficult.

Baculoviruses infect insects but not mammals. After infection, host cells form a crystalline protein (polyhedrin), which may constitute > 50 % of the insect cell. The polyhedrin promoter is therefore useful for the heterologous expression of proteins, using cell cultures of Spodoptera (a butterfly). An advantage of this system is that posttranslational glycosylations in this system resemble those of mammalian cells. Scale-up of this system is, however, limited, rendering it most useful for laboratory experiments.

Forms

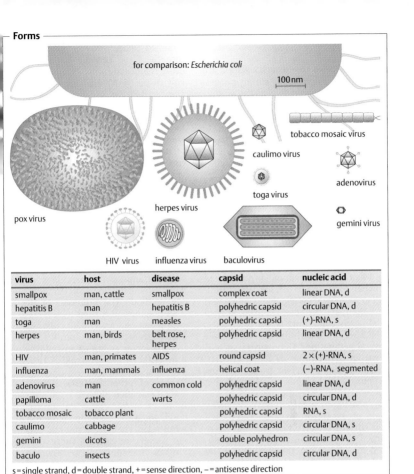

for comparison: *Escherichia coli*

100 nm

tobacco mosaic virus

caulimo virus

adenovirus

toga virus

gemini virus

herpes virus

pox virus

HIV virus   influenza virus   baculovirus

virus	host	disease	capsid	nucleic acid
smallpox	man, cattle	smallpox	complex coat	linear DNA, d
hepatitis B	man	hepatitis B	polyhedric capsid	circular DNA, d
toga	man	measles	polyhedric capsid	(+)-RNA, s
herpes	man, birds	belt rose, herpes	polyhedric capsid	linear DNA, d
HIV	man, primates	AIDS	round capsid	2 × (+)-RNA, s
influenza	man, mammals	influenza	helical coat	(−)-RNA, segmented
adenovirus	man	common cold	polyhedric capsid	linear DNA, d
papilloma	cattle	warts	polyhedric capsid	circular DNA, d
tobacco mosaic	tobacco plant		polyhedric capsid	RNA, s
caulimo	cabbage		polyhedric capsid	circular DNA, s
gemini	dicots		double polyhedron	circular DNA, s
baculo	insects		polyhedric capsid	circular DNA, d

s = single strand, d = double strand, + = sense direction, − = antisense direction

Propagation cycle of a retrovirus

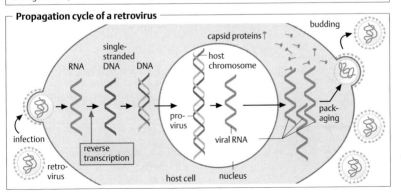

budding

capsid proteins ↑

single-stranded DNA

RNA   DNA

host chromosome

infection

pro-virus

reverse transcription

viral RNA

pack-aging

retro-virus

host cell   nucleus

177

Bacteriophages

General. Viruses that attack bacteria are termed bacteriophages or simply phages. Fermentation processes, e. g., the production of starter cultures, are always endangered by phage infections. As a preventive measure, attempts are usually made to select phage-resistant strains. Phages are useful in genetic engineering, e. g., for the development of cloning vectors or promoters, for DNA sequencing, and for the preparation of gene and protein libraries. Since most cloning experiments are done with *E. coli*, phages specific for this bacterium (λ-, M13-, Qβ-, T-phages) play a key role.

λ Phage. When infecting *E. coli*, λ phage can follow two routes: either its linear double-stranded DNA (ca. 48.5 kbp, ca. 1 % of the *E. coli* genome) is propagated independent of the *E. coli* genome, resulting in lysis (lytic cycle), or it is integrated into the *E. coli* genome, resulting in lysogenic cells containing latent prophages, which replicate with the bacterium over several generations. Upon stress, such as a rise in temperature or UV irradiation, the prophage is excised from the *E. coli* genome and becomes virulent, lysing the host cell. A specific property of λ is its capacity to form cohesive or sticky ends of 12 unpaired nucleotides each (cos sites), which are necessary for circular λ DNA formation and for its integration into the *E. coli* genome. The sticky ends also form the recognition signal for the formation of the viral gene product A, an exonuclease. After replication of the λ DNA into a concatemer of linear λ genomes, endonuclease A cuts at this position, initiating the packaging of progeny into its capsids. Cosmids, an important tool for the construction of large gene libraries, are derived from the λ phage, as is a family of λ plasmids such as λEMBL4, which can be induced by a rise in temperature.

The M13 phage infects *E. coli* according to a different mechanism. It contains single-stranded DNA of ca. 6.4 kb, which after infection directs the synthesis of its complementary strand. The resulting double-stranded phage DNA is not integrated into the *E. coli* genome but is continuously replicated in the cytoplasm, giving rise to up to 1000 phage particles/cell. During division of the host cell, the phage infection is handed on to the daughter cells (ca. 100/cell). Genes that have been cloned into a vector derived from M13 can be obtained as single-stranded DNA – a valuable property for use in DNA sequencing. Prior to the invention of PCR, M13 vectors also served for site-directed mutagenesis of proteins.

T Phages occur in 7 different types. For genetic engineering, two enzymes coded by T phage genomes are useful: the DNA ligase of T4, which links DNA fragments regardless of the quality of their ends (sticky or blunt), and the DNA polymerase of T7, which polymerizes DNA on a single strand DNA matrix; it is used in gene sequencing protocols (Sanger–Coulson method). The promoter of the T7-RNA polymerase is used in several *E. coli* expression vectors. T7-RNA polymerase is used to transcribe DNA into RNA, which in turn serves as mRNA in cell-free protein synthesis, based on mRNA, tRNAs, ribosomes, and ATP.

Phages of other bacteria. Among the > 1000 classified phages, > 300 are specific for enterobacteria, > 230 for bacteriococci, and > 150 each for Bacilli and Actinomycetes. Their structure and function are closely related to those of other viruses, including those specific for *E. coli*. Some of them can be either virulent or lysogenic, similar to the λ phage. Lactobacilli-specific phages are a major problem in the manufacture of milk products. Resistant bacterial strains resist infection by preventing adsorption or replication of these phages. Among the 5 groups of *Bacillus* phages, ø105 and SPO2 are often used in transformation experiments, and PBS1 has been used in construction of the *B. subtilis* genome sequence map. Phage D3112 is the preferred vector for the transformation of Pseudomonads, and SH3, SH5, SH10, or øC31 are preferred for the genetic engineering of Streptomyces.

E. coli phages (selection)

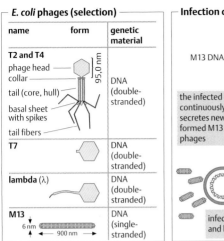

name	form	genetic material
T2 and T4 phage head collar tail (core, hull) basal sheet with spikes tail fibers	95,0 nm	DNA (double-stranded)
T7		DNA (double-stranded)
lambda (λ)		DNA (double-stranded)
M13 6 nm ← 900 nm →		DNA (single-stranded)

Infection cycle of M13

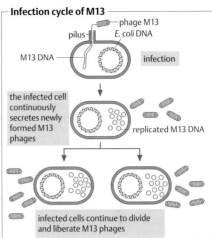

phage M13
E. coli DNA
pilus
M13 DNA
infection

the infected cell continuously secretes newly formed M13 phages

replicated M13 DNA

infected cells continue to divide and liberate M13 phages

Infection cycle of the lambda (λ) phage

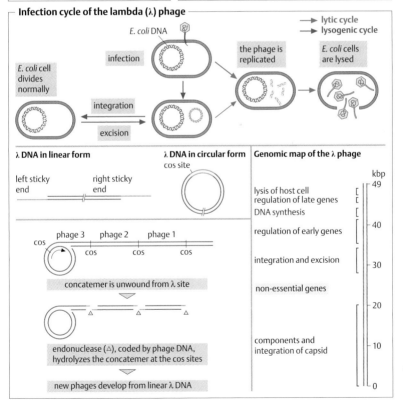

→ lytic cycle
→ lysogenic cycle

E. coli DNA

E. coli cell divides normally

infection

integration

excision

the phage is replicated

E. coli cells are lysed

λ DNA in linear form

left sticky end   right sticky end

λ DNA in circular form
cos site

Genomic map of the λ phage

kbp

lysis of host cell
regulation of late genes
DNA synthesis

regulation of early genes — 40

integration and excision — 30

non-essential genes

components and integration of capsid — 10

— 49

— 20

— 0

phage 3   phage 2   phage 1
cos   cos   cos   cos

concatemer is unwound from λ site

endonuclease (△), coded by phage DNA, hydrolyzes the concatemer at the cos sites

new phages develop from linear λ DNA

Microorganisms

General. Microorganisms play a key role in the chemical cycles on earth. They are involved in the biodegradation of many compounds; these processes occur not only in the environment, but also in symbiosis with other organisms (e.g., lichens, intestinal and rumen bacteria). Some microorganisms are parasites or pathogens, impairing the health or life of other organisms. In biotechnology, nonpathogenic microorganisms are used to produce various products such as citric and glutamic acid, antibiotics, xanthan, and enzymes; for the aerobic and anaerobic treatment of wastewater, sludges, soils, and air; and as host organisms for the manufacture of recombinant proteins. Due to their unicellular structure, well established methods for creating and selecting mutants, and their short generation time, they serve as model organisms for understanding the biochemical, genetic, and physiological mechanisms of life. Based on some fundamental differences, prokaryotic and eukaryotic microorganisms can be distinguished; the former are further subdivided into eubacteria and archaebacteria (ca. 6000 different fully characterized strains).

Eubacteria are unicellular organisms that propagate by cell division. Their cell diameter is usually on the order of 1 μm. They have no cell nucleus, and their chromosomal DNA is formed into a tangle, the nucleoid. Frequently, part of their genetic makeup occurs on nonchromosomal genetic elements, the plasmids. Plasmids are often horizontally transferred to other bacteria – a useful mechanism, from the human perspective, for evolving biodegradation pathways for xenobiotic compounds in the environment and sewage plants, but a very dangerous capacity with respect to the evolution of antibiotic resistances. The cell wall, made of peptidoglycan, is more complex in Gram-negative microorganisms and often covered with a slimy layer from which flagella may protrude, which ensure mobility. In the cytoplasm, storage chemicals such as polyhydroxybutyric acid polyphosphate, cyanophycin, or others may be deposited. Eubacteria have a wide potential for variations in the basic forms of metabolism and thus can grow in a much wider range of habitats than higher organisms. Such highly specialized species often surprise us by their unique proteins and cofactors. Thus, the purple membrane of the halobacteria is a unique functional unit of this genus, exhibiting some analogies to photosynthesis and the chemistry of vision in higher organisms.

Archaebacteria (archaea) are believed to resemble the oldest forms of life on earth. Their footprints have been detected in geological formations many hundreds of millions of years old. They often live anaerobically and are usually specialized for growth in unique biotopes. As just one example, the methanobacteria form the most important group of sludge consortia, reducing acetic acid to methane. They differ from the eubacteria in structural and genetic properties, e.g., in the construction of their cell membrane from ether lipids instead of phospholipids. The function of their enzymes is adapted to their often extreme habitats and have been used in biotechnology. For example, a DNA polymerase from a deep-sea bacterium, *Pyrococcus furiosus*, is often used for PCR reactions with particular high fidelity.

Yeasts and fungi are eukaryotic organisms and so far constitute the largest group of cultivatable microorganisms: about 70 000 different strains have been taxonomically classified. In contrast to the prokaryotes, they contain a cell nucleus and other subcellular functional units, and their cell wall is made of chitin, sometimes also from cellulose. Most yeasts and fungi live aerobically. Their wide differences in reproduction and life cycles provide the most useful basis for their taxonomic classification. The vegetative body of fungi is composed of a hairy network, the mycelium, which can propagate sexually or asexually. Asexual reproduction usually proceeds by spore formation, occasionally by budding. Sexual reproduction of the lower fungi (*Phycomycetes*) proceeds via gametes, of the higher fungi via fruiting bodies (asci) which have the form of a sac (*Ascomycetes*) or a club (*Basidiomycetes*).

Microorganisms

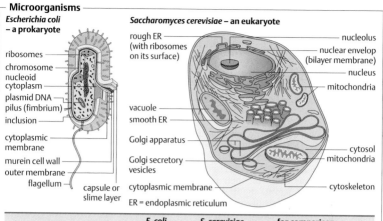

Escherichia coli – a prokaryote

- ribosomes
- chromosome
- nucleoid
- cytoplasm
- plasmid DNA
- pilus (fimbrium)
- inclusion
- cytoplasmic membrane
- murein cell wall
- outer membrane
- flagellum
- capsule or slime layer

Saccharomyces cerevisiae – an eukaryote

- rough ER (with ribosomes on its surface)
- nucleolus
- nuclear envelop (bilayer membrane)
- nucleus
- mitochondria
- vacuole
- smooth ER
- Golgi apparatus
- cytosol
- mitochondria
- Golgi secretory vesicles
- cytoplasmic membrane
- cytoskeleton

ER = endoplasmic reticulum

	E. coli	S. cerevisiae	for comparison: plant and animal cells
cell nucleus, organelles	no	yes	yes
diameter [μm]	~ 1	~ 10	~ 100
volume [μm³]	~ 1	~ 1000	>10000
respiration [μL O_2/mg TS · h]	1000	100	10
generation time [h]	0.3	1.5	>20
genes	~4300	~6000	>30000

Position of the microorganisms in evolution

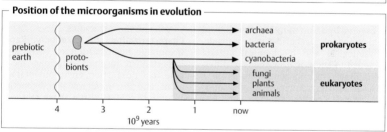

prebiotic earth — proto-bionts →

- archaea
- bacteria — **prokaryotes**
- cyanobacteria
- fungi
- plants — **eukaryotes**
- animals

4   3   2   1   now

10^9 years

Archaea, eubacteria, and lower eukaryotes

	archaea	eubacteria	fungi, yeasts
cell type	prokaryote	prokaryote	eukaryote
cell wall	heteropolysaccharide or glycoprotein	peptidoglycan	glucan, chitin
membrane lipids	ether lipids from isoprenoid building blocks	phospholipids	phospholipids
initiator tRNA	methionine	formyl methionine	methionine
genetic material	small circular chromosome, plasmids, histone-type proteins	small circular chromosome, plasmids	complex nucleus with >1 chromosome and linear DNA, histones
RNA polymerase	complex	simple	complex
size of ribosomes	70S	70S	80S

181

Bacteria

General. Bacteria can be classified by a variety of morphological, biochemical, and genetic methods, as well as by their nutrient requirements. The International Code of Nomenclature of Bacteria (ICNB) presently contains ca. 6000 strains. The analysis of taxonomically relevant sequences of DNA isolated from soil seems to indicate, however, that the number of bacterial species that have not yet been cultured is much larger.

Eubacteria. The oldest method of classifying eubacteria was based on their morphology. Under a simple light microscope, rods, cocci, and spirilli can be seen, some of them forming multicellular aggregates (filaments, colonies) and exhibiting structural details such as spores or flagella. Staining provided further differentiation. Thus, staining according to H. C. Gram's method allows for a classification according to cell wall structure: Gram-positive bacteria have only one cell membrane, covered by a thick murein cell wall, whereas Gram-negative bacteria have two cell membranes, enclosing a periplasmic space. The outer membrane is covered by a thin murein cell wall from which lipopolysaccharides may protrude. Physiological and biochemical criteria have led to additional methods of differentiation. Some important features are:

Response to oxygen: microorganisms can be subdivided according to their ability to grow under aerobic, anaerobic, or both conditions,

Form of energy generation: energy can be generated by photosynthesis (phototrophs), respiration, or fermentation (chemotrophs),

Preferred electron donors: organotrophic microorganisms use organic compounds, and lithotrophic microorganisms use inorganic compounds such a H_2, NH_3, H_2, S, CO, or Fe^{2+}.

Carbon source: autotrophic microorganisms can fix CO_2; heterotrophic microorganisms obtain carbon from organic compounds,

Relation to other organisms: saprophytic microorganisms are autonomous; parasitic microorganisms depend on a host organism.

Phage typing: the susceptibility to phages can also be used for taxonomic identification,

Adaptation to environment: mesophilic microorganisms grow under ordinary conditions, whereas extremophiles are adapted to extreme conditions of temperature, pressure, pH, or electrolyte concentration.

Cell inclusions, pigments, chemical components of the cell wall and cell membrane (fatty acid composition), immunological differentiation of the cell surface (serology), and susceptibility to antibiotics provide further possibilities for phenotype differentiation. Recently, genotyping of bacteria has become more and more important. For example, the GC content of bacterial DNA enables a rough classification. Complete sequencing of microbial genomes enables the most precise differentiation. A particularly useful method for taxonomy, discovered in 1972, is sequencing the DNA coding for the RNA of the 16S and 23S ribosomes. This DNA contains sequences that were highly conserved throughout evolution, and analyses of the sequences suggest three families of living organisms: archaebacteria, eubacteria (the prokaryotes), and the eukaryotes.

Characterization and taxonomy. Rapid taxonomic identification of bacteria is important in hospitals, veterinary medicine, food production, environmental hygiene, and also in microbial and genetics laboratories. Most of the above methods are used, e. g., microscopy, staining procedures, determining the "analytical profile index API" (based on growth on various substrates), fatty acid composition of the membrane, or DNA analysis of taxonspecific sequences coding for the 16S or 23S rRNA. If DNA is isolated from environmental samples, and sequences coding for 16S or 23S rRNA are compared to those of microorganisms deposited in culture collections, there is less than 5% identity, suggesting that > 95% of all microorganisms contained in these samples have not yet been cultivated ("metagenome analysis"). Precise classification of microorganisms is often far from trivial and requires the consideration of a wide range of experimental data; it is usually done by laboratories that archive culture collections.

Forms of unicellular bacteria

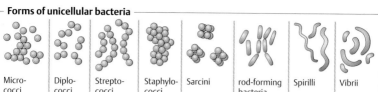

| Micro-cocci | Diplo-cocci | Strepto-cocci | Staphylo-cocci | Sarcini | rod-forming bacteria | Spirilli | Vibrii |

Cell wall composition and Gram-staining

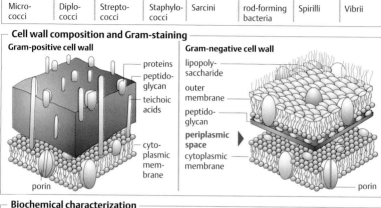

Gram-positive cell wall

- proteins
- peptido-glycan
- teichoic acids
- cyto-plasmic mem-brane
- porin

Gram-negative cell wall

- lipopoly-saccharide
- outer membrane
- peptido-glycan
- **periplasmic space**
- cytoplasmic membrane
- porin

Biochemical characterization

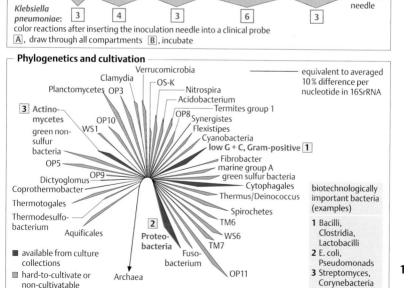

A →

| glucose gas | lysine | ornithine | indole | adonit | lactose | arabinose | sorbit | dulcit phenylalanine | urea | citric acid |

→ B

sterile in-oculation needle

biocode of **Klebsiella pneumoniae:**

| 2+1 | 4 + 2 + 1 | 4 + 2 + 1 | 4 + 2 + 1 | 4 + 2 + 1 |
| 3 | 4 | 3 | 6 | 3 |

color reactions after inserting the inoculation needle into a clinical probe A, draw through all compartments B, incubate

Phylogenetics and cultivation

equivalent to averaged 10 % difference per nucleotide in 16SrRNA

Verrucomicrobia
Clamydia
OS-K
Planctomycetes OP3
Nitrospira
Acidobacterium
Termites group 1
3 Actino-mycetes
OP8
OP10
Synergistes
green non-sulfur bacteria
WS1
Flexistipes
Cyanobacteria
low G + C, Gram-positive 1
OP5
Fibrobacter
marine group A
Dictyoglomus
OP9
green sulfur bacteria
Coprothermobacter
Cytophagales
Thermotogales
Thermus/Deinococcus
Thermodesulfo-bacterium
Spirochetes
Aquificales
TM6
2
WS6
Proteo-bacteria
TM7
Fuso-bacterium
Archaea
OP11

- available from culture collections
- hard-to-cultivate or non-cultivatable

biotechnologically important bacteria (examples)

1 Bacilli, Clostridia, Lactobacilli
2 E. coli, Pseudomonads
3 Streptomyces, Corynebacteria

183

Some bacteria of importance for biotechnology

General. Some bacteria are especially important in biotechnology. Examples are *Escherichia coli*, *Pseudomonas putida*, *Bacillus subtilis*, *Streptomyces coelicolor*, and *Corynebacterium glutamicum*.

Escherichia coli is a saprophyte in the large intestine of mammals and belongs to the Enterobacteriaceae group. It forms rods that carry flagella. The cell wall stains Gram-negative: it encloses two membranes that include a periplasmic space. Under anaerobic growth conditions, *E. coli* generates energy by fermentation and forms acids. In the presence of O_2, energy is supplied through the respiratory chain. Under optimal conditions, its doubling time is ca. 20 min. The *E. coli* genome is ca. 4.6 Mbp in size, the G+C content is 51%. Although *E. coli* is among the best understood microorganisms and its genome has been sequenced, the function of about ⅓ of its gene products is not completely understood. In biotechnology, *E. coli* is used as a host organism for the expression of nonglycosylated proteins, e.g., insulin, growth hormone, and antibody fragments. Since *E. coli* grows in the human large intestine, it is classified in safety group S2; as a consequence, attenuated *E. coli* strains are used, in which all risk factors were eliminated and which can be handled under normal microbiological safety conditions as group S1 organisms (e.g., *E. coli* K12). They are also used for cloning experiments. Various plasmid vectors have been developed for cloning foreign genes in *E. coli*, for example, the BAC cloning vector is used to construct genomic libraries.

Pseudomonas putida is an aerobic rod with polar flagella which lives aerobically in water. The cell wall contains two membranes that enclose a periplasmic space and stains Gram-negative. The *P. putida* genome contains ca. 6.1 Mbp, its G+C content is 61%. Pseudomonads have a wide genetic potential for the degradation of aromatic compounds, which can be horizontally transferred through plasmids. In biotechnology, they are used predominantly in environmental studies.

Bacillus subtilis is a rod without flagella that lives aerobically in soil. Under unfavorable conditions, it forms dormant, thermoresistant spores. Its cell wall stains Gram-positive and encloses only one membrane. Energy is generated via the electron transport chain. Doubling time, under optimal growth conditions, is ca. 20 min. The genome of *B. subtilis* contains ca. 4.2 Mbp and has been completely sequenced; its G+C-content is 44%. In biotechnology, *B. subtilis* is the preferred microorganism for producing extracellular enzymes, e.g., proteases and amylases.

Streptomyces coelicolor is another soil bacterium from the genus Actinomycetes. It propagates in the form of a mycelium and forms aerial hyphae, from which spore-forming conidia are constructed. The cell wall stains Gram-positive and encloses just one membrane. Like most other Streptomyces strains, *S. coelicolor* degrades cellulose and chitin. Its large linear genome has been completely sequenced and contains ca. 8.7 Mbp, nearly twice the number in *E. coli*; its G+C content is 72%. The ca. 8000 structural genes code mainly for enzymes that are required for the formation of secondary metabolites, e.g., for antibiotics.

Corynebacterium glutamicum is a member of the coryneform bacteria which grow in many habitats and include some pathogenic species such as *C. diphteriae*. The club-shaped cells grow aerobically and stain Gram-positive. The *C. glutamicum* genome contains ca. 3.1 Mbp and has been completely sequenced; its GC content is 56%. Deregulated mutants of *C. glutamicum* are important production strains for L-glutamic acid and L-lysine.

Genome sequencing. Including the above organisms, the genomes of more than 50 bacteria have been sequenced, including many pathogens such as *Helicobacter pylori*, *Mycobacterium tuberculosis*, *Neisseria meningitidis*, *Streptococcus pneumoniae*, *Vibrio cholerae*, *Treponema pallidum*, *Yersinia pestis*, and several archae. By the end of 2002, 73 microbial genomes were completed, and sequencing is in progress for a further 140, providing a huge database for comparative microbial genomics and functional analysis (www.tigr.org).

Some important bacteria in biotechnology

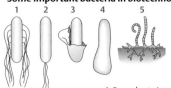

	1	2	3	4	5
flagellation	+	+	–	–	–
Gram-staining	–	–	+	+	+
spore formation	–	–	+	–	+
aerobic growth	+	+	+	+	+
G + C content	51	61	44	56	72
genome size (Mbp)	4,6*	4,2	4,2*	3,1*	8,7*

1 *Escherichia coli*
2 *Pseudomonas putida*
3 *Bacillus subtilis* (germinating from spore)
4 *Corynebacterium glutamicum*
5 *Streptomyces coelicolor* (with conidia)

*genome sequences have been completed

Proteins identified in *E. coli* K12

total	4288	not classified or unknown*	1632
transporter, binding proteins	288	putative regulation proteins	133
putative transporters	146	for DNA replication and modification	115
putative enzymes	251	for cofactors and prosthetic groups	103
for energy metabolism	243	phages, transposons, plasmids	87
for adaptation, defense	188	for nucleotide biosynthesis and catabolism	58
for central intermediary metabolism	188	for transcription and RNA synthesis	55
for structural building blocks	182	for fatty acids and phospholipids	48
for putative structural building blocks	42	with regulatory function	45
for translation or posttranslational modification	182	diverse known gene products	26
for amino acid biosynthesis and catabolism	131	putative membrane proteins	13
for carbon catabolism	130	putative chaperones	9

*immediately after complete sequencing, 1997. Since then, more gene functions have been elucidated

E. coli K12 host strain

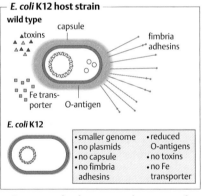

wild type

toxins
capsule
fimbria
adhesins
Fe transporter
O-antigen

E. coli K12
- smaller genome
- no plasmids
- no capsule
- no fimbria adhesins
- reduced O-antigens
- no toxins
- no Fe transporter

BAC cloning vector

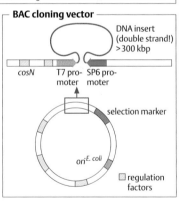

DNA insert (double strand!) >300 kbp

cosN
T7 promoter
SP6 promoter

selection marker

ori E. coli

□ regulation factors

Some completely sequenced genomes of prokaryotes

	disease	genome size (Mbp)
Haemophilus influenzae	childhood meningitis	1,8
Helicobacter pylori	ulcer	1,7
Mycoplasma pneumoniae	bacterial pneumonia	0,8
Mycobacterium tuberculosis	lung tuberculosis	4,4
Treponema pallidum	syphilis	1,1
Mycobacterium leprae	leprosy	3,3

Fungi

General. Fungi play a key role in the carbon catabolism of the biosphere, e. g., in the decomposition of wood and the formation of humic acids. Mycorrhizal fungi are associated with plant roots and assist in the uptake of nutrients, but other fungi, such as mildews, are dangerous plant pathogens. In biotechnology, they have an important role in the decay of food, but also in the preparation of fermented food products. Some fungi produce antibiotics or valuable enzymes. Among ca. 70 000 fungal species that have been classified, the Ascomycetes comprise ca. 20 000 species, forming the largest subgroup, which includes *Penicillium notatum* and *Aspergillus niger*. Among the lower fungi (Zygomycetes), Rhizopus- and Mucor species have the greatest importance in biotechnology. Some of the ca. 12 000 stand mushrooms (Basidiomycetes) are edible (e. g., champignons, shiitake, chanterelles, ceps), and others participate in the degradation of wood (white and red rot fungi). Approximately 300 fungal species are pathogenic to humans. All fungi live heterotrophically. Their cell wall is composed of chitin and glucans.

Reproduction forms. The reproduction of fungi follows highly diverse patterns, which are described here using the Ascomycetes as an example. The cell mass (thallus) consists of a mycelium that is made up of hyphae. During asexual reproduction, the conidiophores, which form at the top of the mycelium, divide and form spores (conidia), which grow into a new mycelium. Like most fungi, Ascomycetes can also propagate by a sexual mechanism. This results in a different phenotype (dimorphism). In this case, their hyphae form male and female sexual organs (antheridia and ascogonia). They fuse, during plasmogamy, into dikaryotic hyphae, which develop into an ascocarp ("fruiting body"). In the terminal cells of the dikaryotic hyphae, the dikaryotic nuclei are fused into a diploid zygote (karyogamy). Meiosis transforms the zygote into 8 haploid ascospores (or 4 basidiospores, in Basidiomycetes), which again grow into a mycelium.

Penicillium notatum grows as a mycelium which forms fruiting bodies liberating spores (conidia) for asexual reproduction. Fungi like Penicillium, which have lost the capacity for sexual reproduction, are termed *Fungi imperfecti*. Consequently, if recombination is required during breeding in the laboratory, protoplast fusion among different types of nuclei (heterokaryosis) must be used. *P. notatum* and the related fungus *Acremonium chrysogenium* are important industrial organisms, since they synthesize the lactam antibiotics. Other Penicillium species such as *Penicillium camembertii* play an important role in the maturation of cheese. The genome of *P. notatum* contains ca.32 Mbp and is sequenced only to a minor extent.

Aspergillus nidulans differs from Penicillium in the form of its conidia. Its genome contains 12.6 Mbp and, although it has been sequenced, data are not yet publicly available. *A. oryzae* is used for industrial production of extracellular enzymes and is a favorite host organism for producing recombinant enzymes from other eukaryotic organisms. Various Aspergillus strains play a traditional role in Asian countries for the manufacture of food products such as soy sauce, miso, and sake and are also used for the production of extracellular enzymes such as proteases or amylases. *A. niger* is the preferred production organism for citric and gluconic acid. Similar to Penicillium, strain improvement is often based on protoplast fusion and selection.

Rhizopus oryzae, a zygomycete, grows on rice, and *R. nigricans* is the black mold on bread. Its hyphae grow rapidly and bore their way through their substrates. Asexual reproduction proceeds by the formation of spores in differentiated mycelium (sporangia). Rhizopus and the closely related Mucor species can also grow on decaying organic materials and synthesize numerous extracellular hydrolases for this purpose. As a result, they have become important organisms for the manufacture of extracellular enzymes. The *R. oryzae* genome sequence has not yet been fully studied.

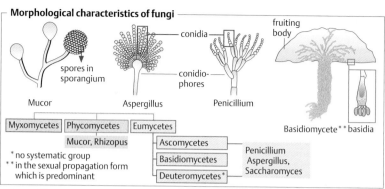

Morphological characteristics of fungi

conidia

fruiting body

spores in sporangium

conidio-phores

Mucor

Aspergillus

Penicillium

Basidiomycete** basidia

Myxomycetes	Phycomycetes	Eumycetes

Mucor, Rhizopus

Ascomycetes

Basidiomycetes

Deuteromycetes*

Penicillium Aspergillus, Saccharomyces

* no systematic group
** in the sexual propagation form which is predominant

Aspergillus niger, an ascomycete

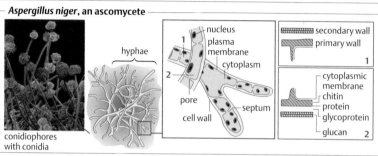

hyphae

nucleus
plasma membrane
cytoplasm

pore

septum

cell wall

conidiophores with conidia

secondary wall
primary wall
1

cytoplasmic membrane
chitin
protein
glycoprotein
glucan
2

Propagation cycle of an ascomycete

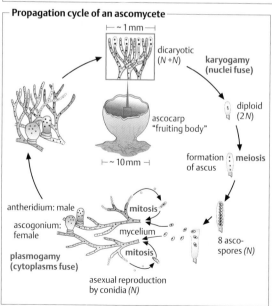

⊢ ~ 1 mm ⊣

dicaryotic (N +N)

karyogamy (nuclei fuse)

ascocarp "fruiting body"

⊢ ~ 10 mm ⊣

diploid (2N)

formation of ascus **meiosis**

antheridium: male
ascogonium: female

plasmogamy (cytoplasms fuse)

mycelium

mitosis

mitosis

asexual reproduction by conidia (N)

8 asco-spores (N)

Parasexual breeding
e.g., of Aspergillus

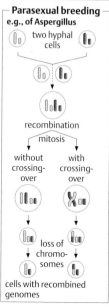

two hyphal cells

recombination

mitosis

without crossing-over

with crossing-over

loss of chromo-somes

cells with recombined genomes

187

Yeasts

General. Yeasts are a subgroup of the Ascomycetes. Because they propagate by budding, they are also termed budding fungi. They grow heterotrophically, preferring acidic media (pH 3.5–5.0) and usually do not form mycelia. Their cell wall is made of chitin. Yeasts of importance for biotechnology are *Saccharomyces cerevisiae*, *Candida utilis*, and *Candida albicans*, *Schizosaccharomyces pombe*, *Hansenula polymorpha*, and *Pichia pastoris*.

Saccharomyces cerevisiae (synonyms: bakers' yeast, brewers' yeast, yeast) can propagate in either a haploid or diploid manner, thus providing an excellent organism for genetic investigations. Haploid laboratory strains belong to one out of two mating types (*MATa* or *MATα*), which can only mate reciprocally. Asexual reproduction proceeds by forming conidia, followed by immigration of either a diploid or a haploid nucleus. Sexual propagation occurs by the fusion of two haploid gametes, followed by meiosis and formation of 4 haploid ascospores, whose phenotype can be separately observed, allowing for simple genetic analysis of the observed traits (tetrad analysis). Due to the simple cultivation of both haploid and diploid cells, the completed genome sequence (12 Mbp, on 16 chromosomes), the general absence of introns, and the short doubling time (90 min), *S. cerevisiae* has become a preferred model organism for the molecular genetics of a simple eukaryote. Another advantage is that yeast occurs with a natural plasmid, termed 2μm (60–100 copies in the cell nucleus), and that a second extrachromosomal element, the killer virion, is also available for recombination experiments. Many cloning vectors have been developed for yeast transformation, which either allow the replication of foreign genes outside the yeast chromosome (YRP = yeast replicating plasmids or YEP = yeast episomal plasmids) or integration of the foreign gene into the chromosome (YIP = yeast integrating plasmids). Artificial yeast chromosomes (YAC = yeast artificial chromosomes) allow for the cloning of large DNA fragments of 600–1400 kbp; they have been widely used for preparing genome libraries, but have a tendency to recombine and thus have been mostly replaced by BACs. The ca. 6000 genes of yeast, located on 16 linear chromosomes, often show a surprisingly high homology to human genes. Thus yeasts have widely served as a simple model system for metabolic and regulation studies. In biotechnology, yeasts are used in the preparation of many food products such as beer, wine, and bread. It is also used in the manufacture of industrial ethanol. Recombinant yeasts have become important host organisms for the manufacture of products such as insulin, interferons, and vaccines (e.g., hepatitis B surface antigen). Unlike *E. coli*, yeast as a host organism allows for the posttranslational modification of gene products, in particular for glycosylation.

Candida utilis differs from Saccharomyces by forming a mycelium, but it propagates solely asexually by budding. Some Candida genes show noncanonical codon usage (e.g., CUG for serine instead of leucine), which has retarded their heterologous expression. Candida strains have been used in biotechnology for production of extracellular enzymes and generation of digestible biomass. They can be grown on unconventional substrates such as sulfite suds or alkane fractions. Some Candida strains, such as *Candida albicans*, are pathogenic to humans.

Pichia pastoris and **Hansenula polymorpha** are both methylotrophic yeasts, which can grow on methanol as their sole carbon source. Isolated and studied in the context of the manufacture of single-cell protein, they are used today as attractive host organisms in cloning experiments. Thus, diverse proteins such as lipases, β interferon, and antibody fragments have been functionally expressed in *P. pastories* in yields of several grams of recombinant products/L of culture broth.

Schizosaccharomyces pombe was first isolated from an East African beer variety (Swahili: pombe, beer). The genome of this ascomycete has been fully sequenced. It contains ca. 12.6 Mbp, which is similar to the *S. cerevisiae* genome. Its ca. 5400 genes are located on just 3 chromosomes.

Morphology

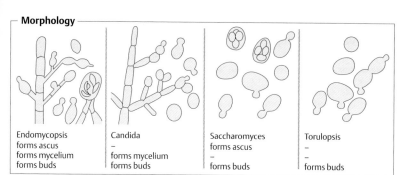

Endomycopsis
forms ascus
forms mycelium
forms buds

Candida
–
forms mycelium
forms buds

Saccharomyces
forms ascus
–
forms buds

Torulopsis
–
–
forms buds

Molecular genetics

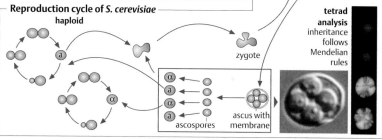

	size of haploid genome [Mbp]	number of chromosomes
Saccharomyces cerevisiae	15	16
Candida utilis	14 – 18	8
Schizosaccharomyces pombe	12.6	3
Pichia pastoris	n. d.	6 – 8
Hansenula polymorpha	n. d.	4 – 6

diploid

Reproduction cycle of *S. cerevisiae*

haploid

a

α

zygote

α
a
α
a

ascospores

ascus with membrane

tetrad analysis
inheritance follows Mendelian rules

Yeast vectors

YEP vector
(yeast episomal plasmid)

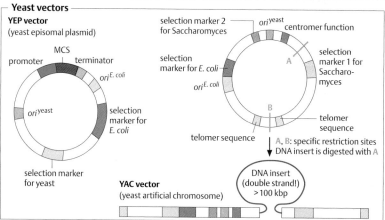

MCS

promoter   terminator

ori^E. coli^

ori^yeast^

selection marker for E. coli

selection marker for yeast

selection marker 2 for Saccharomyces

ori^yeast^

centromer function

selection marker for E. coli

selection marker 1 for Saccharomyces

ori^E. coli^

A

B

telomer sequence

telomer sequence

A, B: specific restriction sites
DNA insert is digested with A

DNA insert (double strand!) >100 kbp

YAC vector
(yeast artificial chromosome)

Microorganisms: isolation, preservation, safety

General. For most experiments with microorganisms, pure cultures are used. In biotechnology, most strains have additionally been optimized for a specific application, using rounds of mutation and selection. Microorganisms are maintained and conserved in culture collections. They are propagated on solid or liquid nutrient media under sterile conditions. Most microorganisms used in biotechnology grow aerobically on organic substrates (heterotrophic growth). Photosynthetic microorganisms are cultured under light, anaerobic bacteria under the exclusion of oxygen.

Pure cultures are obtained from culture collections or from their natural habitats (soil, water, food, other organisms) using enrichment cultures. The preferred method for obtaining a pure culture is the streak plate method, in which a mixed culture is spread over the surface of a sterile nutrient agar (a crosslinked polysaccharide isolated from marine algae) with a sterile wire loop (plating). Usually, growth conditions are chosen that favor the microorganism one wants to isolate (selection): for example, excluding oxygen and working under light with CO_2 as the sole source of carbon and N_2 as the sole nitrogen source leads to enrichment in cyanobacteria. A sugar medium at slightly acidic pH enriches fungi, incubating at elevated temperatures favors thermotolerant microorganisms, and when casein is the sole nitrogen source, protease-secreting microorganisms have a selective advantage. Based on 16S-rRNA analysis, however, it is believed that $< 5\%$ of all naturally occurring microorganisms can be isolated by these methods.

Culture collections are used to conserve pure cultures. The identity, viability, and metabolic functions of conserved cultures must be tested upon reactivation. The conventional method for conservation consists in transferring a pure culture at regular time intervals to a new agar plate or slant. This method may lead, however, to degeneration. Important type or production strains are therefore preserved under either of the following conditions: 1) under metabolically inert liquids such as mineral oil (suitable for hyphae-forming fungi); 2) freezing at $-196\,°C$ in liquid N_2 or at $-70\,°C$ in a deep-freezer; freezing and thawing must be done rapidly and in the presence of glycerol to prevent cell destruction by ice crystals (this method is mainly used for bacteria and yeasts); 3) vacuum drying of cell suspensions on a carrier (sand, silica gel) and in the presence of a mild emulsifier (skim milk, serum) and preservation at $-70\,°C$. In all cases, it must be verified that the conserved strains can be reactivated. Most nations operate large public culture collections from which pure cultures can be ordered. They are either universal for all types of microorganisms (e. g., the American Type Culture Collection, ATCC, or CABRI, Common Access to Biological Resources and Information, a European consortium which includes general resource collections, e. g., the German *Deutsche Sammlung für Mikroorganismen und Zellkulturen*, DSMZ, and specialized collections for particular groups of microorganisms, such as the Dutch *Centralbureau voor Schimmelkultuuren* CBS). All industrial companies that produce biotechnological products, and many hospitals, have their own culture collections.

Safety. Each study using microorganisms must comply with biological safety rules, because dangerous pathogens may occur in all microbial isolates (examples: *Bacillus subtilis:* harmless producer of technical enzymes, *Bacillus anthracis*: anthrax pathogen; *Aspergillus oryzae*: used for soy sauce production, *Aspergillus flavus*: forms highly hepatotoxic and carcinogenic aflatoxins). For safety considerations, microorganisms are classified into four risk groups. Both the construction and the equipment of a laboratory and the operating rules must be adapted to the relevant risk group. Risk group 1 (generally safe) includes microorganisms that have been used in food production for centuries, e. g., *Saccharomyces cerevisiae* and *Aspergillus oryzae*. Most microorganisms used in biotechnology fall into risk group 1.

Pure cultures

1 streak plate method using nutrient agar

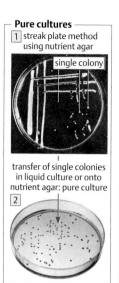

single colony

transfer of single colonies in liquid culture or onto nutrient agar: pure culture

2

Enrichment cultures (examples)

bacteria		energy source, nutrients
phototropic		
Rhodospirilla	●	light, H_2 or organic acids, CO_2
Cyanobacteria	■	light, CO_2, N_2 as nitrogen source
chemolithotrophic		
Nitrosomonas	●	NH_4^+ as H donor, O_2 as H acceptor
Thiobacillus	●	H_2S, S or $S_2O_3^{2+}$ as H donor
methane formers	■	H_2 as H donor, CO_2 as H acceptor
heterotrophic		
Pseudomonads	■	2% KNO_3 as H acceptor, organic acids
Clostridia	■	starch, NH_4^+, pasteurized inoculate
Enterobacteria	■	glucose, NH_4^+
lactic acid bacteria	■	glucose, yeast extract, pH 5
Bacilli	●	starch, NH_4^+
Streptomycetes	●	mannitol, NH_4^+
enzyme secretors		
protease-forming strains	●	glucose, NH_4^+, casein
lipase-forming strains	●	glucose, NH_4^+, tributyrin

● aerobic or ■ anaerobic growth conditions

Diversity of microorganisms

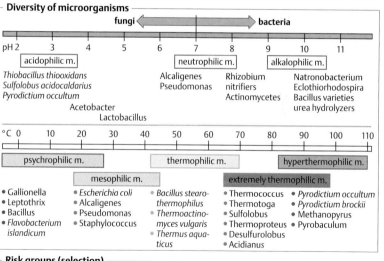

fungi ⟷ bacteria

pH 2　3　4　5　6　7　8　9　10　11

| acidophilic m. | | neutrophilic m. | | alkalophilic m. |

Thiobacillus thiooxidans
Sulfolobus acidocaldarius
Pyrodictium occultum

Alcaligenes
Pseudomonas

Rhizobium
nitrifiers
Actinomycetes

Natronobacterium
Eclothiorhodospira
Bacillus varieties
urea hydrolyzers

Acetobacter
Lactobacillus

°C 0　10　20　30　40　50　60　70　80　90　100　110

| psychrophilic m. | | thermophilic m. | | hyperthermophilic m. |

mesophilic m.　　　extremely thermophilic m.

- ● Gallionella
- ● Leptothrix
- ● Bacillus
- ● *Flavobacterium islandicum*

- ● *Escherichia coli*
- ● Alcaligenes
- ● Pseudomonas
- ● Staphylococcus

- ● *Bacillus stearo-thermophilus*
- ● *Thermoactino-myces vulgaris*
- ● *Thermus aquaticus*

- ● Thermococcus
- ● Thermotoga
- ● Sulfolobus
- ● Thermoproteus
- ● Desulfurolobus
- ● Acidianus

- ● *Pyrodictium occultum*
- ● *Pyrodictium brockii*
- ● Methanopyrus
- ● Pyrobaculum

Risk groups (selection)

risk group 1	risk group 2	risk group 3
Acetobacter acetii, Agrobacterium tumefaciens, Bacillus subtilis, Lactobacillus casei	*Acinetobacter calcoaceticus, Escherichia coli, Pseudomonas aeruginosa*	*Bacillus anthracis, Mycobacterium tuberculosis, Yersinia pestis*
Penicillium notatum, Rhizopus oryzae, Aspergillus niger, Candida tropicalis	*Aspergillus flavus, Candida albicans, Trychophyton rubrum, Histoplasma capsulatum*	*Histoplasma capsulatum*

bacteria　　　fungi, yeasts

Microorganisms: strain improvement

General. Microorganisms isolated from environmental samples rarely exhibit all the properties that are required in a technical application. Thus, they are usually optimized by a series of mutation and selection steps. The targets of strain improvement are usually: 1) to increase the yield of the desired product; 2) to remove undesired by-products; and 3) to improve general properties of the microorganism during fermentation (e. g., reduced fermentation time, no interfering pigments formed, resistance to bacteriophages). A great advantage in dealing with microorganisms is their short doubling time (often < 1 h): it allows a very large number of mutants to be produced and screened in a short time. In eukaryotic organisms, e. g., fungi, recombination events must also be taken into account. With increasing knowledge of microbial metabolism, its regulation and its coding by the genome, genetic methods that delete or amplify defined metabolic steps in a targeted way are on the increase (metabolic engineering).

Mutation. The spontaneous mutation frequency (changes in DNA sequence due to natural mutation events and errors during replication) is on the order of 10^{-7} for a gene (1000 bp) of normal stability. Most mutations remain silent or they revert genetically or functionally or by DNA repair mechanisms to the original state. Thus, for industrial strain improvement harsher mutation conditions are required: the use of UV radiation or of mutagenic chemicals are methods of choice, and, depending on the experimental goals, conditions are chosen to achieve a mortality rate of 90 % to > 99 %. Survivors exhibiting the desired properties are then selected according to their phenotypes.

Selection using surface cultures. Phenotype selection is often synonymous to the selective isolation of mutants with high productivity. A key requirement for such experiments is the availability of an indicator reaction. For example, the resistance of a mutant to antibiotics, inhibitors, or phages can be identified if the mutant can grow on a nutrient agar that contains one of these agents. Replica plating first on a nutrient-rich agar, followed by plating on a selection medium, may yield very useful information. An enrichment step in a penicillin-containing agar (penicillin inhibits only growing cells) can help to identify auxotrophic mutants, which depend on the presence of a given metabolite for growth. If mutants that form a biologically active metabolite (e. g., an antibiotic or an enzyme) in higher yields are to be isolated, the size of inhibition or lysis plaques can be used as an indicator. The great advantages of such selection procedures are 1) high flexibility in the choice of the selection criterion and 2) high number of mutants that can be visually screened (several hundred on a single agar plate). Due to the random method of mutagenesis, however, the strains obtained by this kind of selection are usually defective in several genes and must be tested for their robustness as production strains in separate experiments. To this end, they are subjected to further selection with respect to growth, productivity, and other features using shake flasks and then small bioreactors under conditions resembling the production process. The best candidates may then be backcrossed with wild-type or less mutated strains to reduce the negative effects arising from many passages of random mutation.

Selection in submersed culture. Continuous fermentation has also been used for selecting microorganisms. A pure culture of a microorganism is grown in a chemostat in the presence of a mutagenic agent and subjected to selective pressure, e. g., by gradually replacing a good carbon source with a poor one. During continuous growth, those mutants that are better adapted to the altered growth conditions prevail. This method cannot be used, however, for selecting mutants that form a desired metabolite in higher concentrations.

Strain improvement of microorganisms

mutagens	mechanism	applications
physical agents		
ionizing radiation (x-ray)	leads to single- and double-strand DNA breakage	major genetic alterations
UV light (254 nm)	thymidine and cytidine form dimers	point mutations
chemical agents		
nitrite	deaminates adenine to hypoxanthine, cytidine to uridine	point mutations
alkylating agents	alkylate purines	point mutations
base analogs	are incorporated into replicated DNA	major genetic alterations
acridine orange	intercalates into DNA	major genetic alterations
biological agents		
transposons	transfer DNA elements within a chromosome	gene markers

Replica plating and penicillin technique

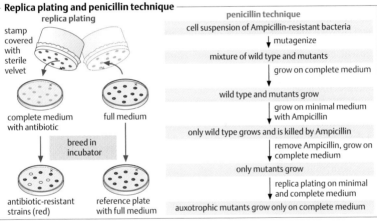

replica plating

stamp covered with sterile velvet

complete medium with antibiotic

full medium

breed in incubator

antibiotic-resistant strains (red)

reference plate with full medium

penicillin technique

cell suspension of Ampicillin-resistant bacteria

↓ mutagenize

mixture of wild type and mutants

↓ grow on complete medium

wild type and mutants grow

↓ grow on minimal medium with Ampicillin

only wild type grows and is killed by Ampicillin

↓ remove Ampicillin, grow on complete medium

only mutants grow

↓ replica plating on minimal and complete medium

auxotrophic mutants grow only on complete medium

Selection media

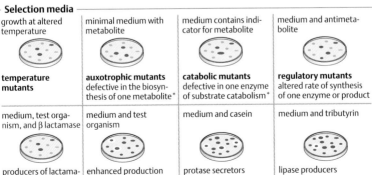

growth at altered temperature	minimal medium with metabolite	medium contains indicator for metabolite	medium and antimetabolite
temperature mutants	**auxotrophic mutants** defective in the biosynthesis of one metabolite*	**catabolic mutants** defective in one enzyme of substrate catabolism*	**regulatory mutants** altered rate of synthesis of one enzyme or product
medium, test organism, and β lactamase	medium and test organism	medium and casein	medium and tributyrin
producers of lactamase resistant antibiotic	enhanced production of antibiotics	protase secretors	lipase producers

*using penicillin technique and replica plating

Growing microorganisms

General. Microorganisms are cultivated either on solid nutrients (surface cultivation) or in liquid culture (submerged cultivation). In laboratory experiments, agar plates or shake flasks prevail. Under industrial conditions, bioreactors are the cultivation vessels of choice. The composition of the culture medium is of key importance for product formation. In most cases, contamination by undesired microorganisms is excluded by using sterile cultivation conditions.

Shake flasks. The standard vessels are Erlenmeyer flasks with baffles (filled with ca. 50–500 mL liquid), containing a sterile liquid nutrient solution. Oxygen saturation is assured by shaking the flasks on a thermostatted reciprocal or gyrating shaker. When anaerobic bacteria are to be grown, oxygen is removed from the nutrient medium by boiling and degassing, followed by addition of thiogycolate. Further handling is done under an O_2-free sterile hood.

Bioreactors (fermenters) are closed reactors with a capacity of 1 L to > 500 m³. In the standard, stirred bioreactor, mass transfer and air distribution are accomplished with a stirrer. Bioreactors can be operated as batch cultures, as batch cultures with subsequent addition of substrates (fed-batch culture), or as continuous cultures. In industrial practice, batch and fed-batch cultures are preferred; whereas in fundamental studies, continuous culture is of great importance because it allows cells to be kept at a constant specific growth rate for many days or even weeks. In many microbial fermentations, product formation begins only in the late logarithmic phase of culture growth. If at this point more nutrients are added in fed-batch mode, the production phase of the fermentation can be prolonged, and the yield of the end product is increased. Another reason to use fed-batch fermentations is to prevent substrate inhibition: often, microorganisms produce less product in the presence of high glucose concentrations (catabolite repression).

Medium optimization. Most microorganisms that are used in biotechnology are heterotrophic and grow aerobically. They require organic compounds as a source of carbon and energy, in addition to inorganic or organic nitrogen, salts, and trace elements. The nutrient medium is usually optimized in shake flasks, using product yield and substrate cost and availability as the major parameters (in some fermentations, such as for ethanol or citric acid production, the cost of the C source may exceed 50 % of the production costs). For cost reasons, most industrially used nutrient media are composed of components that are not very well defined, such as corn starch hydrolysate, molasses, or soy meal (complex media), whereas in the research laboratory, defined media components such as glucose or mixtures of amino acids are preferred.

Sterilization. Autoclaves are best for sterilizing nutrient media in the laboratory prior to inoculation. Autoclaving for 15 min at 121 °C is sufficient to kill even the spores of thermophilic microorganisms (test organism: *Bacillus stearothermophilus*). Heat-sensitive medium components such as glucose and vitamins are usually added via sterile filtration to the autoclaved medium after cooling. If a bioreactor exceeds ca. 10 L in volume, it is usually autoclaved in place with steam at 1.4–3 bar. The method is time-consuming (heating and cooling cycles of several hours) and, due to the long exposure to heat, may lead to changes in the medium composition. Thus, continuous sterilization is preferred in industry. In this procedure, the nutrient broth is exposed to steam at 140 °C for ca. 2–3 min (holding time). Using countercurrent heat exchangers, the formation of vapor condensate is prevented, and ca. 90 % of the introduced energy is recovered. Air must be purified by filters before it enters the bioreactor: 1 m³ of air may contain up to 2000 colony-forming units (cfu), including up to 50 % fungal spores and 40 % Gram-negative bacteria). For a bioreactor of 100 m³ working volume, operating at an aeration rate of 1 vvm (volume air/volume liquid · min), 6000 m³ of sterile air is required per hour.

A shake flask

air-per-meable cotton or wool plug

rotating culture medium

Stirred bioreactor

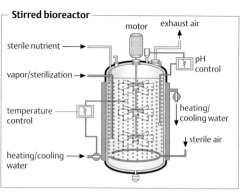

motor | exhaust air

sterile nutrient →

vapor/sterilization →

temperature control

heating/cooling water

pH control

heating/cooling water

sterile air

Nutrient components

component	origin	composition/remarks
complex carbon sources		
sugar beet molasses	sugar production	~ 48% saccharose,
sugar cane molasses		~ 33% saccharose, ~ 22% invert sugar,
corn steep liquor	corn milling	~ 1–3% glucose, 11–13% lactose
distillers´ solubles	alcohol production	variable
starch and dextrins	corn starch hydrolysis	variable
sulfite suds	paper production	2–4% hexoses and pentoses
whey	dairies	3–5% lactose
hydrocarbons	petrochemical plants	aliphatic alkanes >C-5
defined carbon sources		
glucose		can be metabolized by most organisms
mannitol		good carbon source for Streptomycetes
methanol		can be metabolized by many bacteria and yeasts
complex nitrogen sources		
soymeal, peanut meal, wheat germ meal, corn steep liquor, whey powder, yeast extract		content of raw protein between 20% and 6%, contain vitamins and trace elements
defined nitrogen sources		
ammonia salts, nitrates, urea, amino acids		
vitamins, trace elements		
thiamin, riboflavin, pyridoxin, nicotinic acid, nicotinamide, pantothenic acid, cyanocobalamin, folic acid, biotin, α-lipoic acid, purines, pyrimidines, heme		

macro minerals (10^{-3}–10^{-4} M): salts of K, Ca, Mg, Fe, S,P
micro minerals (10^{-6}–10^{-8} M): salts of Mn, Mo, Zn, Cu, Co, Ni, V, B, C, Na, Si

Prices for some carbon sources

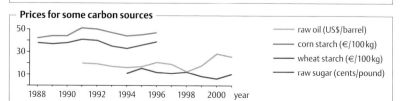

— raw oil (US$/barrel)
— corn starch (€/100 kg)
— wheat starch (€/100 kg)
— raw sugar (cents/pound)

Growth kinetics and product formation

General. The rules governing the growth of microorganisms are well defined for single-cell organisms but not for myceliar organisms (Streptomyces, *fungi*). Several varieties of fermentation can be distinguished, depending on the kinetics of product formation.

Growth kinetics of unicellular microorganisms. Most microorganisms and yeasts are unicellular. They propagate by cell division, and an increase in the number of cells can be monitored continuously by optical methods, e.g., turbidity. In a static culture, e.g., in a small shake flask or a batch reactor, a lag phase (when the formation of enzymes important for biosynthesis is induced) is followed, after a short transition phase, by a phase of logarithmic growth (log phase) having first-order kinetics. The following transition phase II is reached if one substrate becomes limiting or one product becomes inhibiting. This is followed by the stationary phase, where substrate limitation, excess population density, limited oxygen transfer, or the accumulation of toxic metabolites have terminated growth. A death phase characterized by decreasing cell number may ensue. To characterize a growth curve, important parameters are 1) the *lag phase* (dimension: [h]), which depends on the microorganism, the physiological conditions of the inoculation material, and the nutrient composition; 2) *the specific growth rate* μ (dimension: h^{-1}), which allows the rate of cell formation to be correlated with cell concentration during the exponential phase. When written as an equation $\mu = \mu_{max} \cdot S/(K_S + S)$ (the Monod equation), μ enables the experimental determination of the velocity of cell growth; 3) the *saturation constant K_S* of this equation relates to the substrate concentration (in $mg\,L^{-1}$), at which 50% of the maximum growth rate has been reached. From a formal point of view, K_S is equivalent to

the K_M of enzyme kinetics, the Michaelis constant. 4) Growth rates are linked to the *generation* or *doubling time*. This parameter indicates in [h], how fast a bacterial culture doubles under exponential conditions. 5) The yield coefficient Y_S is a measure of biomass formation per consumed substrate. Various yield coefficients can be defined, because the formation of cell mass depends on both chemical (pO_2, C/N ratio, phosphate content) and physical parameters (e.g., temperature). If complex nutrient media are used, two log phases separated by an intermediate lag phase may be observed (diauxic growth). This is explained by the lag time required to induce new enzymes after the first carbon source is exhausted.

Growth kinetics of mycelium-forming microorganisms. Fungi, and also mycelium-forming prokaryotes such as Streptomycetes, do not grow only by cell division but also by longitudinal growth of the mycelium. Determination of growth is usually done by weighing dried biomass and leads to complex kinetics.

Product formation in most fermentation processes is either coupled or not coupled to growth. Only very few processes do not belong to either of these two basic types. In the traditional classification, growth-coupled product formation is termed type I; this includes e.g., cell mass formation (baker's yeast, SCP, algae) and alcohol fermentation. In type III fermentations, product formation is decoupled from growth and occurs at the end of the logarithmic phase; the product does not arise from the primary, but rather from the secondary metabolism. Examples are: antibiotics and extracellular enzymes. In type II fermentations, the product originates from side paths of the primary metabolism and is synthesized in parallel to the growth of cell mass. Examples are citric acid and amino acids. Since the formation of most type II products is also coupled to growth, classification today is usually restricted to type I and type III fermentations.

Growth in a batch reactor

growth curve of microorganisms

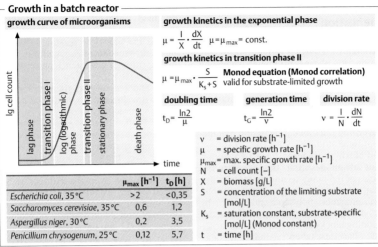

growth kinetics in the exponential phase

$$\mu = \frac{1}{X} \cdot \frac{dX}{dt} \qquad \mu = \mu_{max} = \text{const.}$$

growth kinetics in transition phase II

$$\mu = \mu_{max} \cdot \frac{S}{K_s + S}$$ **Monod equation (Monod correlation)** valid for substrate-limited growth

doubling time	generation time	division rate
$t_D = \dfrac{\ln 2}{\mu}$	$t_G = \dfrac{\ln 2}{\nu}$	$\nu = \dfrac{1}{N} \cdot \dfrac{dN}{dt}$

ν = division rate [h^{-1}]
μ = specific growth rate [h^{-1}]
μ_{max} = max. specific growth rate [h^{-1}]
N = cell count [–]
X = biomass [g/L]
S = concentration of the limiting substrate [mol/L]
K_s = saturation constant, substrate-specific [mol/L] (Monod constant)
t = time [h]

	μ_{max} [h^{-1}]	t_D [h]
Escherichia coli, 35 °C	>2	<0,35
Saccharomyces cerevisiae, 35 °C	0,6	1,2
Aspergillus niger, 30 °C	0,2	3,5
Penicillium chrysogenum, 25 °C	0,12	5,7

Product formation

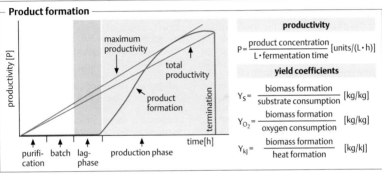

productivity

$$P = \frac{\text{product concentration}}{L \cdot \text{fermentation time}} \ [\text{units}/(L \cdot h)]$$

yield coefficients

$$Y_S = \frac{\text{biomass formation}}{\text{substrate consumption}} \ [kg/kg]$$

$$Y_{O_2} = \frac{\text{biomass formation}}{\text{oxygen consumption}} \ [kg/kg]$$

$$Y_{kJ} = \frac{\text{biomass formation}}{\text{heat formation}} \ [kg/kJ]$$

Fermentation types

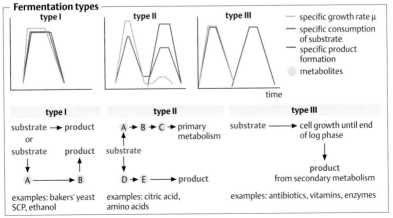

— specific growth rate μ
— specific consumption of substrate
— specific product formation
● metabolites

type I

substrate → product
or
substrate   product

A ─────→ B

examples: bakers' yeast
SCP, ethanol

type II

A → B → C → primary metabolism

substrate

D → E ─────→ product

examples: citric acid, amino acids

type III

substrate ─────→ cell growth until end of log phase

product
from secondary metabolism

examples: antibiotics, vitamins, enzymes

Fed-batch and continuous fermentation

General. In fed-batch fermentations, the production phase is prolonged by feeding nutrient medium to the fermentation. This protocol is the preferred procedure in industry. Continuous fermentations are less practical, but of great fundamental importance, since they allow the laws governing microbial growth and metabolism to be studied.

Fed-batch procedures have two important advantages: first, they increase the yields of the many secondary metabolites that are industrially produced (antibiotics, enzymes, polysaccharides, etc.) by providing fresh medium or intermediary building blocks at the end of the logarithmic phase just when secondary metabolism takes off. Second, substrate inhibition can be prevented by carefully limiting the glucose level in the medium. Glucose is the most widely used carbon and energy source in fermentation, but an excess represses product formation by catabolite repression, e. g., during production of antibiotics. Bakers' yeast production is another example of catabolite repression: higher sugar concentrations lead to an increased specific growth rate μ; however, the biomass yield coefficient Y_S decreases strongly because an increasing amount of glucose is converted to ethanol (Crabtree effect). Thus, sugar is added in fed-batch mode to the fermentation broth. Similar requirements have sometimes been found for nitrogen and phosphorous sources.

Continuous fermentations. A batch reactor is usually considered a closed system, though strictly speaking there is gas exchange with the environment. During continuous fermentations, there is not only gas exchange with the environment, but the bioreactor is an open system to which sterile nutrient broth is continuously fed and from which culture medium is continuously removed. Three varieties of continuous fermentation modes are usually distinguished: two are chemostats, where nutrient levels are held constant, and turbidostats, where cell mass is held constant. The third is the plug-flow reactor, in which the culture medium flows without backmixing through a tubular reactor, while the cell mass is recovered at the reactor outlet and returned at the entrance of the reactor. In such a system, conditions along the direction of flow, e. g., medium composition, biomass concentration, and product concentration resemble the conditions that are obtained in a batch reactor over time. In a continuous fermentation under equilibrium conditions, cell loss from the effluent is balanced by the specific growth rate μ of the microorganism; and the substrate concentration S and the rate of product formation Q_X remain unchanged. Under such conditions, Q_X depends as a first approximation solely on the flow rate. Concepts for continuous fermentations are much harder to develop for secondary metabolites (type III fermentations), since cell growth and product formation are not directly coupled. Continuous fermentations are useful for 1) optimizing cell growth and product formation, and 2) analyzing the limiting nutrient components. In industrial practice, however, continuous fermentations are rarely used. Some few exceptions are the aerobic and anaerobic treatment of wastewater, newer process variants for the production of beer, and the manufacture of human insulin using recombinant yeast strains. In most industrial fermentation processes, however, it is held that 1) continuous processes, compared with batch-fed fermentations, show break-even economics only after 500–1000 h of continuous operation – an extremely difficult condition to meet, from the point of view of sterile operation and anti-infection management; 2) it is difficult to keep the composition of nutrient media constant over such a long period of time; and 3) the genetic stability of recombinant host organisms cannot be assured over so many generations of cell division.

Continuous fermentations

reactor types

chemostat	turbidostat	plug-flow reactor

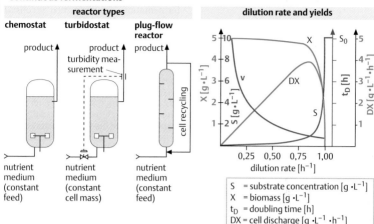

chemostat
product ▲

nutrient medium (constant feed)

turbidostat
product ▲
turbidity measurement

nutrient medium (constant cell mass)

plug-flow reactor
product ▲

cell recycling

nutrient medium (constant feed)

dilution rate and yields

dilution rate [h⁻¹]

S = substrate concentration [g·L⁻¹]
X = biomass [g·L⁻¹]
t_D = doubling time [h]
DX = cell discharge [g·L⁻¹·h⁻¹]

biomass and substrate balances

biomass

$$\frac{dX}{dt} = \mu \cdot X - D \cdot X \quad (1)$$

substrate

$$\frac{dS}{dt} = D(S_0 - S) - \frac{\mu \cdot X}{Y_S} \quad (2)$$

for the stationary state it follows:

$$\frac{dX}{dt} = \frac{dS}{dt} = 0 \quad (3) \quad \text{and it further follows that } \mu = D$$

using (3) with (1) and (2):

$$X = Y_S(S_0 - S) \qquad D = \mu_{max}\frac{S}{K_S + S} \qquad S = \frac{D \cdot K_S}{\mu_{max} - D}$$

Y_S = yield coefficient [kg/kg]
K_S = Monod constant [g·L⁻¹]
S_0 = substrate concentration in feed [g·L⁻¹]
S = substrate concentration in bioreactor [g·L⁻¹]
D = dilution rate [h⁻¹]
X = biomass [g·L⁻¹]
μ = specific growth rate [h⁻¹]
μ_{max} = maximum specific growth rate [h⁻¹]

Fed-batch process

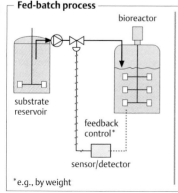

bioreactor

substrate reservoir

feedback control*

sensor/detector

*e.g., by weight

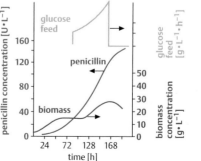

penicillin production with glucose feed

penicillin concentration [U·L⁻¹]

glucose feed

penicillin

biomass

glucose feed [g·L⁻¹·h⁻¹]

biomass concentration [g·L⁻¹]

time [h]

Fermentation technology

General. To produce biotechnological products at acceptable costs, bioprocess engineering, a discipline shaped by engineers, is as important as the biosciences developed by biologists and biochemists. Key objectives of industrial process engineering are the operational safety of a process and the minimization of both investment and process costs. Important aspects of these tasks are 1) optimized mass transfers; 2) technical solutions to keep temperature constant; and 3) optimization of aeration (for aerobic processes).

Mixing in a bioreactor is achieved by stirrers or pumps, resulting in a turbulent current in the immediate neighborhood of the stirrer, which is characterized by its Reynolds number Re. In aerobic processes, aeration also contributes to mixing. A factor in the numerical calculation of the Reynolds number is viscosity, which depends on the concentration of microorganisms, on their physical shape (e. g., mycelia in fungal fermentations), and on the type of product (e. g., xanthan). In an ideally mixed bioreactor, turbulence in the reaction zone is distributed in a homogenous manner. This target, however, is only approximated, because the sensitivity of the biological materials is usually limiting. For example, stirrer speed is limited by the shear sensitivity of a mycelium. Various factors, such as the geometry of the stirrers, their form and number, the position of mechanical units such as baffles, the position of pumps (in unstirred reactors), and the configuration and positioning of aeration plates and air ejectors (in stirred and unstirred reactors). The power number Ne describes the energy requirement of stirred reactors and is correlated, in an ungassed state, with the Reynolds number. For industrial use, various stirrers have been developed, e. g., disc, turbine, MIG, and InterMIG impellers, which support good mixing and O_2 transfer. Mixing and O_2 transfer parameters are measured according to the volumetric transfer coefficient $k_L a$.

Temperature control. For optimal results, fermentations are performed at constant temperature. After an initial heating as required for cell growth, fermentation reactors are usually cooled, because both microbial growth and stirrer mechanisms produce heat, which must be removed. To calculate the heat produced in a process, the heat transfer number and the exchange area of the fermenter must be considered. It is usually sufficient to remove heat with a water cooling system that surrounds the bioreactor, but when yield coefficients are very high (low Y_{kJ} values), as in, e. g., alkane fermentations by yeasts, additional internal heat exchangers must be used.

Aeration. The growth of aerobic cultures is usually limited by the oxygen content of the culture solution. To optimize the oxygen content, several biological and technical factors must be considered. For example, the optimal oxygen transfer in a bioreactor is correlated with the specific maximum oxygen uptake rate $q_{O_2}^{max}$ of the microorganism. In addition, oxygen is transported in a ternary-phase system comprising the gas and liquid phases and the microorganism. For oxygen transfer, several phase boundaries must be overcome: 1) from the gas bubble to the phase boundary surface; 2) through the phase boundary surface into the liquid; 3) through the liquid to the boundary surrounding the microorganism; and 4) into the cell. In single-cell organisms, phase boundary surface 2 is often limiting, whereas cell associations or mycelia often show limitations in 4. O_2 transfer also depends on a wide range of technical conditions such as reactor dimensions, hydrostatic pressure (filling height), stirrer performance, aeration system and rate, chemical and physicochemical parameters such as nutrient type, medium density and viscosity, temperature, surface pressure (antifoams), and finally on biological factors, e. g., the growth form of the microorganism. An important parameter for the characterization of oxygen transfer in a bioreactor is the volumetric transfer coefficient $k_L a$, which can be determined experimentally.

Mixing in a stirred reactor

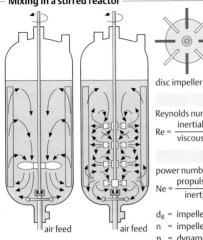

MIG impeller

InterMIG impeller

disc impeller   turbine impeller

characterization of flow

Reynolds number Re

$$Re = \frac{inertial\ force}{viscous\ force} = \frac{d_R^2\, n\, \rho}{\eta}\ [-]$$

energy requirement

power number Ne

$$Ne = \frac{propulsion}{inertia} = \frac{P_o}{d_R^5 n^3 \rho}\ [-]$$

d_R = impeller diameter [m]
n = impeller revolutions [s^{-1}]
η = dynamic viscosity [Pa·s]

P_o = impeller performance [W]
ρ = density [kg/m^3]

air feed   air feed

Aeration

sparger   sieve plate   injector   self-aspiring stirrer

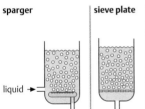

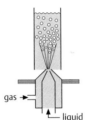

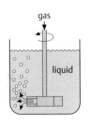

gas

liquid

liquid ←   gas ←

↑ gas   ↑ gas   ↑ liquid

specific O₂ requirement Q_{O2}

$$Q_{O_2} = Xq_{O_2} = k_L a\,(C_{O_2}^* - C_{O_2})\ [mol \cdot L^{-1} \cdot h^{-1}]$$

$q_{O_2}^{max}$ = maximum O$_2$ resorption rate [mol·g^{-1}·h^{-1}]
q_{O_2} = specific O$_2$ resorption rate [mol·g^{-1}·h^{-1}]
$k_L a$ = volumetric transfer coefficient [h^{-1}]
X = biomass concentration [g·L^{-1}]

O₂ resorption rate r

$$q_{O_2} = q_{O_2}^{max} \frac{C_{O2}}{K_{O_2} + C_{O2}}\ [mol \cdot g^{-1} \cdot h^{-1}]$$

C_{O_2} = dissolved O$_2$ concentration [mol·L^{-1}]
$C_{O_2}^*$ = saturation concentration O$_2$ [mol·L^{-1}]
K_{O_2} = Michaelis constant O$_2$ [mol·L^{-1}]

O₂ transfer: k_La value

$$k_L a = k \left(\frac{P}{V_R}\right)^\alpha (u_G^0)^\beta\ [h^{-1}]$$

k, α, β = constants [–]
P = impeller performance [W]
V_R = reactor volume [m^3]
u_G^0 = gas superficial velocity [m/s]

aeration rate B

$$B = \frac{volume\ air}{bioreactor\ volume}\ min\ (vvm)$$

aeration number N_B

$$N_B = \frac{V_G}{nd_R^3}\ [-]$$

V_G = gas volume current [m^3/s]
n = impeller revolutions [s^{-1}]
d_R = impeller diameter [m]

Fermentation technology: scale-up

General. During the transfer of processes from a development to a production scale (scale-up), changes in several parameters must be considered. Depending on the process and the desired production volume, several types of bioreactors can be used, the stirred reactor being the most popular version. Traditionally, scale-up is done in decimal steps (30 L to 300 L to 3000 L to production scale).

Scale-up. Even on a pilot scale, bioreactors are equipped with impellers, turbines, baffles, pumps, or aeration modules to allow good mixing. For results at the pilot scale to be transferred to production scale, one must take into account that mixing time strongly increases with volume and that fast mixing in reactor volumes > 150 m^3 is not only difficult to achieve, but entails prohibitive energy costs. This fact also plays against plasmids that carry a λ promoter: the rapid rise in temperature required for induction is completely out of scope in a large production fermenter. Apart from mixing, mechanical shear stress also sets limits to fermentations using fungal or Streptomyces strains. Similar arguments are valid for the distribution of air bubbles during aeration and for heat removal.

Bioreactor types. Surface bioreactors for the manufacture of citric acid are mostly of historical importance, although the trickling filter in aerobic wastewater treatment is still a widely-used example of this reactor type. Surface reactors are simple to operate but their volume-space-yield of products is rather limited. Today, stirred bioreactors are the equipment of choice. They are temperature- and pH-controlled, equipped with stirrers and aeration systems, and have sterile feed and sampling valves. Stirrers are usually of a multi-stage type, complemented by baffles; in some reactors, one-stage stirrers with overflow and conduit tubes are still being used. For acetic acid production and aerobic wastewater treatment, self-aspiring agitators are in use. Research-type bioreactors range from 1 to ca. 300 L, and stirred bioreactors with a working volume up to 500 m^3 can be found in industrial manufacturing. At larger volumes, energy requirements for fast mixing and heat transfer quickly increase. Thus, loop or airlift reactors are preferred for large-volume fermenters with a volume up to 1500 m^3; in these systems, mammoth pumps or injectors serve as mixing devices. Single-cell protein manufacture and aerobic waste water treatment have been mentioned in this book as typical examples in which airlift reactors are being used industrially.

Measurement and control are most important for the economic optimization of a bioprocess, but also for the operational safety of a plant. Routinely measured parameters include reactor weight, temperature, pH value and O$_2$ content of the nutrient broth, and revolutions and energy use of the stirrer. Usually, feed and exhaust air are analyzed for CO$_2$ (by IR spectroscopy) and O$_2$ (by paramagnetic resonance), providing the respiratory quotient (RQ), which yields important clues about the growth and condition of a culture. Consumption of substrate and formation of product are usually determined outside the bioreactor, after sterile sampling. In view of the high value of a bioreactor filling (even if based on a market value of just $10 \, € \, kg^{-1}$ product and an end-product concentration of $100 \, g \, L^{-1}$, the value of product after fermentation in a 100 m^3 bioreactor is 100 000 €), methods for reliable operational and sterility control are continuously being updated.

Foam breakdown. Foam forms during aeration of proteinaceous solutions and is a frequent impediment in fermentation processes. It is usually counteracted by a mechanical foam breaker (thermal foam destruction or foam centrifuge), which is located on top of the impeller shaft. If foam formation is very heavy, chemical antifoam agents such as erucic acid or silicones have been used. Their disadvantage is, however, that they may wind up in the end product of fermentation and may be difficult to remove.

Bioreactor types

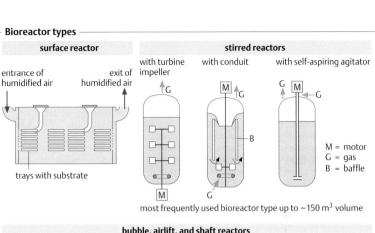

surface reactor

entrance of humidified air   exit of humidified air

trays with substrate

stirred reactors

with turbine impeller   with conduit   with self-aspiring agitator

M = motor
G = gas
B = baffle

most frequently used bioreactor type up to ~150 m³ volume

bubble, airlift, and shaft reactors

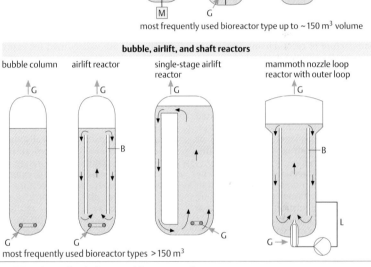

bubble column   airlift reactor   single-stage airlift reactor   mammoth nozzle loop reactor with outer loop

most frequently used bioreactor types > 150 m³

Mixing time - dependence on bioreactor volume

volume [L]	3	9	100	300	1000	3000	24000
impeller speed [Upm]	750	2000	230	350	200	180	30
mixing time [s]	5	3	6,6	5	25	20	66

Measurement and control in bioreactor technology

physical parameters	chemical parameters	biological parameters
temperature	**pH value**	enzyme activities
pressure	dissolved oxygen	ATP content
power input	O_2 and CO_2 in exhaust gas	NADH content
viscosity	redox potential	protein content
flow rate of air	**substrate concentration**	
nutrient feed	product concentration	
turbidity	ion concentration	
bioreactor weight		

measured parameters **regulated parameters**

203

Cultivation of mammalian cells

General. Mammalian cell cultures are preferentially used 1) to produce vaccines, and 2) to manufacture therapeutic and diagnostic proteins that cannot be obtained from recombinant microorganisms. These include proteins that contain numerous disulfide bridges, that are effective only after complex posttranslational modifications (e. g., glycosylations), or which, due to a wrong glycosylation pattern, lead to an immune response upon long-term administration. Examples are therapeutic antibodies, factor VIII, erythropoietin, and tPA. The manufacture of recombinant proteins in animal cell culture is technologically demanding and expensive. As a recent alternative technology, attempts are being made to produce these type of proteins in transgenic animals or plants. Human tissue cultures are also being investigated for the manufacture of tissue replacements for transplantation medicine or for the testing of drugs and cosmetics (tissue engineering).

Human cell cultures. Cell populations taken from human tissue and expanded in nutrient media have been used for a long time for the identification and propagation of human-specific viruses, in part also for the manufacture of virus vaccines. Human cells can be stored in the gas phase of cryocontainers at $-120\,°C$, and uniform cell cultures can be obtained from such stocks over a long period of time. The life span of these primary cells, however, is limited to ca. 50 cell divisions, and they absolutely require a solid surface for growth. As a result, the yield of cell material obtained by this method is limited. In contrast, continuous cell lines have the capacity to expand (propagate) indefinitely. Examples of continuous cell lines are HeLa cells and the Namalva cell line, obtained from human cervix cancer and lymphoma, respectively. Embryonic stem cells, although not cancerous, are also continuous cells, since they can divide indefinitely.

Cell lines in biotechnology. Continuous cell lines are useful for the production of therapeutic proteins. Cell lines used in industry originate from animal tumors. They divide without limit and exhibit these properties: 1) short doubling time of 20 – 30 h; 2) uncomplicated culture conditions; 3) ability to grow in suspension to a high cell density with adequate stability to shear stress, which allows their cultivation in large bioreactors; and 4) availability of vectors for transformation. Today, the following cells are mainly used for this purpose: 1) mouse hybridoma cells for the manufacture of monoclonal antibodies (preferentially for diagnostic applications); 2) Chinese hamster ovary fibroblasts (CHO cells); and 3) tumor cells from Syrian baby hamster kidneys (BHK cells). CHO or BHK cells are widely used for expressing recombinant proteins, e. g., tPA, EPO, or factor VIII. They lead to products whose post-translational modification, in particular their glycosylation pattern, is very similar to the native human proteins.

Cloning vectors. Several types of vectors are used for preparing genetically stable recombinant animal cells. They all integrate into the genome of the target cell and may carry a sequence for external induction. Usually, shuttle vectors are used that can also transform *E. coli*, which is used as a host to optimize the vector. In laboratory experiments, typical markers applied for the selection of transformed cells are proteins conferring resistance to toxic medium components such as neomycin or Cd salts. For industrial applications, such agents cannot be used. Instead, dihydrofolate reductase (DHFR) in combination with *dhfr*-deficient host cells (e. g., CHO-K1) is the marker of choice. DHFR is competitively inhibited by the folic acid analog methotrexate, disturbing thymidine biosynthesis. Thus, transformed *dhfr*-deficient cells auxotrophic for thymidine can grow on a minimal medium in the presence of metothrexate after transfection with a DHFR vector. They even amplify the foreign *dhfr* gene, ligated to the cDNA of the desired gene product. Two vectors used for the industrial production of factor VIII and tPA in CHO cells are shown on the opposite page.

	origin	applications
Namalva	human lymphoma	interferon
BHK	baby hamster kidney	factor VIII
CHO	Chinese hamster ovary	tPA, EPO, FMD virus vaccine
SP 2/0 mouse hybridoma	mouse myeloma	monoclonal antibodies
BS-C-1 vero cells	primary monkey kidney	human vaccines

Animal cells used for protein production

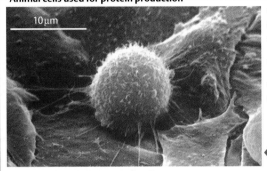

10 µm

desired properties

- immortal
- simple and stable transformation
- propagate to high density in suspension culture
- high duplication rate
- undemanding culture conditions

◀ CHO cell (attached to a surface)

Examples of CHO expression vectors

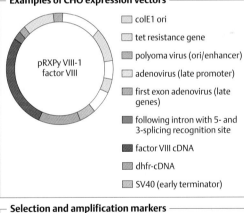

pRXPγ VIII-1
factor VIII

- ☐ colE1 ori
- ☐ tet resistance gene
- ☐ polyoma virus (ori/enhancer)
- ☐ adenovirus (late promoter)
- ☐ first exon adenovirus (late genes)
- ☐ following intron with 5- and 3-splicing recognition site
- ☐ factor VIII cDNA
- ☐ dhfr-cDNA
- ☐ SV40 (early terminator)

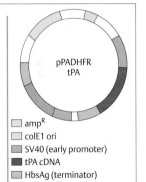

pPADHFR
tPA

- ☐ amp^R
- ☐ colE1 ori
- ☐ SV40 (early promoter)
- ☐ tPA cDNA
- ☐ HbsAg (terminator)
- ☐ dhfr cDNA

Selection and amplification markers

marker gene on expression vector	component in nutrient medium	selection/ amplification	comments
neomycin phosphotransferase	neomycin	selection of neomycin-resistant cells	unsuitable for production media
metallothionein	Cd ions	selection of Cd-resistant cell lines	research use
dehydrofolate reductase (DHFR)	thymidine and methotrexate (a DHFR inhibitor)	selection of dhfr-complemented cell lines, amplification by methotrexate	frequently used for production

Mammalian cell bioreactors

General. In recent decades, numerous protocols for the cultivation of mammalian cells on a laboratory scale have been developed. In particular, nutrient media have been optimized. Because several recombinant proteins such as human or humanized antibodies, factor VIII, and some forms of tPA can be produced in sufficient quality and quantities only in mammalian cell culture, manufacturing processes based on mammalian cells have been scaled up to the > 10 000 L scale. In this context, typical aspects of bioprocess engineering such as mixing and aeration play a key role. Isolation and purification of products obtained from animal cells are extremely demanding, in view of undesired protein constituents, as well as the need to detect even traces of viral DNA or retroviral RNA.

Nutrient media. Besides a good supply of oxygen and CO_2, a sufficient supply of nutrient components is of utmost importance. Glucose is preferred as the carbon source. Amino acids, vitamins, nucleotides, proteins, fatty acids, inorganic salts, and growth-promoting substances must also be included. Most older nutrient mixtures contained fetal calf serum as a complex source of growth promoters. The serum-free media used today are less expensive and more acceptable with respect to animal welfare concerns. They contain supplements of bovine serum albumen, insulin, transferrin, ethanolamine, and selenite (e. g., the ITES medium). To regulate cellular pH, bicarbonate is added, or experiments are conducted under a CO_2 atmosphere.

Laboratory procedures. In the laboratory, cells are preferentially grown in roller or T flasks (tissue flasks). For larger scales, up to 10 L, spinner flasks are used. Cell concentrations in these flasks reach 1–2 million cells mL^{-1}.

Cell reactor. For scaling up experiments to a cell reactor, one must be aware that toxic metabolites may enrich during batch or fed-batch cultivation and inhibit formation of the desired product. In per-fusion cultures, spent medium is replaced with fresh nutrient solution. For industrial fermentations, however, fed-batch protocols are preferred, since they reduce the risk of infection. Because cells are in suspension, oxygen supply and shear stress stability are important but opposing requirements that must be optimized. The k_La values for the oxygen supply of animal cells are on the order of 2.2 h^{-1}, or 1–2 magnitudes lower than the k_La values for microorganisms. A wide range of indirect oxygen supply systems has been investigated, e. g., semipermeable silicon membranes. For example, in a 1000-L bioreactor with a rotating membrane system, an oxygen transfer efficiency of 120 g O_2 m^{-3} was achieved, applying an internal membrane pressure of up to 6 bar. Another important task is full exploitation of the expensive nutrients. In perfusion experiments, the cells are retained by a microporous membrane filter. Once the desired cell density has been reached, continuous workup of the product is initiated. Cultures of this type have been held at equilibrium for > 900 h. With cells fixed on microbead carriers with a large internal surface, e. g. made from porous silicon, a stable production yield was obtained for over 30 d. In modern production techniques, which have been optimized to the > 10 000 L scale, only fed-batch procedures are used, and cells are suspended in a serum-free medium.

Purification of products. In contrast to the recovery of most microbial products, recovery of mammalian-cell products does not require breaking the cells, since the desired products are secreted into the culture medium. However, lengthy purification protocols must be followed to achieve highly pure products, and quality control is a key issue. Recovery steps may include affinity purification. Identity of the protein product is proved by using peptide maps, terminal amino acid sequencing, MALDI-TOF, and other methods; the absence of oncogenic or transformable DNA must be proven to a concentration of 100 pg (10^{-10} g) per dose of the pharmaceutical end product; and no retroviral RNA must be present.

Nutrient media

serum-containing media	serum-free media
glucose, glutamine	glucose, Na pyruvate, glutamine
amino acids	amino acids
fetal calf serum	bovine serum albumen, insulin, transferrin, ethanolamine, selenite (ITES medium)
minerals	minerals
CO_2- and O_2 supply	CO_2- and O_2 supply

Cultivation in the laboratory and pilot plant

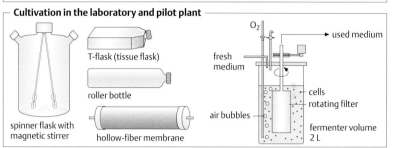

spinner flask with magnetic stirrer

T-flask (tissue flask)

roller bottle

hollow-fiber membrane

O_2

used medium

fresh medium

cells
rotating filter

air bubbles

fermenter volume 2 L

Methods of cultivation

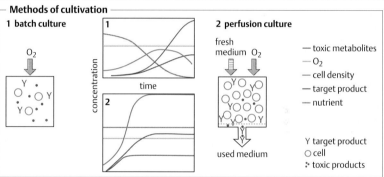

1 batch culture

O_2

Y
O
Y O
Y O

concentration

time

2 perfusion culture

fresh medium O_2

used medium

— toxic metabolites
— O_2
— cell density
— target product
— nutrient

Y target product
O cell
❖ toxic products

Scheme of a fed-batch culture using CHO cells

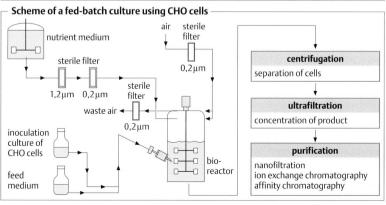

nutrient medium

air   sterile filter

sterile filter

$1,2\,\mu m$ $0,2\,\mu m$

sterile filter

$0,2\,\mu m$

waste air

$0,2\,\mu m$

inoculation culture of CHO cells

feed medium

bio-reactor

centrifugation

separation of cells

ultrafiltration

concentration of product

purification

nanofiltration
ion exchange chromatography
affinity chromatography

207

Enzyme and cell reactors

General. Biocatalysts used in industrial processes are usually immobilized on a carrier, both for stability and cost savings. For one-step biotransformations, enzymes are the biocatalysts of choice. If multistep biotransformations are required, or if intracellular enzymes are difficult to purify to a sufficient extent, immobilization of whole microbial cells is preferred. Sometimes the cells are inactivated under conditions where the desired enzymatic activity is preserved.

Chemistry of immobilization. Adsorption of enzymes or cells onto charged surfaces may often suffice: to this end, a wide range of ion-exchange materials is available. If covalent immobilization of an enzyme is preferred, 3 main methods have been used: 1) crosslinking of surface ε-amino groups of lysine with glutardialdehyde, leading to azomethines which can be further stabilized by hydrogenation with sodium borohydride; 2) crosslinking with diisocyanates; and 3) binding to polymeric epoxides (oxiranes). Numerous inorganic and organic carriers have been investigated as matrix materials. For the inclusion of cells within a carrier material, prepolymers are mixed with cells, and the mixture is subjected to radical or photochemical polymerization. For example, cells (or enzymes) can be incorporated into a polyacrylamide gel when they are mixed with acrylamide, and then N,N′-methylene bisacrylamide and potassium persulfate are added. Urethane prepolymers and other materials have been used for photochemical polymerization. For microencapsulation of enzymes or cells, polymerization of suitable prepolymers may be carried out at the boundary phase of water, which contains the enzyme or cells, and an organic solvent that does not mix with water. Water-soluble polymers such as polyethylene glycol may be used for immobilization of detergent enzymes. Another popular method uses ionotropic gels such as alginate, which form a gel in the presence of Ca^{2+}. Weaving of enzymes and cells into fibers, e. g., into cellulose derivatives or collagen, has also been described.

Properties of immobilized biocatalysts. Upon immobilization, biocatalysts may change their properties, because catalytic efficiency is not influenced only by the catalyst itself, but also by mass transfer, which depends on the properties of the immobilization matrix used. Thus, empirical rules have been elaborated that are useful for optimizing the immobilized catalyst as well as the packing of bioreactors and their volume-time yields. Besides the intrinsic properties of the catalyst such as V_{max} and K_M, these rules take into account both particle size and mass-transfer parameters. A key property of an immobilized biocatalyst is its operational stability. In favorable cases, it may remain active for several months (aspartase, glucose isomerase, penicillin acylase).

Reactor type and process technology. Enzyme and cell reactors have been engineered as discontinuous or continuous stirred reactors; as fixed-bed, fluidized-bed, or hollow-fiber column reactors; and as membrane reactors. In addition to biocatalyst stability and optimization of mass transfer, upstream and downstream processing are most important for good economy. These processes include preparation of substrates, suppression of side reactions, and development of a recovery protocol that is adequate for the intended use of the product. The overall scheme must be robust, simple to operate, and optimized for minimal investment and operational costs. Measurement and control are also key issue. For large-volume products such as isomerose (high-fructose corn syrup, HFCS) or 6-aminopenicillanic acid, continuously operating plants are preferred. For this, several bioreactors are used in parallel but with consecutive catalyst fillings; thus, each module with exhausted catalyst activity can be exchanged without decreasing the overall productivity of the plant. For products with lower tonnage such as L-aspartic acid, a single-cell reactor of immobilized *E. coli* may suffice for a production run.

Materials used for immobilization

inorganic carriers		natural and synthetic polymers		
Al$_2$O$_3$	Ca$_2$PO$_3$ gels	activated charcoal	oxiranes	collagen
bentonite	porous ceramic materials	polyacrylamide	agarose	polyamide
porous glass		carboxymethyl cellulose	alginate	cellulose
		polyurethanes	dextrans	

Immobilization onto glass surfaces

$$\begin{array}{c} \quad\quad\quad OR \\ \rule{3mm}{0.5mm}OH + H_5C_2-O-\underset{\underset{OR}{|}}{\overset{\overset{|}{OR}}{Si}}-CH_2-CH_2-CH_2-NH_2 \xrightarrow{-C_2H_5OH} \rule{3mm}{0.5mm}O-\underset{\underset{OR}{|}}{\overset{\overset{|}{OR}}{Si}}-CH_2-CH_2-CH_2-NH_2 \end{array}$$

$$\rule{3mm}{0.5mm}NH_2 + OHC-(CH_2)_3-CHO + \boxed{H_2N-\text{E}} \longrightarrow \rule{3mm}{0.5mm}N{=}CH-(CH_2)_3-CH{=}N-\boxed{\text{E}}$$

glutardialdehyde

▨ enzyme with ε-amino groups of lysine   █ glass surface

Reactor types

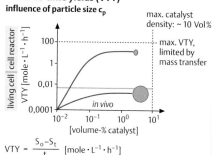

S = substrate
P = product
⊶ enzyme
—— membrane

fixed bed reactor with circulation   fixed bed reactor   fluidized bed reactor   slurry reactor   enzyme membrane reactor

Volume-time yields (VTY)

influence of particle size c_p

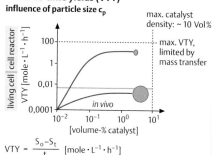

max. catalyst density: ~ 10 Vol%
max. VTY, limited by mass transfer

VTY [mole · L^{-1} · h^{-1}]
living cell || cell reactor
in vivo
[volume-% catalyst]

$$VTY = \frac{S_o - S_t}{t} \quad [\text{mole} \cdot L^{-1} \cdot h^{-1}]$$

S_o = starting concentration of substrate [mol · L^{-1}]
S_t = concentration of substrate at time t [mol · L^{-1}]

influence of enzyme immobilization

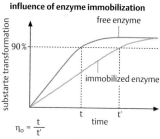

substrate transformation
free enzyme
90%
immobilized enzyme
t   time   t'

$$\eta_o = \frac{t}{t'}$$

η_o = degree of utilization of catalyst
t   = time for defined substrate turnover for free (t$_{12}$) or immobilized (t'$_{12}$) biocatalyst

Influence of mass transfer

inner mass transfer: **Thiele modulus** φ

$$\varphi = \frac{\text{reacting substrate}}{\text{diffused substrate}}$$

$$\varphi = \frac{d_p^2}{4} \cdot \frac{\delta_{max}}{K_m \cdot D^e}$$

external mass transfer: **Sherwood number** Sh

$$Sh = \frac{\text{mass transfer}}{\text{diffusion}} \quad\quad Sh = \frac{k_f \cdot d_p}{D_0} \quad [-]$$

D^e effective diffusion coefficient [m^2 · s^{-1}]
k_f mass transfer coefficient [m · s^{-1}]
d_p particle diameter [m]
D_0 molecular diffusion coefficient [m^2 · s^{-1}]
δ_{max} maximum catalyst density
K_m Michaelis constant

Recovery of bioproducts

General. Bioproducts obtained during a fermentation process are either cellular (cell mass, intracellular proteins, inclusion bodies) or extracellular (amino acids, antibiotics, enzymes). In classical fermentations, the product concentration is often low (< 10%, often < 1%). Using genetically engineered microorganisms, higher product yields are usually obtained (e.g., up to 50% protein per wet cell mass). For isolating purified bioproducts, the sequence of concentration and purification steps should be appropriate for the intended use of the product. Thus, pharmaceutical and diagnostic preparations require complex purification protocols, but technical enzymes are enriched with fewer and simpler processing steps. Product loss during purification may exceed 50%, indicating the importance of good protocols for economic production. Safe and inexpensive disposal of waste fractionsis another task of economic and ecological importance.

Cell mass. The manufacture of baker's yeast is a good example of the preparation of microbial biomass. After fermentation is complete, cells are separated using centrifuges or centrifugal separators, followed by washing and filtration through a pressure leaf or vacuum rotating filter; the moist product is then packaged. Dried baker's yeast is produced from moist yeast with a cyclone and can be stored much longer. Diafiltration or cross flow filtration are additional methods used in cell separation.

Intracellular products. The target products are usually intracellular proteins, and the separated cells must be disrupted without inactivating proteins. Mechanical methods such as high-pressure homogenizers or ball mills are often used for this step. For the production of yeast autolysate, intracellular proteases of yeast are activated and lead to cell destruction. On a laboratory scale, ultrasonication in cooled baths is often used. Lysozyme or other lytic enzymes can also be used in combination with mild surfactants. During fermentations using recombinant *E. coli* cells, target proteins often form inclusion bodies, in which disulfide bridges have been formed wrongly. Depending on the leader sequences used, inclusion bodies may form in the periplasmic space or in the cytosol. They are isolated after mild cell breakage by differential centrifugation. Reduced forms of inclusion bodies (from the cytoplasm) are first oxidized by oxidative sulfitolysis; the resulting S-sulfonates are then reduced with thiol reagents (as are oxidized forms of inclusion bodies from the periplasmic space), and the proteins are denatured with urea or similar reagents that break hydrogen bonds. In a following dialysis step under oxidizing conditions, urea is removed, and part of the protein folds into the correct conformation. Overall yields of functional protein rarely exceed 20% with this process.

Extracellular products. Low-molecular-weight bioproducts such as citric acid and amino acids are usually precipitated from the broth after removal of the cell mass. They are purified further by dissolution and precipitation steps, often leading to crystalline products. Antibiotics are usually isolated by multi-stage solvent extractions using n-pentanoyl- or n-butyl acetate. The isolation of extracellular proteins is usually initiated with an ultrafiltration step, followed by salting out with ammonium or sodium sulfate. Alternatively, low concentrations (2 – 10%) of cooled organic solvents such as 2-propanol can be used. Technical enzymes are often used in crude form, either as liquid concentrates or in spray-dried form.

Integrated procedures. Numerous attempts to simplify workup protocols have been made. In expanded-bed adsorption, cell separation and product concentration occur simultaneously, by using a density gradient within a fluidized bed of ion-exchange or affinity adsorbents. Another procedure is based on two-phase systems and consists of an aqueous salt phase and a second phase composed of water-soluble polymers. Since these phases do not mix, they allow cell fragments and proteins to be enriched in different phases and have been fine-tuned for integrated product recovery.

Survey

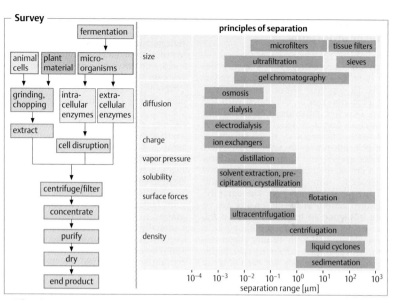

	principles of separation
fermentation	

Process flow (left):
- animal cells / plant material / micro-organisms
- fermentation
- grinding, chopping → extract
- intracellular enzymes / extracellular enzymes
- cell disruption
- centrifuge/filter
- concentrate
- purify
- dry
- end product

Separation principles chart:

- size: microfilters, tissue filters, ultrafiltration, sieves, gel chromatography
- diffusion: osmosis, dialysis, electrodialysis
- charge: ion exchangers
- vapor pressure: distillation
- solubility: solvent extraction, precipitation, crystallization
- surface forces: flotation
- density: ultracentrifugation, centrifugation, liquid cyclones, sedimentation

10^{-4} 10^{-3} 10^{-2} 10^{-1} 10^{0} 10^{1} 10^{2} 10^{3}

separation range [µm]

Filtration
vacuum rotary filter

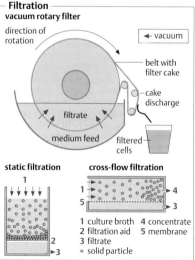

- direction of rotation
- ← vacuum
- belt with filter cake
- cake discharge
- filtrate
- medium feed
- filtered cells

static filtration / **cross-flow filtration**

1 culture broth   4 concentrate
2 filtration aid   5 membrane
3 filtrate
∘ solid particle

Centrifugation

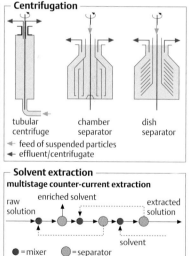

tubular centrifuge   chamber separator   dish separator

▷ feed of suspended particles
◀ effluent/centrifugate

Solvent extraction
multistage counter-current extraction

raw solution → enriched solvent → extracted solution

solvent

● = mixer   ⬤ = separator

Membrane-based procedures

	reverse osmosis	ultrafiltration
principle	transport by diffusion	separation by molecular size
retained particles	$M_R < 500 – 1\,000$	$M_R > 1\,000$
osmotic pressure	$0.8 – 10\,\text{mPa}$	very small
working pressure	$1 – 15\,\text{mPa}$	$< 1\,\text{mPa}$

Recovery of proteins: chromatography

General. Chromatography is a very important step in most purification protocols. The optimal layout of a chromatographic step can be calculated with the van Deemter equation. Column filling materials are classified according to the separation principle: 1) in gel chromatography, substances are separated by their molar masses and shapes; 2) in absorption chromatography, hydrophilic or hydrophobic interactions predominate; 3) in ion-exchange chromatography, charged amino acid side chains are of key influence; 4) in chromatofocussing, the isoelectric point of a protein governs separations; and 5) in affinity chromatography, separation occurs by specific interactions with ligands. For each of these methods, a wide variety of commercial adsorbents is available. A valuable piece of equipment for selecting the most suitable adsorbent and elution protocol within a few hours is the ÄKTA™ system. In the laboratory, many chromatographic separation steps are done at medium or high pressure (e.g., FPLC = fast protein liquid chromatography). Procedures scaled to the production level, however, do not include high-pressure chromatographic steps.

Gel chromatography. Important commercially available carrier materials are gels formed from dextrans or agarose, whose pore size can be modulated by cross-linking. After partial alkylation of hydroxyl groups by alkylating reagents, they can also be used with organic solvents.

Adsorption chromatography. A hydrophilic material in wide use is hydroxyapatite. Hydrophobic chromatography is usually carried out with Sepharose gels that are derivatized with butyl, octyl, or phenyl groups. They enable purification of hydrophobic proteins through their interaction with the hydrophobic adsorbent material.

Ion-exchange chromatography. Ion exchangers are frequently used in protein purification, since they are efficient and can be scaled up easily. Sulfonated or carboxylated polymers are often used as cation exchangers. Anion exchangers are often based on polymers containing quaternary amino groups. Polysaccharides or synthetic polymers are used as a polymeric support. The separation principle rests on the difference in net charge of different proteins. Elution is done by raising the salt concentration or changing the pH.

Affinity chromatography. This elegant method relies on the interaction of proteins with a specific ligand that is coupled to the carrier material. For purification of dehydrogenases, for example, dextran-coupled pigments that fit into the NADH binding pocket of these proteins have been used. Immunochromatography, a very expensive technique, is occasionally used in industrial purification of pharmaceutical proteins such as factor VIII. The carrier is loaded with monoclonal antibodies specific for this protein. Elution from the matrix is achieved with low-molecular-weight competitive ligands, or the column is eluted by raising the salt concentration or lowering the pH. Often, on the small scale, recombinant proteins are purified by using affinity ligands that have been purposely introduced into the protein by genetic engineering. Fusion proteins with an easily purified helper protein such as streptavidin (purified over a carrier carrying biotin ligands) is one example. Another useful technique is the addition of a polyhistidine tail (his_n, $n = 4–6$) to the N or C terminus of a protein. Proteins modified with such a residue can be purified with high selectivity on a metal-containing carrier matrix (IMAC = immobilized metal affinity chromatography), and preferentially Ni^{2+}, but also Cu^{2+}, Zn^{2+}, Co^{2+}, Fe^{2+}, or Fe^{3+} are used as matrix-bound metal chelators. If cleavage sites highly specific for a protease such as factor X are introduced adjacent to the his tag, the tag can be selectively removed after purification. Other peptide tags have been genetically introduced into proteins, e.g., a sequence binding to the biotin site of streptavidin, allowing purification over a column modified with streptavidin. The fused protein or peptide tag is usually linked in a manner that allows for its removal by a specific protease after purification.

van Deemter equation

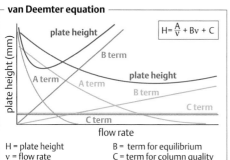

$$H = \frac{A}{v} + Bv + C$$

plate height

B term

plate height

A term

A term

plate height

B term

C term

C term

flow rate

H = plate height
v = flow rate
A = diffusion term

B = term for equilibrium
C = term for column quality

—— high molecular-weight compounds (proteins)
—— low molecular-weight compounds

Polymer matrix
example: cationic polysaccharide

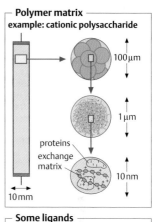

100 μm

1 μm

proteins
exchange
matrix

10 nm

10 mm

Chromatography types

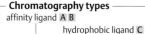

affinity ligand **A B**

hydrophobic ligand **C**

weak cation exchanger **D**

polymer
matrix

strong cation exchanger **E**

weak anion exchanger **F**

strong anion exchanger **G**

gel permeation chromatography **H**

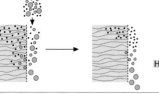

H dextrans as
molecular
sieve

Some ligands

hydrophobic interaction

C butyl, octyl, phenyl

cation exchanger

D carboxymethyl
E sulfopropyl

anion exchanger

F diethylaminoethyl
G trimethylaminomethyl

affinity chromatography

A group-specific ligands
B immunoreagents

Immunoaffinity chromatography

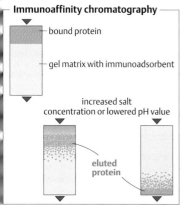

bound protein

gel matrix with immunoadsorbent

increased salt
concentration or lowered pH value

eluted
protein

his-tag chromatography

genetically modified
protein with terminal
poly-histidine tag (n = 4–6)

nitrilotriacetic
acid (NTA) as
ligand

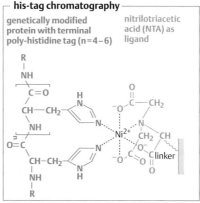

213

Economic aspects of industrial processes

General. For industrial applications, biotechnological processes are optimized with economic considerations in view. The key goals are usually 1) to develop a simple and robust process; 2) to keep the personnel and investment costs low; 3) to obtain high yields in a short time, using inexpensive, readily available raw materials of constant quality and keeping energy costs low; and 4) to keep environmental costs (disposal of waste, wastewater treatment) low.

Simple and robust processes. During multistage processes, which are usual in biotechnology, the elimination of even one process step may offer significant advantages. Thus, the so-called Westphalia decanter became very popular for production of antibiotics because it combines the two steps of cell separation and solvent extraction. Robust processes are necessary because biotechnological processes take a long time and are often carried out by unskilled workers in shift operations. Infections during a bioreactor run pose a severe and permanent problem and may lead to losses of hundreds of thousands of euros or dollars.

Low personnel and capital costs. Between 10% and 40% of biotechnology production costs are for labor. A typical fermenter run plus recovery takes about one week, and the process must be controlled by workers' shifts around the clock. The capital investment for a bioreactor production facility is on the order of 10^7 to 10^8, depending on the product. Depreciation and insurance may amount to 10% or more of the manufacturing costs.

Energy balance and raw materials. Energy costs are mainly determined by the costs of sterilization, cooling, and stirring. The large bioreactors used in industrial processes are sterilized in a continuous mode or by steam injection (140 °C for 4 min). During cell growth, about half the energy content of the carbon source is dissipated as heat. In production fermenters, cooling jackets and helical cooling coils are used to remove this heat. With heat exchangers, ca.90% of the energy used for sterilization and obtained from cooling is recovered. In classical fermentation processes such as citric acid production, raw materials, in particular the carbon source, may account for between 30 and 6% of the total manufacturing cost. Complex, inexpensive feedstocks are mostly used, whose composition may vary from lot to lot. Thus, standardization and quality control of raw materials is an important requirement in manufacturing.

Cell mass disposal and wastewater treatment. After all fermentations, cell mass and used nutrient media accumulate in large volumes; and their BOD_5 is too high for direct disposal. For example, the production of citric acid in a 300 m^3 bioreactor leads to 15 t of wet cell mass of *Aspergillus niger* mycelium as a filter cake. For environmental reasons, these residues are incinerated – an expensive procedure for wastes with high water content. At the end of a fermentation process, the desired product is dissolved in a highly diluted aqueous solution; and even in well optimized protocols, yields rarely exceed 10%. During product purification, nutrient-rich waste streams are produced, which may contain salts or organic solvents; their treatment in sewage plants leads to wastewater treatment costs that must be included in the price calculations.

Sensitivity analysis. Each bioprocess can be divided into single unit operations that can be subjected to economic scrutiny. This procedure helps to analyze which process step is cost-intensive and where optimization measures will result in the largest savings. Table calculation programs are mostly used for these analyses, since they facilitate viewing the influence of individual factors in a networked manner. Thus, material costs for the production of recombinant human insulin (opposite side) immediately demonstrate that the price of the chemicals used in recovery exceeds the price of the fermentation raw materials by a factor of nearly 10, with guanidine HCl (for unfolding of inclusion bodies) and carboxypeptidase B (for cleavage of a fusion protein) being the two major cost factors. As a result, optimization of process economy will focus on protocols where these reagents are not required.

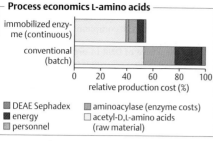

Process economics L-amino acids

immobilized enzyme (continuous)

conventional (batch)

relative production cost (%): 0 20 40 60 80 100

- ▦ DEAE Sephadex
- ■ energy
- ▧ personnel
- ▨ aminoacylase (enzyme costs)
- ☐ acetyl-D,L-amino acids (raw material)

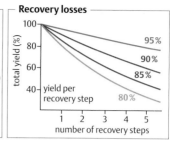

Recovery losses

total yield (%): 40 60 80 100

95%
90%
85%
80% yield per recovery step

number of recovery steps: 1 2 3 4 5

Aspects of recovery during industrial production of citric acid

feed solution	process step	isolated or waste product
	fermenter solution, 284000 L	
wash water	sedimentation tank	wet biomass, 3400 kg dry weight
	filter	
~22000 kg lime [Ca(OH)₂]	precipitation tank, separator	calcium citrate
	heat to 80–90 °C, then 95 °C	
	rotary filter; in filter caken 40000 L calcium citrate suspension	waste (filtrate)
~35000 kg 95% sulfuric acid, mother liquors	acid tank	
wash water	filter	~6000 kg calcium sulfate
	evaporator: 67% citric acid in residues	
	ion exchanger, activated charcoal	
	crystallizer	
wash water	centrifuge	mother liquors
hot air	dryer	
	packaging	40456 kg citric acid, anhydrous

Material costs for production of a recombinant protein

(h-insulin in *E. coli*, old Hoechst process, 35 m³-fermenter, batch-fed, 24 h, 25 g/L dry cell mass productivity 1 t/year)

type of cost	quantity (kg/year)	price (US$/kg)	value (US$/kg/year)	specific cost (%)
glucose	432640	0,69	298520	68,3
corn steep liquor	652800	0,12	78336	17,9
salts	~12000		~40000	8,4
antifoam agent	2448	4,86	11897	2,7
tetracycline	163,2	55	8965	2,7
fermentation raw materials			437000	100 rel%
guanidine HCl (for unfolding of inclusion bodies)	1007	2,15	2165100	56,4
carboxypeptidase B (for cleavage of fusion protein)	0,8	1023,00/g	818400	21,3
formic acid	262280	1,25	327850	8,5
bromcyan	22848	11,00	251330	6,5
all other chemicals			28000	7,3
recovery materials			3839000	100 rel%

DNA: structure

General. During cell division, the genetic information of a cell is transferred from a parental cell to daughter cells (in prokaryotic cells: during cell division; in eukaryotic cells: after fusion of two haploid parent cells). The chemical substance containing this information is deoxyribonucleic acid (DNA), a supermolecular double helix of M_R up to 10^9 Da, composed of 2 single molecules. Most living beings store their genetic information in DNA molecules. This makes its heterologous transfer among different, unrelated species in principle possible, although such events are rare, e. g., during viral infection. Since the early 1970s, technical methods have been developed that allow the transfer of genetic information among different organisms (genetic engineering). These new techniques have led to a revolution in cell biology and to major advances in biotechnology.

Chemical structure of DNA. The individual building blocks of DNA are termed nucleotides. Their structure has 2 components: a deoxyribose-5′-phosphate moiety and one of the 4 bases adenine (A), guanine (G), thymine (T) or cytosine (C), which are glycosidically linked via their N1 nitrogen to position 1′ of the sugar moiety. In DNA, nucleotides are linked together as a sugar phosphodiester polymer by a phosphate bridge between the 5′-C atom of one and the 3′-C atom of a second nucleotide. Such polymers can hybridize in a highly specific manner to a supramolecular double helix, if the base sequence allows for the sequential formation of either 2 hydrogen-bonded AT or 3 hydrogen-bonded GC base pairs. As a consequence, only two single-strand polymers of DNA which are (largely) complementary in their nucleotide sequence may form a double helix. DNA isolated from organisms has the following properties: 1) its molar mass is extremely high, 2) the genetic information is stored in its linear sequence of nucleotides, 3) the two single-strand polymers are unidirectional, i. e., they each contain one 3′- and one 5′-terminus, and 4) both strands may serve as a template for transferring its sequence information to a copy with a complementary nucleotide sequence.

Structure of DNA in organisms. The total DNA of an organism is termed its genome. The large size of the DNA molecule and its important function in the storage and inheritance of genetic information requires special subcellular structures. In higher organisms DNA is usually distributed among several chromosomes whose number does not depend on genome size. Thus, bakers' yeast (*Saccharomyces cerevisiae*) contains 16, the fruit fly (*Drosophila melanogaster*) has 4 chromosomes, and the much larger quantity of DNA in humans is distributed among 23 chromosomes. The chromosomes are localized in the cell nucleus, where they form chromatin [a complex of DNA with basic proteins (histones)]. The length of DNA is usually given as the number of base pairs (bp). Human chromosome 3, for example, contains 2 DNA strands of ca. 160 million bases each. Since 3×10^6 bp have a calculated length of 1 mm, the extended length of the DNA double helix of chromosome 3 would be ca. 5 cm. The human DNA of all 23 chromosomes in a single haploid egg or sperm cell has a combined mass of ca. 3×10^9 bp, corresponding to a molar mass of $M_R = 1.8 \times 10^{12}$ and a calculated length of ca. 1 m. The ca. 10^{13} standard diploid cells of our body contain a double set of 46 chromosomes, equivalent to ca. 2 m of DNA per cell. During each cell division, all 46 double strands are replicated and packaged again into 46 chromosomes in the daughter cells. The DNA of prokaryotes exhibits simpler packaging. Because prokaryotes do not contain a cell nucleus, their DNA molecule is contained in a subcellular region of the cytoplasm, the nucleoid. Even the genome of *Escherichia coli*, a single circular DNA double helix of M_R 2.4×10^9, would have an extended length of 1.5 mm. The replication of this huge molecule in *E. coli* under favorable growth conditions does not take more than 20 min, the doubling time of this organism. It proceeds with extremely high fidelity (error rate ca. 10^{-7} per gene of 1000 bp).

From a chromosome to the double helix

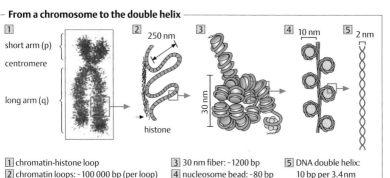

| 1 | short arm (p), centromere, long arm (q) | 2 | 250 nm, histone | 3 | 30 nm | 4 | 10 nm | 5 | 2 nm |

1 chromatin-histone loop
2 chromatin loops: ~100 000 bp (per loop)
3 30 nm fiber: ~1200 bp
4 nucleosome bead: ~80 bp
5 DNA double helix: 10 bp per 3.4 nm

Structure of DNA

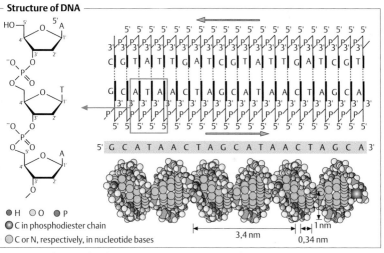

```
5' 5' 5' 5' 5' 5' 5' 5' 5' 5' 5' 5' 5' 5' 5' 5' 5' 5' 5' 5'
3' 3' 3' 3' 3' 3' 3' 3' 3' 3' 3' 3' 3' 3' 3' 3' 3' 3' 3' 3'
C  G  T  A  T  T  G  A  T  C  G  T  A  T  T  T  G  A  T  C  G  T
G  C  A  T  A  A  C  T  A  G  C  A  T  A  A  C  T  A  G  C  A
3' 3' 3' 3' 3' 3' 3' 3' 3' 3' 3' 3' 3' 3' 3' 3' 3' 3' 3' 3'
5' 5' 5' 5' 5' 5' 5' 5' 5' 5' 5' 5' 5' 5' 5' 5' 5' 5' 5' 5'
```

5' G C A T A A C T A G C A T A A C T A G C A 3'

● H ○ O ● P
◉ C in phosphodiester chain
○ C or N, respectively, in nucleotide bases

3,4 nm   1 nm   0,34 nm

Purine and pyrimidine bases

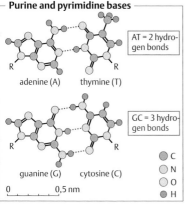

AT = 2 hydro-gen bonds

adenine (A)   thymine (T)

GC = 3 hydro-gen bonds

guanine (G)   cytosine (C)

○ C
○ N
○ O
○ H

0   0,5 nm

Bacterial DNA

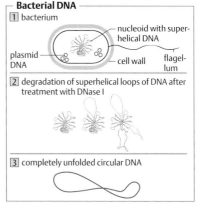

1 bacterium

nucleoid with super-helical DNA
plasmid DNA
cell wall
flagel-lum

2 degradation of superhelical loops of DNA after treatment with DNase I

3 completely unfolded circular DNA

217

DNA: function

General. The information stored in DNA codes for the biosynthesis of proteins. In prokaryotic organisms, protein biosynthesis occurs in two steps: transcription of a DNA segment (usually a gene) into messenger RNA (mRNA), a polyribonucleic acid, and translation of mRNA into a protein sequence, using the machinery of the ribosomes. In eukaryotic organisms, this process is more complex. First, only part of the DNA, the exons, codes for the synthesis of proteins. In higher organisms, the noncoding sequences of the DNA (introns) may exceed exon DNA by a factor of 10 or more in length; the function of introns is still unclear. Within the cell nucleus, the DNA is transcribed into mRNA (primary transcript), followed by removal of the intron part of the transcript by splicing. The spliced, mature mRNA now leaves the nucleus and attaches to the ribosomes, which are located in the cytoplasm and on the endoplasmic reticulum. At the ribosomes, the protein chains are assembled according to the sequence information encoded in the mRNA. In some cases, the mRNA also encodes leader sequences, which direct a protein into a special compartment of the cell. Protein properties can be further modified through differential splicing during translation, removing one or several exons, or through posttranslational modification (e. g., glycosylation or phosphorylation). Such modifications in higher organisms may lead to a roughly 50-fold greater diversity of proteins than the number of coding genes; in humans, 30 000 – 40 000 genes may code for ca. 2 million protein variants, which differ among the types and age of cells. Such variations are analyzed by proteomics techniques.

The genetic code used by living organisms to translate DNA sequences into protein structures is nearly universal. Each triplet of nucleotides of coding DNA, through the transcribed mRNA sequence, leads to highly selective incorporation of one distinct amino acid into the growing peptide chain. Due to the universal character of the genetic code, it is in principle possible to transfer genetic information from one organism to another. The genetic transformation of a host organism with DNA of foreign origin is the basis of genetic engineering technology. In gene-transfer experiments, donor and host organism may show different preferences for some triplet codons. This problem relates to the fact that the genetic code is degenerate: there are more unique triplet codes (A, T, G and C: $4^3 = 64$) than amino acids (20), leading to a redundancy of up to 6 triplet sequences all coding for a single amino acid (e. g., UUA, UUG, CUU, CUC, CUA, and CUG all code for leucine). As shown in this example, the third base of the triplet contributes least to codon specificity (wobble hypothesis). However, the actual usage of a code type (in more precise words: the amount of each triplet-specific tRNA) differs among organisms and may lead to severe problems in gene transfer experiments. Due to the rapid advances in genome sequencing, however, the codon usage of more and more organisms has become available.

Synthetic oligonucleotides are required for many experiments in genetic engineering. They are used as probes for hybridization experiments and as primers (starter sequences for DNA polymerase) in polymerase chain reaction (PCR) experiments and for site-directed mutagenesis of proteins. A well-selected probe or primer hybridizes with a highly specific sequence of the DNA isolated from an organism or a cell and allows for the detection, cloning, and amplification of an individual gene sequence. If no DNA sequence of the protein to be cloned is available, a putative sequence may be derived by sequencing the purified protein. In such cases, the fact that the genetic code is degenerate necessitates the chemical synthesis of many putative primers ("degenerate probe/primer") for the cloning experiment. If the probe or primer mixture leads in fact to hybridization or amplification (preferred method of identification: Southern blot or gel electrophoresis after amplification), the desired DNA has been isolated and can be sequenced to derive its real nucleotide sequence.

Functions of DNA

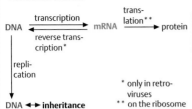

transcription and translation in a prokaryotic cell

* only in retroviruses
** on the ribosome

The genetic code

amino acid	codon	amino acid	codon	amino acid	codon	amino acid	codon
phenylalanine (F)	UUU	serine (S)	UCU	tyrosine (Y)	UAU	cysteine (C)	UGU
	UUC		UCC		UAC		UGC
leucine (L)	UUA		UCA	stop	UAA	stop	UGA
	UUG		UCG	stop	UAG	tryptophan (W)	UGG
leucine (L)	CUU	proline (P)	CCU	histidine (H)	CAU	arginine (R)	CGU
	CUC		CCC		CAC		CGC
	CUA		CCA	glutamine (Q)	CAA		CGA
	CUG		CCG		CAG		CGG
isoleucine (I)	AUU	threonine (T)	ACU	asparagine (N)	AAU	serine (S)	AGU
	AUC		ACC		AAC		AGC
	AUA		ACA	lysine (K)	AAA	arginine (R)	AGA
methionine (M)	AUG		ACG		AAG		AGG
valine (V)	GUU	alanine (A)	GCU	aspartic acid (D)	GAU	glycine (G)	GGU
	GUC		GCC		GAC		GGC
	GUA		GCA	glutamic acid (E)	GAA		GGA
	GUG		GCG		GAG		GGG

Primer design

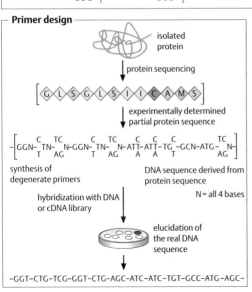

isolated protein

protein sequencing

experimentally determined partial protein sequence

synthesis of degenerate primers

DNA sequence derived from protein sequence

N = all 4 bases

hybridization with DNA or cDNA library

elucidation of the real DNA sequence

–GGT–CTG–TCG–GGT–CTG–AGC–ATC–ATC–TGT–GCC–ATG–AGC–

Codon usage

amino acid	codon	frequency		
Glu, E	CAG	0,30	0,31	0,59
	CAA	0,70	0,69	0,41
Lys, K	AAG	0,24	0,43	0,60
	AAA	0,76	0,57	0,40
Pro, P	CCG	0,55	0,12	0,11
	CCA	0,20	0,42	0,27
	CCU	0,15	0,31	0,29
	CCC	0,10	0,15	0,33

■ E. coli
■ S. cerevisiae
■ man

Genetic engineering: general steps

General. Although genetic engineering has a wide range of applications, only a few fundamental steps are required to transfer and express foreign DNA in a host cell. These include 1) manipulation of DNA, in particular its isolation, amplification, enzymatic modification, characterization, sequencing, and chemical synthesis, and 2) cloning and expression of the DNA in pro- and eukaryotic cells.

Purification, enzymatic modification, and amplification. DNA is found in each cell in very small quantities (just one molecule in a prokaryotic cell and in each chromosome of a eukaryotic cell), but it can still be isolated in pure form by extraction. For the next steps required in genetic engineering, intact DNA is too large. Fortunately, enzymes have been found that selectively cut, modify, ligate, or amplify DNA. Other enzymes help to translate DNA sequences into mRNA in a test tube (in vitro). Shorter DNA fragments (< 100 bp) can be chemically synthesized, using a robotic device. For most tasks, the amount of DNA fragments that can be obtained from a cell is not enough. It is therefore very important that DNA segments up to a length of 1000 bp or more be amplified using the polymerase chain reaction (PCR). This procedure allows, for instance, recombining DNA fragments of different origins (e. g., from different organisms or from chemical synthesis).

Characterization and sequencing. DNA fragments can be characterized according to 1) its melting behavior (a DNA double helix containing a high amount of GC base pairs, with 3 H bonds, melts at a higher temperature than AT-rich DNA, with 2 H-bonds), 2) its molar mass, which is usually determined by gel electrophoresis, 3) its nucleotide sequence, 4) biological characterization of its functional elements, and 5) mapping of those sequences that can be enzymatically cut by diverse but selective restriction enzymes.

220 Cloning. A DNA sequence contained in the genome of a prokaryotic organism, usually a gene which codes for a protein, can be directly cloned. If a functional gene from a eukaryotic cell is to be cloned, the introns must first be removed from the DNA. This need is usually circumvented by starting with mature mRNA, i. e., mRNA after splicing, and transcribing it in vitro into complementary double-stranded DNA (cDNA), using the enzyme reverse transcriptase (RT), which is mostly isolated from mammalian retroviruses. Heat-stable DNA polymerase from the thermophile *Thermus thermophilus* can also be used in this reaction. In the presence of Mn^{2+}, this enzyme functions like a reverse transcriptase and accepts RNA as well as DNA as a template. The mRNA-DNA hybrid that results from this reaction is then degraded to single-stranded DNA using RNase and serves as a template in a standard PCR reaction to generate double-stranded DNA. Another feature of eukaryotic mRNA isolated from a cell can be exploited if one wants to amplify it in an unspecific manner: because all eukaryotic mRNA molecules carry 3′ sequences of polyadenine (polyA), a complementary polythymidine oligomer (polyT) can be used as a primer for enzymatic synthesis of the mRNA-DNA hybrid.

Expression. Several methods are available, and are described later, for cloning, replicating, and expressing (transcribing and translating) foreign DNA or cDNA in a host cell. Several types of bacteria are favored as host cells for the following reasons: 1) their genome consists of a single DNA double helix, 2) their molecular genetics has been thoroughly studied, 3) phages or plasmids are available that can be used as vectors, 4) they propagate rapidly and can be bred in large quantities in a bioreactor. The preferred species for the cloning and expression of foreign DNA are attenuated strains of *Escherichia coli*, e. g., *E. coli* K12. However, other bacteria, such as *Bacillus subtilis*, Lactobacilli, various Pseudomonas and Streptomyces strains, have also been successfully used. Higher organisms such as fungi, yeasts, mammalian or insect cells, whole animals or plants can also be transformed by foreign DNA.

Fundamental steps in genetic engineering

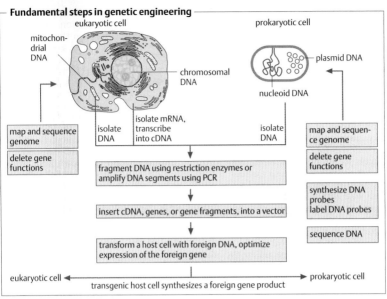

eukaryotic cell

prokaryotic cell

mitochondrial DNA

chromosomal DNA

plasmid DNA

nucleoid DNA

| map and sequence genome |
| delete gene functions |

isolate DNA

isolate mRNA, transcribe into cDNA

isolate DNA

| map and sequence genome |
| delete gene functions |

fragment DNA using restriction enzymes or amplify DNA segments using PCR

↓

insert cDNA, genes, or gene fragments, into a vector

↓

transform a host cell with foreign DNA, optimize expression of the foreign gene

| synthesize DNA probes |
| label DNA probes |

| sequence DNA |

eukaryotic cell ◄——— transgenic host cell synthesizes a foreign gene product ———► prokaryotic cell

Cloning of eukaryotic genes

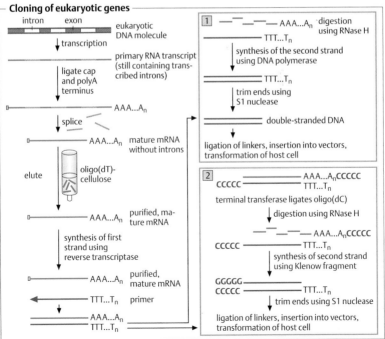

intron exon   eukaryotic DNA molecule

↓ transcription

primary RNA transcript (still containing transcribed introns)

↓ ligate cap and polyA terminus

AAA...A$_n$

↓ splice

AAA...A$_n$ mature mRNA without introns

oligo(dT)-cellulose

elute

AAA...A$_n$ purified, mature mRNA

↓ synthesis of first strand using reverse transcriptase

AAA...A$_n$ purified, mature mRNA

TTT...T$_n$ primer

AAA...A$_n$
TTT...T$_n$

1

— — — — AAA...A$_n$ digestion using RNase H
——————— TTT...T$_n$

↓ synthesis of the second strand using DNA polymerase

═══════════ TTT...T$_n$

↓ trim ends using S1 nuclease

═══════════ double-stranded DNA

↓

ligation of linkers, insertion into vectors, transformation of host cell

2

————— AAA...A$_n$CCCCC
CCCCC ————— TTT...T$_n$

terminal transferase ligates oligo(dC)

↓ digestion using RNase H

— — — AAA...A$_n$CCCCC
CCCCC ————— TTT...T$_n$

↓ synthesis of second strand using Klenow fragment

GGGGG
CCCCC ————— TTT...T$_n$

↓ trim ends using S1 nuclease

ligation of linkers, insertion into vectors, transformation of host cell

Preparation of DNA

General. The preparation, modification, and characterization of DNA from living cells is usually the first step in genetic engineering. Restriction endonucleases are an essential tool in such experiments. They are also used for constructing physical gene maps.

Preparation of DNA. Depending on the organism or cell type, DNA may occur in different forms, requiring various protocols for its isolation. Prokaryotic DNA is not contained in a cell nucleus. It can be isolated after enzymatic lysis of the cell wall with a lysozyme/detergent mixture (usually sodium dodecylsulfate, SDS), and denaturation of the cell proteins by phenol/chloroform. If this mixture is centrifuged, DNA can be precipitated from the supernatant by the addition of cold ethanol. Frequently, bacteria contain not only a single molecule of genomic DNA, but also plasmid DNA, which is of much lower molecular mass and is very important in genetic experimentation. If cell lysis is followed by the addition of NaOH and detergent, proteins are precipitated and chromosomal DNA is partially hydrolized. Under suitable conditions, plasmid DNA is preserved as a circular double strand and can be isolated, from the supernatant of a centrifugate, by centrifugation in a density gradient (sucrose or CsCl), by ethanol precipitation, or, much more simply, by anion-exchange chromatography. Depending on the amount of isolated DNA, this last procedure is termed miniprep (ca. 20 µg DNA) or maxiprep (several mg). Phage DNA is enriched from infected bacterial cultures by removing the lysed bacteria by centrifugation, precipitating the phage particles with polyethylene glycol, removing the phage capsid by phenol extraction, and precipitating the phage DNA from the supernatant with ethanol. Eukaryotic DNA is distributed among several chromosomes that are contained in the cell nucleus. Total DNA from animal cells is obtained by lysing the cells with detergent, digesting proteins and RNA with proteinase K

and RNase, removing detergent and proteins by salt precipitation, and, finally, precipitating DNA from the supernatant with ethanol. In many genetic experiments on eukaryotes, spliced mRNA is used as a source of genetic information. mRNA is isolated from organs or cell cultures by lysing the cells, removing the cell nuclei by centrifugation, removing cytoplasmic proteins by phenol extraction, and precipitating the mRNA from the supernatant. Mitochondria and the plastids of eukaryotic photosynthetic organisms contain their own DNA, which replicates independently of the chromosomal DNA. It can be isolated by analogous procedures from these organelles after their isolation by differential centrifugation.

Restriction enzymes (restriction endonucleases) are synthesized by bacteria as protection against foreign DNA (e. g., phage DNA) that may enter their cytoplasm. They are widely used in genetic engineering to selectively cut the long native DNA strands into smaller, well defined fragments. They bind to a recognition sequence of DNA that is 4–10 nucleotides long and then hydrolyze the DNA double strand within or near the recognition site so as to leave even (blunt) ends or partially single-stranded (sticky or cohesive) ends. If a purified DNA molecule is hydrolyzed by different types of restriction enzymes and the size of the fragments from each experiment is recorded by gel electrophoresis or DNA sequencing, a restriction map of the DNA is obtained, which can be used as a strategic basis for inserting and expressing foreign DNA. The frequency of cleavage sites for a given restriction enzyme can be statistically calculated from the length of the recognition sequence; thus, an enzyme with a 4-bp recognition site (e. g., *Alu* I, AGCT) will make cuts at $(\frac{1}{4})^4 = (\frac{1}{256})$, or once every 256 bp. However, the sequence information in DNA is not randomly distributed, so the observed cleavage frequency differs from this calculation. Recognition also depends on the G+C content of a DNA and thus can be used for characterizing an organism.

Isolation of bacterial DNA

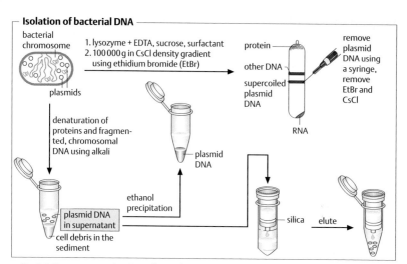

bacterial chromosome

1. lysozyme + EDTA, sucrose, surfactant
2. 100 000 g in CsCl density gradient using ethidium bromide (EtBr)

protein

other DNA

supercoiled plasmid DNA

RNA

remove plasmid DNA using a syringe, remove EtBr and CsCl

plasmids

denaturation of proteins and fragmented, chromosomal DNA using alkali

plasmid DNA

plasmid DNA in supernatant

cell debris in the sediment

ethanol precipitation

silica

elute

Restriction enzymes (endonucleases)

enzyme	organism	recognition sequence (5' → 3')	3'-end
Not I	*Nocardia otitidis-caviarum*	GCGGCCGC	sticky
Eco RI	*Escherichia coli*	GAATTC	sticky
Bam HI	*Bacillus amyloliquefaciens*	GGATCC	sticky
Bgl II	*Bacillus globigii*	AGATCT	sticky
Pvu I	*Proteus vulgaris*	CGATCG	sticky
Pvu II	*Proteus vulgaris*	CAGCTG	blunt
Hind III	*Haemophilus influenzae* R$_d$	AAGCTT	sticky
Hinf I	*Haemophilus influenzae* R$_f$	GANTC	sticky
Sau 3A	*Staphylococcus aureus*	GATC	sticky
Alu I	*Arthrobacter luteus*	AGCT	blunt
Taq I	*Thermus aquaticus*	TCGA	sticky

DNA complex of a restriction enzyme

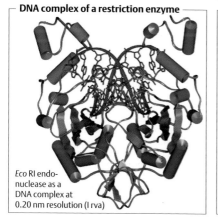

Eco RI endo-nuclease as a DNA complex at 0.20 nm resolution (I rva)

Cleavage frequency

cleavage of λ -DNA using	number of cleavage sites	
	calculated	found
Bgl II	12	6
Bam HI	12	5
Sal I	12	2

all 3 restriction enzymes recognize sequences of 6 base pairs. This leads to a statistical cleavage frequency of one every 4^6 or 4 096 base pairs, corresponding to 12 cleavage sites for the ~49 000 bp of λ phage. However, significantly fewer cleavage sites are observed experimentally, because the sequences required for hydrolysis are not randomly distributed over the phage genome

223

Other useful enzymes for DNA manipulation

General. For manipulation of DNA in vitro, the following enzymes have found wide use: 1) hydrolases that cut DNA at specific sequences, 2) lyases, which ligate DNA fragments to each other, 3) synthetases that polymerize a DNA double strand on a single-strand DNA template, 4) hydrolases and transferases that modify DNA at the 3' or 5' end, and 5) topoisomerases, which introduce or remove DNA supercoils.

Hydrolases. The most important hydrolases are the restriction endonucleases, which were discussed above. A wide choice of enzymes is available commercially, differing in their recognition sequences and in their ability to generate blunt or sticky ends (i.e., ends with a single-strand overhang a few nucleotides long). Another frequently used hydrolase is nuclease S1, obtained from *Aspergillus niger*: it cleaves single-stranded DNA and double-stranded DNA at single-stranded gaps.

Lyases, transferases. The most important enzyme of this group is DNA ligase. It functions in the cell as a repair enzyme: it repairs gaps that may have occurred in a double-stranded DNA molecule during replication, recombination, or accidental events. This important enzyme, which occurs in all cells, is used in genetic engineering to recombine DNA fragments in vitro (e. g., to insert foreign DNA into a DNA vector). DNA ligase can ligate both blunt and sticky ends. However, because the ligation of sticky ends is much more effective (because the single-stranded sequence facilitates hybridization with the complementary sequence to be ligated), blunt ends are often transformed into sticky ends before ligation. This can be done by using linkers or adapters or by adding polymer tails (tailing) in the presence of the enzyme terminal deoxynucleotidyl transferase. A detailed description of all individual steps used in such protocols can be found in the pertinent literature. DNA ligase is usually isolated from *E. coli* cells that were infected with bacteriophage T4 (T4 DNA ligase), terminal transferase is obtained from calf spleen.

Synthetases. The two most important synthetases are DNA polymerase I and reverse transcriptase. DNA polymerase I is obtained from *E. coli*. In the presence of a single-stranded DNA primer, it adheres to single-stranded DNA and synthesizes the second strand in the 5'→3' direction. In addition to its polymerase activity, it also shows 3'→5' and 5'→3' exonuclease activity. It also recognizes short stretches of single-stranded DNA (gaps) in an otherwise double-stranded DNA molecule, using them as a template for polymerization. Removal of the first 323 amino acids of this enzyme results in loss of the 5'→3' nuclease activity, but the resulting enzyme fragment (Klenow fragment) still has polymerase and 3'→5' nuclease activity and can fill in single-stranded gaps in an otherwise double-stranded DNA. The Klenow fragment is often used to introduce radioactive labels into DNA, a useful technique for visualizing traces of DNA by autoradiography (by exposure to an x-ray film). In labeling experiments, the preferred method is to use sticky 5' ends of the DNA fragment as a template, complementing it with ^{32}P-labelled deoxy nucleotides in the presence of Klenow fragment. DNA polymerases from thermophilic microorganisms are widely used in PCR reactions. Reverse transcriptase (RT) is a key enzyme in retroviruses. It uses viral RNA as a template in synthesizing complementary DNA in a host cell. This property can be used in genetic engineering to synthesize cDNA from mRNA. RT is isolated from animal cells that have been infected with retroviruses such as the monkey myoblastoma virus (MMV) or the Moloney mouse leukemia virus (MMLV).

Endgroup-modifying enzymes. It is often necessary during cloning experiments to add or remove a terminal phosphate group at the 5' position of a DNA fragment. For this purpose, alkaline phosphatases and polynucleotide kinase are commercially available.

Topoisomerases are an interesting type of enzyme as they are able to modify the conformation of circular DNA by introducing or removing supercoils; at present, however, they are of little practical use in genetic engineering.

Enzymes for the manipulation of DNA

enzymes used in	function
DNA degradation	
restriction enzyme, endonuclease, [1]	cleaves internal phosphodiester bonds
nuclease S1 [2]	cleaves single-stranded DNA or single-stranded gaps
synthesis or ligation of DNA	
DNA ligase [3]	repairs single-stranded breakages, ligates two DNA molecules
DNA polymerase I [4]	synthesizes double strand, fills gap in single strand
Klenow fragment [5]	fills gaps in single strand, no 5'–3' exonuclease activity
reverse transcriptase [6]	synthesizes DNA on a RNA matrix
modification of end groups	
alkaline phosphatase	removes phosphate group from 5' end
polynucleotide kinase	adds phosphate group to 5' end
terminal deoxynucleotide transferase	adds phosphate group to 3' end

Enzymes for DNA cleavage

[1] restriction enzymes

```
–N–N–A–G–C–T–N–N–        AluI      –N–N–A–G        C–T–N–N–
–N–N–T–C–G–A–N–N–      ------->    –N–N–T–C        G–A–N–N–
„N" = A, G, C or T                             blunt ends
```

```
–N–N–G–A–A–T–T–C–N–N–      EcoRI      –N–N–G              A–A–T–T–C–N–N–
–N–N–C–T–T–A–A–G–N–N–    ------->    –N–N–C–T–T–A–A  ↑        G–N–N–
                                                  ↖ sticky ends
```

[2] nuclease S1

single-stranded gaps

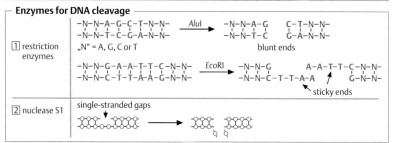

Enzymes for DNA ligation

[3] DNA ligase

repair of single-stranded nicks — DNA ligase → or

ligation of blunt ends — DNA ligase →

ligation of sticky ends — DNA ligase →

Enzymes for polymerization of DNA on a matrix

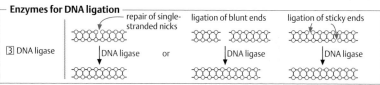

[4] DNA polymerase I

```
5'–A–T–G–C–A–A–T–G–C–A–T–ᶟ'      5'–A–T–G–C–A–A–T–G–C–A–T–³'
                G–C–T–A–        3'–T–A–C–G–T–T–A–C–G–T–A–⁵'
DNA matrix   primer  3'   5'              newly synthesized strand
```

```
–A–T–G–C–A–A–T–G–C–A–T–        –A–T–G–C–A–A–T–G–C–A–T–
–T–A–C–G          G–T–A–        –T–A–C–G–T–T–A–C–G–T–A–
        single-stranded gap    nucleotides are replaced   gap is
                               (exonuclease activity)     filled in
```

[5] Klenow fragment

```
–A–T–G–C–A–A–T–G–C–A–T–        –A–T–G–C–A–A–T–G–C–A–T–
–T–A–C–G          G–T–A–   →   –T–A–C–G–T–T–A–C–G–T–A–
                              only the gap is filled (no 5', 3'-exonuclease activity)
```

[6] reverse transcriptase

```
–A–T–G–C–A–A–T–G–C–A–T–   →   –A–T–G–C–A–A–T–G–C–A–T–
              G–T–A–     RNA  –T–A–C–G–T–T–A–C–G–T–A–
RNA template                 DNA    new DNA strand
```

PCR: general method

General. The polymerase chain reaction (PCR) is one of the most important practical developments in genetic engineering, and its inventor, Kary Mullis, was honored with a Nobel Prize. Using PCR, any short sequence of DNA, e. g., a gene or gene fragment, can be amplified many times by using DNA polymerase. This technique is most valuable for the identification of genes and for their manipulation.

Method. The standard protocol requires two oligonucleotides (primers), which bind to either end of the DNA target sequence that is to be amplified (one for each DNA strand). This implies that these DNA sequences are either already known or that they can be determined from a protein sequence (if so, the ambiguities of the degenerate genetic code must be considered). Besides the DNA template and the two primers, a mixture of the 4 deoxynucleotides and Taq DNA polymerase is required. The PCR reaction then proceeds in three steps: 1) at 94 °C, the DNA double strand is melted, forming two single-stranded DNA molecules (denaturation), 2) after lowering the temperature to 40–60 °C, the two primers each hybridize to a DNA strand (annealing), and 3) after increasing the temperature to 72 °C, two new complementary strands are formed (extension). If heating to 94 °C is repeated, the newly formed DNA strands come off their DNA template, and the reaction cycle is repeated after cooling. With an automatic thermocycler, this cycle is repeated 25–40 times (a few seconds to several minutes per cycle, depending on the template), leading to amplification of the original DNA to between 2^{25} and 2^{40} copies within just a few hours. A condition for the reaction is that the DNA polymerase used must tolerate the high temperature required for melting the two DNA strands without becoming inactive. DNA polymerase from thermophilic microorganisms, e. g.. from $Thermus\ aquaticus$, $Pyrococcus$ $furiosus$, or $Thermotoga\ maritima$ (Taq, Pfu, or Tma-polymerase) have such properties.

The error rate (mutation frequency per bp per amplification cycle) of Taq polymerase is about 8×10^{-6}. The two other polymerases are even more precise, because they also have proofreading activity. The molar mass and yield of PCR products are determined by gel electrophoresis or, in real time, by including reporter groups such as SYBR Green (light cyclers) – a fluorescent dye that binds to double-stranded DNA. As the amount of DNA produced in a PCR reaction increases, the amount of fluorescence from the dye increases proportionally, and the orginal amount of DNA can be obtained by extrapolation.

Practical applications. Using PCR, defined sequences of DNA can rapidly be cloned and sequenced. Because even single DNA molecules can be amplified by PCR, as has been shown by the amplification of DNA from a single sperm cell, the method has been applied in archeology and paleontology. In clinical diagnostics, it can be advantageously used, once a correlation between a DNA sequence and a disease has been established. This is already true for many infectious and, increasingly, also for genetically determined diseases. In the fields of food and environmental analysis, PCR methods help to identify traces of transgenic plant materials or of pathogens. Once the consensus sequences of a protein family are known, primers can be designed to help identify still unknown members of the family (reverse genetics). By using modified primers or deliberately increasing the error rate of a PCR reaction, defined or statistical mutations can be introduced into a protein. RNA is also amenable to PCR analysis after it has been transcribed into cDNA. It can be amplified (RT-PCR), and then used to, for example, determine 1) the load of RNA viruses in a cell (e. g., HIV virus), or 2) the relative quantities of mRNA in a cell. PCR techniques have been successfully miniaturized to proceed in micromachined capillaries ("PCR on a chip"). Using appropriate microdevices, a desired sequence of DNA may be amplified by a factor of 2^{20} in < 1 h, to be further used for diagnostic assays, e. g., in a DNA array.

Polymerase chain reaction (PCR)

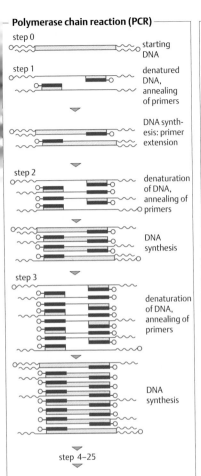

step 0 — starting DNA

step 1 — denatured DNA, annealing of primers

DNA synthesis: primer extension

step 2 — denaturation of DNA, annealing of primers

DNA synthesis

step 3 — denaturation of DNA, annealing of primers

DNA synthesis

step 4–25

analysis of PCR products	
method	**analysis**
polyacrylamide gel electrophoresis	ethidium bromide staining (UV light, image analysis)
agarose gel electrophoresis	hybridization with labeled probe (Southern blot) incorporation of radioactive tracer, e. g., ^{32}P phosphate silver staining
HPLC	UV analysis
gel electrophoresis or HPLC after degradation with restriction enzymes	see above

Functional principle of a thermocycler

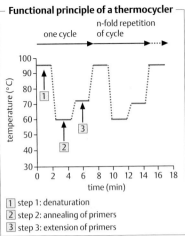

one cycle → n-fold repetition of cycle

1 step 1: denaturation
2 step 2: annealing of primers
3 step 3: extension of primers

Thermocycler

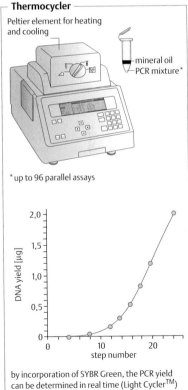

Peltier element for heating and cooling

mineral oil
PCR mixture*

* up to 96 parallel assays

by incorporation of SYBR Green, the PCR yield can be determined in real time (Light Cycler™)

227

PCR: laboratory methods

General. PCR is a key protocol for a wide range of molecular genetic experiments. Only a limited number of applications is discussed here, namely: 1) incorporation of functional elements into DNA, 2) amplification of RNA (RT-PCR), 3) fusion of two DNA fragments, 4) incorporation of new DNA fragments into a gene, 5) removal of a DNA fragment from a gene, 6) site-directed mutagenesis of a gene, and 7) multiplex PCR. PCR with degenerate primers is a standard procedure for isolating a gene whose sequence is not precisely known.

Incorporation of functional elements. Functional elements of DNA include cloning sites (recognition sequences for restriction enzymes), tags (sequences encoding, for example, an N- or C-terminal oligo histidine, allowing for rapid purification of the translated protein by metal affinity chromatography), and start- or stop-codons.

Amplification of mRNA (RT-PCR). mRNA can be amplified by PCR after its (partial) sequence is known directly or from the corresponding protein sequence. A pertinent primer is synthesized, annealed to mRNA isolated from a cell, and translated into the first single strand of cDNA using reverse transcriptase and a nucleotide mixture.

Fusion of two DNA fragments. If two fragments of genes are to be fused, the desired sequences are amplified in two separate PCR steps, using a set of two primers. This results in PCR products that contain identical sequences at the desired fusion positions. In a third PCR reaction, the two PCR products are used as template, adding the terminal primers, which leads to hybridization of the complementary strands of the identical sequences and amplification of the fusion product. In this protocol it is important that the reading frame for the desired triplet is correctly chosen. It also may be necessary to insert a spacer between the two coding genes, e. g., a sequence coding for polyalanine. Such spacers may help preserve the free mobility of each of the two fused proteins (single-chain antibodies are a relevant example).

Insertion or removal of a gene segment. By analogy to gene fusion, a skilful choice of primers for internal or terminal sequences may lead to truncated DNA (and proteins), from which a desired segment has been deleted.

Site-directed mutagenesis is a very useful technique, for example, for elucidating enzyme mechanisms or for the targeted modification of an enzyme's substrate specificity. An older method for site-directed mutagenesis is based on introducing mutations into the single-stranded DNA of the M13 phage. It has been completely replaced by PCR protocols. Because DNA fragments can hybridize even when there is a mismatch between single nucleotides, a modified triplet code leading to the desired amino acid substitution can be introduced into any position of the DNA under study and can be amplified by PCR. Another method uses two complementary oligonucleotides, carrying the mutation, and a double-stranded plasmid, composed of permethylated DNA as a template. Using suitable primers and Pfu polymerase, the complete plasmid is amplified in vitro. The methylated template DNA (the DNA amplified in vitro is not methylated) is removed by digestion with the restriction endonuclease *Dpn*I, which hydrolyzes only methylated DNA. The newly synthesized DNA that carries the mutation can be directly transformed into *E. coli*, and time-consuming cloning steps are not longer necessary. Mutation kits based on this principle are commercially available.

Multiplex PCR. It is possible to amplify a few sequences simultaneously in a single PCR, by combining suitable primer pairs.

PCR with degenerate primers. Degenerate primers are families of homologous sequences of a single-strand DNA in which one or several of the nucleotides may be any of the 4 bases. This allows genes whose precise sequence is unknown to be cloned, e. g., if the putative sequence was derived from a protein sequence and the codon usage is uncertain, or if genes of a multigene family are to be cloned. Deoxyinosine pairs with all other bases and thus can also be used as a "universal base" in a degenerate primer.

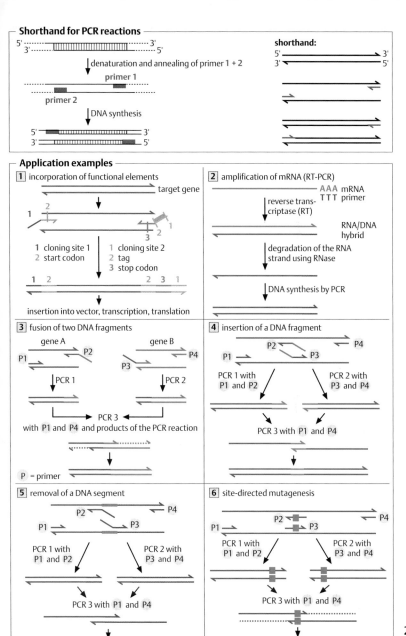

Shorthand for PCR reactions

5' ·········||||||||||||||||·········3'
3' ·········||||||||||||||||·········5'

shorthand:

5' ————————→ 3'
3' ←———————— 5'

↓ denaturation and annealing of primer 1 + 2

primer 1

primer 2

↓ DNA synthesis

5' ████|||||||||||████ 3'
3' ████|||||||||||████ 5'

Application examples

1 incorporation of functional elements

target gene

↓

1 cloning site 1   1 cloning site 2
2 start codon   2 tag
                   3 stop codon

↓

insertion into vector, transcription, translation

2 amplification of mRNA (RT-PCR)

AAA mRNA
TTT primer

↓ reverse trans-
  criptase (RT)

RNA/DNA hybrid

↓ degradation of the RNA
  strand using RNase

↓ DNA synthesis by PCR

3 fusion of two DNA fragments

gene A   gene B

P1   P2   P3   P4

↓ PCR 1   ↓ PCR 2

└———→ PCR 3 ←———┘
with P1 and P4 and products of the PCR reaction

↓

P = primer

4 insertion of a DNA fragment

P2   P4
P1   P3

PCR 1 with   PCR 2 with
P1 and P2   P3 and P4

↓   ↓

PCR 3 with P1 and P4

↓

5 removal of a DNA segment

P2   P4
P1   P3

PCR 1 with   PCR 2 with
P1 and P2   P3 and P4

PCR 3 with P1 and P4

↓

6 site-directed mutagenesis

P2   P4
P1   P3

PCR 1 with   PCR 2 with
P1 and P2   P3 and P4

↓

PCR 3 with P1 and P4

↓

229

DNA: synthesis and size determination

General. Short single-stranded DNA fragments up to ca. 100 bp (oligonucleotides) are chemicals that can be synthesized simply, quickly, and economically in the laboratory. They are useful for various steps in genetic engineering, e. g., as primers for PCR. For molar mass determination of DNA fragments up to ca. 30 kbp, gel electrophoresis is used and standardized with DNA fragments of known M_R.

DNA synthesis. The method of choice is the phosphoamidite method, which is usually carried out in an automated synthesizer. All 4 nucleotide bases (A, C, G, and T) are present as phosphoamidites, in which the 3′ phosphite group is protected by diisopropylamine and a methyl group. The 5′ hydroxy group of the deoxyribose and the amino groups of the purine and pyrimidine bases are also protected. Nucleoside 1 is now bound to an insoluble carrier material. Chemical unblocking of the 5′ hydroxy group leads to a nucleophilic attack on the tetrazol-activated phosphoamidite group of nucleotide 2. The resulting phosphotriester bond is now oxidized to a 5-valence phosphate ester, using iodine. This cycle, which in contrast to the biosynthesis of DNA proceeds from 3′ to 5′, is repeated for each base. After the complete DNA fragment has been synthesized, all protecting groups are removed, and the single-stranded oligonucleotide is purified by gel electrophoresis or HPLC. Even if a 98 % yield is achieved in each reaction cycle, the total yield for a 20mer oligonucleotide is only 67 %, and for a 40mer, only 45 %, resulting in DNA mixtures that are difficult to analyze and purify. For the synthesis of longer DNA segments or whole genes, complex strategies are necessary and are usually based on PCR. Oligonucleotides are mostly used 1) for synthesis of gene fragments or short genes, 2) as probes or primers for the identification or isolation of gene fragments from genomic or cDNA using hybridization or PCR, 3) for site-directed mutagenesis of a gene, and 4) for DNA sequencing. DNA synthesis is usually carried out by specialized laboratories which provide good quality and fast delivery at an acceptable price.

Size determination of DNA. Due to its net negative charge, DNA is easily separated by gel electrophoresis. Gels usually consist of agarose (large pore size), polyacrylamide (small pore size), or mixtures of both materials that allow one to define a mesh size that permits rapid analysis of the distribution of molar masses in a mixture of DNA fragments up to a size of ca. 30 kbp with high precision. In most protocols, denaturing conditions are used (SDS-PAGE): if electrophoresis in a polyacrylamide gel (PAGE) is carried out in the presence of the surfactant sodium dodecyl sulfate (SDS), the mobility of single-stranded DNA depends only on its molecular mass, because the formation of secondary structures and intermolecular aggregates is prevented. Detection of DNA in a gel is done either by staining with ethidium bromide, by autoradiography, using radioactive labels, or by labeling with luminescence markers such as rhodamine:luminol. Ethidium bromide is mostly used; however, it is genotoxic and must be used under appropriate safety conditions. The sensitivity of the ethidium bromide method is limited to > 25 ng DNA. DNA labeled with ^{32}P, using ^{32}P-labeled ATP and nick-translation with DNA polymerase I, can be detected at much lower concentrations, but requires radiation-safety equipment and routine monitoring for contamination. As a result, less demanding protocols based on fluorescent dyes are increasingly being used, e. g., SYBR Green, which is 25–100 fold more sensitive than ethidium bromide and allows a detection limit of > 250 pg DNA. Analysis is carried out in a phosphoimager after excitation with UV light. The M_R of a fragment can be calculated from its migration distance, but usually a set of DNA markers of various M_R is used for this purpose. Analysis of the molar mass of DNA may be important for restriction analysis of unknown DNA fragments, for constructing restriction maps, and for the identification of genes and gene fragments from chromosomal, plasmid, or viral DNA after PCR cloning.

Chemical solid-phase synthesis of DNA

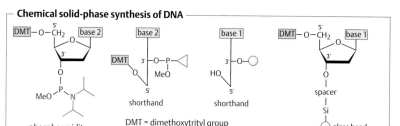

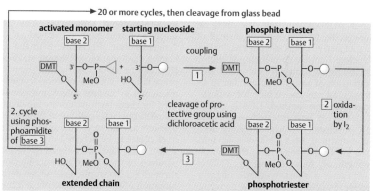

DMT = dimethoxytrityl group
base = adenine, thymine, guanine, cytosine

Determination of DNA molar mass using gel electrophoresis

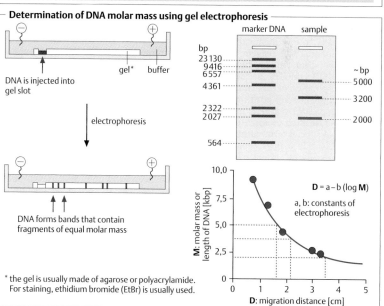

* the gel is usually made of agarose or polyacrylamide.
For staining, ethidium bromide (EtBr) is usually used.

DNA sequencing

General. Two alternative methods can be used to sequence DNA: the Sanger-Coulson and the Maxam-Gilbert procedures. Both permit the sequencing of single-stranded DNA fragments up to a length of ca. 600 bp. The sequence of longer DNA stretches must be derived from overlapping shorter fragments. When very long DNA fragments are sequenced, as in genome sequencing, highly automated methods are used. They rely on base-specific fluorescent dyes instead of the radioactive labels used in classical sequencing. Genome sequencing is highly demanding in terms of computer-based comparison of a very large number of sequences (an exercise in bioinformatics).

Sanger-Coulson method. DNA is cloned into an *E. coli* host infected with phage M13, resulting in phage progeny with single-stranded DNA. It serves as template for sequencing, using the Klenow fragment or, more frequently, T7-DNA polymerase, a short synthetic oligonucleotide as primer, and the 4 deoxynucleotides dATP, dTTP, dGTP, and dCTP as substrates. Double-stranded DNA is synthesized along the single-stranded DNA template. To four identical reaction vessels containing this reagent mixture, one of the four dideoxynucleotides (ddATP, ddTTP, ddGTP, or ddCTP) is added. Incorporation of these nucleotides at their complementary positions causes termination of DNA synthesis in a statistical manner, leading to a mixture of all possible DNA species terminating at each of these nucleotide analogs. Separation of this DNA mixture by gel electrophoresis allows the molar mass of the fragments to be identified, and, implicitly, the DNA sequence. Visualization on the gel is usually accomplished by autoradiography, after addition of a [32]P- or ^{35}S-labeled nucleotide to the reaction mixtures.

Maxam-Gilbert method. This procedure is used less today. It is based on the partial chemical hydrolysis of double-stranded DNA in 4 independent chemical reaction sequences, after the labeling of one terminus. Each base-specific reaction involves several steps (e. g., treatment with formic acid, dimethyl sulfate, hydrazine, etc.) and leads to (partially) selective cleavage at this base in the DNA strand, resulting in a family of terminally labeled DNA fragments which, as in the Sanger-Coulson method, are separated by gel electrophoresis and visualized by autoradiography. This method can also be run automatically, using a solid-phase procedure and labeling the terminal nucleotide with a fluorescent marker.

High-throughput sequencing. The presently preferred method is based on the Sanger-Coulson procedure, with the following modifications: 1) double-stranded DNA can be sequenced by using specific primers in a PCR-type reaction (cycle sequencing), 2) the four dideoxynucleotides used for chain termination are labeled by coupling one of four different fluorescent markers to each base. This allows all four nucleotides to be detected in a single reaction assay and, after time-resolved separation of the DNA fragments by flow-through gel electrophoresis, allows determination of the molar mass of each fragment, leading directly to the DNA sequence. The read length is ca. 900 bp, the duration of one cycle is 13 h plus 2 h of setup time. In commercially available instruments with 96 parallel electrophoresis lanes, the sequencing capacity is thus slightly less than 100 000 b in 15 h. If capillary electrophoresis is used instead of gel electrophoresis, the read length is reduced to ca. 650 b, but the separation time is only 3 h plus 1 h of setup time. Thus, with a commercial capillary sequencer having 96 capillaries, 65 000 nucleotides can be sequenced in 4 h, or ca. 400 000 per day, in one instrument. Sequencers with 384 capillaries are now also available. To correct for reading errors, multiple sequence determinations are usually carried out; still, a modern sequencing factory-lab using dozens or even hundreds of high-performance robots, in combination with supercomputers for sequence alignment, allows the genome sequences of whole organisms to be completed in a short time.

DNA sequencing after Sanger and Coulson

a annealing of the primer

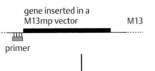

gene inserted in a
M13mp vector   M13

primer

DNA polymerase
dATP, dTTP, dGTP, dCTP   **dideoxy-ATP (ddATP)**

^{32}P- or ^{35}S for autoradiography

b example: synthesis of strands in lane 1
addition of dideoxy-ATP

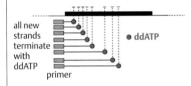

all new
strands
terminate
with
ddATP

primer

• **ddATP**

ddTTP has been added in lane 2,
ddGTP in lane 3, and ddCTP in lane 4.

c gel electrophoresis and autoradiography

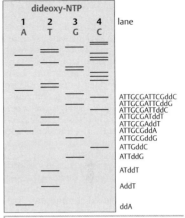

| dideoxy-NTP | | | | lane |
| 1 A | 2 T | 3 G | 4 C | |

ATTGCGATTCGddC
ATTGCGATTCddG
ATTGCGATTddC
ATTGCGATddT
ATTGCGAddT
ATTGCGddA
ATTGCGddG
ATTGddC
ATTddG
ATddT
AddT
ddA

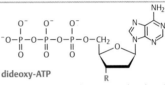

dideoxy-ATP

R = OH   deoxy-ATP (dATP), normal nucleotide,
strand synthesis

R = H   dideoxy-ATP (ddATP), termination of
strand elongation

High-throughput sequencing

DNA single strand by PCR

primer

DNA polymerase,
dATP, dTTP, dGTP, dCTP
ddATP ddTTP ddGTP ddCTP*

DNA strands of various chain lengths
with terminal fluorescence marker

↓

separation by electrophoresis,
time-resolved detection

buffer with
DNA strands

polyacrylamide gel or capillary
electrophoresis, up to 384 lanes
or capillaries in parallel

flow
direction

time-resolved
detection

laser ········· detector

CGTTG
GGCTG
AAGCT

*using dideoxynucleotides with base-specific fluorescence marker for chain termination

time-resolved detector signal

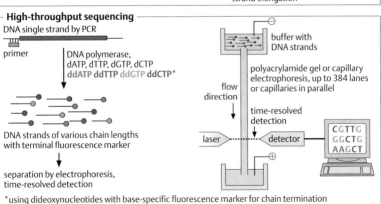

AATACGACTCACTATAGGGCGAATTCGAGCTCGGTACCCGGGGATCCTCTAGAGTCGACCTGCAGGCATGCAAGCTTTTATTCCTCTAGGATAAATGCCCAAGTGTACTCTTGTTGGCTTT
        10   20   30   40   50   60   70   80   90   100   110   120

233

Transfer of foreign DNA in living cells (transformation)

General. In nature, DNA is transferred into living cells in various ways: 1) transfer by plasmids, phages, or viruses (conjugation, transduction, transfection), or 2) direct uptake (transformation). Cells that have incorporated foreign DNA are called transformed cells. In genetic engineering experiments, foreign (heterologous) DNA is usually transferred and expressed in a host cell. The transformation methods used are partially of biological, partially of technical origin.

Plasmids occur nearly exclusively in bacteria. Most are circular double-stranded DNA molecules that replicate independently of the bacterial chromosome, but they can integrate into chromosomal DNA (episomes). Plasmids contain an origin of replication and, nearly always, one or several genes that are advantageous to the bacterium, e. g., a gene coding for antibiotic resistance. Plasmid DNA can be easily separated from chromosomal DNA and, like the latter, can be manipulated with enzymes in vitro. This has led to a wide range of cloning and expression vectors for use in genetic engineering. The most important functional properties of a plasmid vector are 1) an origin of replication (*ori*) for replication in the host organism; 2) optional: origins of replication for other host organisms (shuttle plasmids) – this allows construction of an appropriate vector in an easy-to-handle organism such as *E. coli* before transferring the genetic information to the desired, but more complex, host; 3) unique restriction sequences for inserting a gene only at the desired position (MCS, multiple cloning sites); 4) one or several resistance or auxotrophic markers for selection of positive recipient clones. Reporter genes facilitate the screening of transformed clones. For example, in the "blue-white" screening often used in combination with pUC plasmids, the plasmid-coded *lacZ'* gene complements the chromosomal *lacZ* gene of an *E. coli* host strain that lacks the *lacZ'* gene sequence (deletion lacZΔM15); only

transformed *E. coli* clones can synthesize functional β-galactosidase, which in turn hydrolyzes the leuko dye 5-bromo-4-chloro-3-indolyl-β-D-galactopyranoside (X-gal), forming dark blue 5,5'-dibromo-4,4'-dichloro indigo. Plasmid vectors are usually smaller than 10 kbp, to facilitate manipulation and prevent their elimination from dividing cells due to negative selection pressure. Most plasmid vectors have been developed for *E. coli*, but plasmids useful in cloning experiments have also become available for Bacillus, Pseudomonas, Streptomyces, Lactobacillus, and some other bacteria. Plasmids are rare in eukaryotes. One of the few exceptions is the 2μ plasmid of *Saccharomyces cerevisiae*. The Ti plasmid derived from the soil bacterium *Agrobacterium tumefaciens* has become an important vector for transforming dicotyledon plants.

Bacteriophages and viruses permit the transfer of DNA into a host cell by transfection. Phage and viral vectors are attenuated by removing gene segments that are responsible for cell lysis or other mechanisms of pathogenicity. Specific phages are known for most bacterial species; many of them are used in genetic engineering, e. g., the λ and M13 phages for experimentation with *E. coli*. A small number of vectors based on attenuated viruses are also available for transforming plant and animal cells.

Nonbiological methods comprise chemical and physical procedures. Biolistics was mentioned in the section on transformation of plant cells. For transformation of animal cells or plant protoplasts (both contain no cell wall) DNA can be precipitated as the Ca salt on the surface of the cells, initiating endocytosis. Electroporation uses a short electrical pulse that leads to the transient formation of pores in the cell membrane, resulting in uptake of the DNA. Other procedures include fusion of cells with liposomes containing DNA (lipofection) and microinjection of DNA into the nucleus of eukaryotic cells. With these methods, the number of transformed cells remains small, and the procedure must be optimized for each experimental setup.

Plasmids

	example	size [kbp]	occurrence
f(ertility) plasmids	F-plasmid	95	*Escherichia coli*
r(esistance) plasmids	RP4	54	*Pseudomonas* sp.
toxin (colicin) plasmids	ColE1	6.4	*Escherichia coli*
degradative plasmids	TOL	117	*Pseudomonas putida*
virulence plasmids	Ti plasmid	213	*Agrobacterium tumefaciens*

Cloning vectors

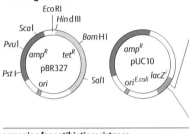

multiple cloning site (MCS)

- *Hin*d III
- *Sph*I
- *Pst*I
- *Sal*I, *Acc*I, *Hinc*II
- *Xba*I
- *Bam*HI
- *Sma*I, *Xma*I
- *Kpn*I
- *Sac*I
- *Eco*RI

two cloning vectors for *E. coli*

ori^{E.coli}	origin of replication for *E. coli*
amp^R	selection marker 1 (ampicillin resistance)
tet^R	selection marker 2 (tetracycline resistance)
lacZ'	gene fragment of β-galactosidase

screening for antibiotic resistance

amp^R*tet*^S-recombinant clone. Gene fragment was inserted into *tet*^R using BamH1 or SalI restriction sites

amp^R*tet*^R: no recombinant clone

blue-white screening

- agar + X-gal + IPTG + ampicillin
- white colony = recombinant clone (*lacZ'* destroyed)
- blue colony = no recombinant clone (*lacZ'* intact)

X-gal: chromogenic substrate for β-galactosidase
IPTG : inducer for β-galactosidase promoter

shuttle vector for gene expression in eukaryotes

ori^{E.coli}
eukaryotic resistance factor

amp^R resistance factor

transcription terminator (polyA signal)

multiple cloning site

ori^{euk}
eukaryotic promoter for expression

ori^{E.coli}	origin of replication for *E. coli*
amp^R	selection marker for *E. coli*
ori^{euk}	origin of replication for eukaryotes (e.g., of 2μ plasmid for *S. cerevisiae*, of SV40 virus for animal cells)

Transformations based on nonbiological methods

endocytosis

calcium phosphate solution

DNA is deposited on the cell surface

monolayer of animal cells

electroporation

cells and foreign DNA

selection agar for transformands

liposome fusion (lipofection)

transfer of DNA into the cell nucleus

fused liposome (DNA in red)

microinjection

DNA solution

cell nucleus

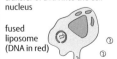

~10 000 V/m²
for ~10 ms

Gene cloning and identification

General. PCR methods are widely used for cloning genes whose sequences are known or can be derived from a protein's partial amino acid sequence. If a gene of unknown protein or gene sequence must be cloned, a genomic library (prokaryotes) or cDNA library (eukaryotes) is prepared and deposited in a set of transformed host organisms (usually *E. coli*). The desired gene must be present in this library if its product is functionally expressed and can be identified by its activity (shotgun cloning), through its transcribed mRNA (but only if it is a major cell product), or by immunological identification of the gene product (Western blot).

Cloning with PCR methods. Once enough sequence information about the desired gene or gene product is available, synthetic primer can be constructed, which allows the gene to be cloned from DNA or cDNA using PCR. Degenerate primers must be used if the precise sequence of the gene is unknown, e. g., if it is derived from an amino acid sequence. Often, PCR cloning is combined with the insertion of a restriction site for later ligation into an expression vector.

Preparation of gene libraries. DNA or cDNA from the donor cells (bacteria, plant, mammalian, or insect cells) is digested by restriction nucleases, and competent host cells (usually *E. coli*) are transformed with this library. Because very many different clones must be analyzed for heterologous gene inserts, efficient selection methods are crucial. Thus, the vectors used for transformation may contain marker genes that confer resistance to antibiotics, e. g., *amp^R* or *tet^R*, so that only transformed cells will survive on the selection agar. In the marker-rescue approach, the auxotrophic mutant of a wild-type strain is transformed with DNA fragments from a gene library containing the gene responsible for the auxotrophic properties: the transformands that are complemented by this gene can grow on minimal medium without additives. Usually two marker genes are used: one for the selection of transformand, the other as a component of the multiple cloning site (MCS) into which the foreign DNA is inserted. The successful integration of a foreign DNA into the MCS can then be recognized by a loss of the marker phenotype, e. g., antibiotic resistance. After these preparatory steps, a gene library can now be analyzed for transformands containing the desired gene.

Detection of genes and gene products. The most important procedures are based on 1) gene-specific hybridization of a DNA or RNA probe, and 2) expression of the gene product. In the first procedure, a gene-specific probe complementary to a sequence in the desired DNA is synthesized and radioactively or otherwise labeled; it is then used for hybridization experiments with single-stranded DNA obtained from the transformands (Southern blot), or directly in colony or phage hybridization. mRNA transcribed by a transformand can be analyzed in an analogous way, using a labeled DNA or RNA probe (Northern blot). If the DNA sequence of the gene product is not known, a Western blot can be tried. Here, the gene library is prepared by using an expression vector. The desired protein may then be found with an immune reaction using labeled antibodies. Even protein fragments may be discovered by this technique – a useful property, since the encoding gene may have been cut during preparation of the gene library, being distributed among 2 or more transformands. If a gene library is to be searched for regulatory elements, e. g., for promoters, vectors that contain a reporter gene (e. g., for luciferase or green fluorescent protein) behind the MCS are used. A promoter isolated from the gene library is then detected by the expression of the reporter gene.

Other detection methods. PCR methods based on specific primers are usually used to monitor the insertion of a gene into chromosomal DNA. Restriction patterns may provide the first indication of the successful cloning of a new gene, if novel restriction patterns appear during gel electrophoresis upon comparison of wild-type and transformand genes.

Cloning of genes

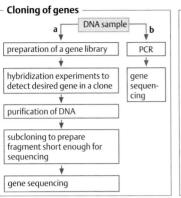

DNA sample

a → preparation of a gene library → hybridization experiments to detect desired gene in a clone → purification of DNA → subcloning to prepare fragment short enough for sequencing → gene sequencing

b → PCR → gene sequencing

Detection methods

Southern blot	detection of DNA	hybridization with labeled DNA or RNA probe
Northern blot	detection of mRNA	hybridization with labeled DNA or RNA probe
Western blot	detection of proteins	immunological detection by labeled antibodies
reporter groups	detection of regulatory elements	expression of genes coding for reporter proteins

Southern blot

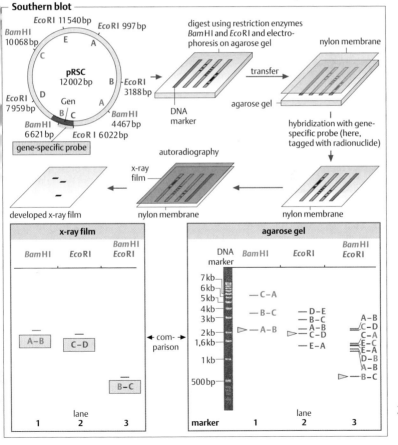

EcoRI 11540bp   EcoRI 997bp
BamHI 10068bp
pRSC 12002bp
EcoRI 3188bp
EcoRI 7959bp
Gen
BamHI 6621bp   EcoRI 6022bp   BamHI 4467bp
gene-specific probe

digest using restriction enzymes BamHI and EcoRI and electrophoresis on agarose gel → transfer → nylon membrane / agarose gel → hybridization with gene-specific probe (here, tagged with radionuclide)

autoradiography ← x-ray film ← nylon membrane ← nylon membrane

developed x-ray film

x-ray film

BamHI	EcoRI	BamHI EcoRI
A–B	C–D	B–C

lane 1   2   3

agarose gel

DNA marker: 7kb, 6kb, 5kb, 4kb, 3kb, 2kb, 1,6kb, 1kb, 500bp

BamHI (lane 1): C–A, B–C, A–B
EcoRI (lane 2): D–E, B–C, A–B, C–D, E–A
BamHI EcoRI (lane 3): A–B, C–D, C–A, E–C, E–A, D–B, A–B, B–C

com-parison

marker   lane 1   2   3

237

Gene expression

General. A main objective of genetic engineering is to express a foreign gene or operon (several coordinated genes) in a host organism. This requires expression vectors which, depending on the choice of the host organism, are either replicated outside the chromosome (as is usual in *E. coli*) or integrated into the chromosomal DNA (the standard in all higher eukaryotic cells). In expression vectors, the foreign gene is often preceded by an inducible promoter that allows the gene to be switched on or off by appropriate external conditions. In higher organisms, it has become possible to direct and express foreign DNA in a desired compartment, e. g., the chloroplast, by using appropriate leader sequences.

Expression vectors for prokaryotes. A typical expression vector for bacteria contains an origin of replication (*ori*), a marker gene to enable selection of transformed clones, and the foreign structural gene or operon (ORF = open reading frame), with its start codon ATG and its terminal stop codon. Several recognition sequences provide the appropriate commands to the transcription and translation machinery of the cell to form the gene product. In *E. coli*, RNA polymerase binds to sequences upstream of the ORF (the so-called −35 and −10 boxes) and transcribes DNA into mRNA. Transcription ends at a transcription terminator region that is located downstream of the ORF, sometimes by forming a stem-loop region in the mRNA. For constructing expression vectors, inducible promoters are usually preferred. For example, the promoter of the lactose operon of *E. coli* can be switched on by adding the inducer isopropyl-β-D-thiogalactoside (IPTG) to the medium; in the presence of this inducer, a repressor protein is removed from an operator sequence, allowing for RNA polymerase to bind to the promoter and to transcribe the gene into mRNA. Many expression vectors are commercially available. They usually contain multiple cloning sites (MCS).

Expression vectors for eukaryotes have a similar structure. The contain a selection marker, often an inducible promoter with consensus sequences (TATA, CCAAT and GC boxes), a start codon (ATG), followed by a multiple cloning site, and a terminator sequence. The mRNA obtained during transcription in an eukaryotic cell is polyadenylated at its 3′-position (polyA tail) and carries a 7-methyl guanosine triphosphate residue at its 5 position ("cap"). Specific signal sequences may lead to expression of the gene product in a desired cellular compartment. Eukaryotic expression vectors for higher organisms rarely replicate autonomously: they are usually inserted into a chromosome of the host organism by recombination. To select clones of transformed animal cells having many copies of the heterologous gene, auxiliary genes are cloned into the vector which provide the cells with resistance to toxic culture-medium components. Thus, coexpression of dihydrofolate reductase (DHFR) or neomycin phosphotransferase with the desired gene product ensures that only cells with the desired gene can survive in a medium that contains high concentrations of methotrexate or neomycin.

Promoters. Strong and weak, and tight and loose, promoters are known. Promoters used in technical processes should be strong and tight, i.e., they should remain switched off in the absence of inducer. Typical promoters for *E. coli*-based processes are the *lac*, *trp* and *tac* promoters, which are induced by adding a reagent to the medium. In contrast, the λP$_L$-promoter is induced by raising the temperature of the medium from 30 to 42 °C. For expressing foreign genes in fungi and yeasts, the GAL10 galactose promoter is often used in *Saccharomyces cerevisiae*, the alcohol oxidase promoter AOX in *Pichia pastoris*, and the glucoamylase promoter in *Aspergillus* strains. The metallothionein promoter is popular for expression in animal cells. For transgenic plants and animals, promoters that are regulated by the host organism are often used. Thus, in transgenic animals the target gene is often cloned behind the strong lactalbumen promoter of the mammary gland; the recombinant protein is then formed in large quantities after induced lactation.

Expression vectors

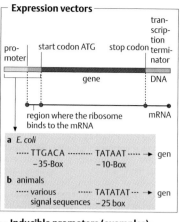

a *E. coli*

····· TTGACA ·········· TATAAT ····· → gen
    −35-Box   −10-Box

b animals

····· various ······ TATATAT ··· → gen
  signal sequences   −25 box

Induction and repression

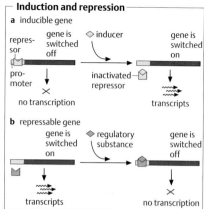

a inducible gene

b repressable gene

Inducible promoters (examples)

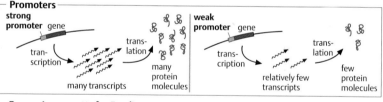

promoter	protein	inducer	host organism
lacZ	β-galactosidase	IPTG	*E. coli*
λ*P_L*	–	temperature rise 30–42 °C	*E. coli*
GAL10	β-galactosidase	galactose	*S. cerevisiae*
AOX	alcohol oxidase	methanol	*Pichia pastoris*
metallothionein	metallothionein	Zn^{2+}	animal cells

Promoters

strong promoter

many transcripts — many protein molecules

weak promoter

relatively few transcripts — few protein molecules

Expression cassette for *E. coli*

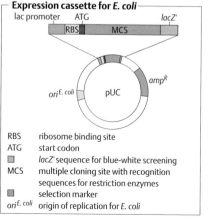

RBS	ribosome binding site
ATG	start codon
◻	*lacZ'* sequence for blue-white screening
MCS	multiple cloning site with recognition sequences for restriction enzymes
◻	selection marker
ori^E. coli	origin of replication for *E. coli*

Expression cassette for *Pichia pastoris*

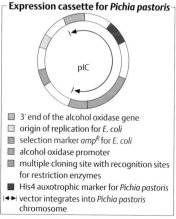

- ◻ 3' end of the alcohol oxidase gene
- ◻ origin of replication for *E. coli*
- ◻ selection marker *amp^R* for *E. coli*
- ◻ alcohol oxidase promoter
- ◻ multiple cloning site with recognition sites for restriction enzymes
- ◼ His4 auxotrophic marker for *Pichia pastoris*
- |◄►| vector integrates into *Pichia pastoris* chromosome

239

Gene silencing

General. The targeted silencing of genes is an important technique in basic molecular biology studies and in biotechnology, for example, to eliminate undesired properties in the breeding of domestic animals and plants, in the development of microbial strains, and in medicine, e.g., for tumor therapy. Unlike random mutagenesis based on chemical mutagens or radiation, genetic techniques have the potential to knock out specific genes. Experimentally, this is done by recombination or RNA-based techniques. Although gene silencing has undoubtedly been successful in some cases, most phenotypes are multigenic, and it is usually very difficult, often impossible, to assign a desired phenotype to the function of a single gene.

Knockout by recombination. DNA replacement vectors used for creating knockout mutants are nearly homologous to the gene or exon to be silenced, but contain a mutation or deletion that results in a nonfunctional gene product after the vector has recombined with the chromosomal DNA. Reliable recombination is ensured only if the length of the inserted fragment exceeds ca. 150 bp. Because recombination events are rare ($< 10^{-3}$), markers are required to select transformed cells. Growth inhibitors are often used for this purpose, e.g., methotrexate, whose inhibitory activity can be overcome by cells that express enough dihydrofolate reductase (DHFR).

RNA-based techniques. In RNA-based techniques, the gene to be inhibited is ligated in the reverse direction into an expression vector which is used for transformation of the host cell. The mRNA that is formed during transcription of this gene (antisense RNA, asRNA) is complementary to the mRNA of the normal gene and prevents synthesis of the gene product. Probably both species form an RNA double strand that either is not bound to the ribosomes or is rapidly degraded by RNases. The antisense technique offers an interesting new concept in medical therapy, which complements gene therapy: if a disease is not due to an erroneous gene product, but to its excessive formation, replacement of the gene is less promising than interference with its translation by antisense techniques. In fact, the cancer-inducing properties of glioblastoma cells in brain tumors, which are due to errors in the formation of insulin-like growth factor, were decreased by expression of asRNA. The as RNA was expressed by means of an episomal vector containing a metallothionein promoter. A completely different concept for using asRNA in medicine is based on its direct injection. Because RNA is quickly degraded in vivo by RNases, RNA analogues such as phosphothionates are being investigated. Several asRNA-based drugs are now under clinical scrutiny, e.g., an oral preparation against Crohn's enteritis and several antiviral preparations. Interfering and small interfering RNAs (RNAi and siRNA) have recently been shown to play a major role in the underlying mechanism. A technically successful example of a plant produced by antisense techniques is the FlavrSavr™ tomato, whose fruits can be left to ripen and form aroma on the vine and have a long shelf life. For its production, part of the gene coding for the enzyme polygalacturonase was linked in reverse to an RNA promoter derived from tobacco mosaic virus and transferred via a Ti plasmid vector into tomatoes. As shown by Southern and Northern blotting, the gene was inserted into the tomato genome, and asRNA was formed. Transformed tomatoes showed a greatly reduced activity of polygalacturonidase in mature fruits, leading to a longer ripening period and less softening. A related technique was used to prepare virus-resistant plants, expressing asRNA coding for the capsid of the plant-pathogenic virus. Plants transformed by such vectors contain marker genes, e.g., coding for antibiotic resistance. Critics have pointed out that horizontal gene transfer of these resistance markers may lead to the spread of antibiotic resistance throughout the ecological system.

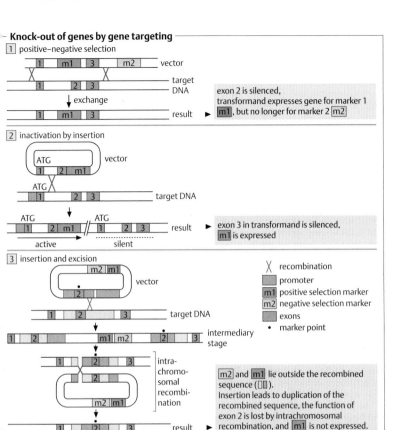

Knock-out of genes by gene targeting

1 positive–negative selection

| 1 | m1 | 3 | m2 | vector |

| 1 | 2 | 3 | target DNA |

↓ exchange

| 1 | m1 | 3 | result

► exon 2 is silenced, transformand expresses gene for marker 1 m1, but no longer for marker 2 m2

2 inactivation by insertion

ATG

| 1 | 2 | m1 | vector

ATG

| 1 | 2 | 3 | target DNA

↓

ATG   ATG

| 1 | 2 | m1 | // | 1 | 2 | 3 | result

◄ active ►   ······ silent ······

► exon 3 in transformand is silenced, m1 is expressed

3 insertion and excision

| m2 | m1 | vector

| 2 |

| 1 | 2 | 3 | target DNA

↓

| 1 | 2 | m1 | m2 | 2 | 3 | intermediary stage

| 1 | 2 | 3 | intra-chromo-somal recombi-nation

| 2 |

| m2 | m1 |

↓

| 1 | 2 | 3 | result

X recombination

☐ promoter

m1 positive selection marker

m2 negative selection marker

☐ exons

• marker point

► m2 and m1 lie outside the recombined sequence (☐☐). Insertion leads to duplication of the recombined sequence, the function of exon 2 is lost by intrachromosomal recombination, and m1 is not expressed.

Eukaryotic selection markers

marker	cell type	indicator
recessive		
adenosine deaminase	CHO mutant	9-β-xylofuranoxyl
dominant		
dihydrofolate reductase	all	methotrexate
neomycin phosphotransferase thymidine kinase fusion protein	all	neomycin sulfate
metallothionein I	all	Cd^{2+}, Zn^{2+}

Antisense technique

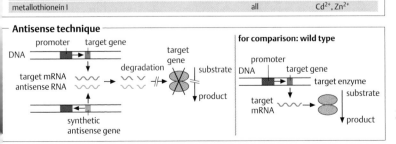

promoter target gene

DNA

target mRNA
antisense RNA → degradation

target gene

substrate

product

synthetic antisense gene

for comparison: wild type

promoter target gene

DNA

target enzyme

target mRNA

substrate

product

RNA

General. It is widely held that the DNA-based genome, the genetic program of today's biosphere, was preceded by a simpler life form whose replication was based on RNA. Regardless, the machinery used by cells for protein synthesis is largely based on ribosomal RNA, transfer RNA and messenger RNA. In biotechnology, RNA-based techniques play a considerable role. Examples are 1) RNA-based aptamers as bioaffinity molecules; 2) mRNA-based procedures to prepare proteins *in vitro*; 3) the capacity of interfering RNA to knock out gene functions; and 4) RNA-based vectors in gene therapy.

Aptamers are artificial nucleic acid ligands that bind with high affinity to hydrophilic molecules such as peptides and drugs. In the SELEX process (systematic evolution of ligands by exponential enrichment), vast combinatorial libraries of synthetic nucleic acids ($10^{14} - 10^{15}$ different molecules) are screened for binding to a target molecule. Those sequences that interact with the target are amplified by RT-PCR and transcribed in vitro, providing after each round a sublibrary with further enhanced binding properties. Aptamers with binding capacities in the lower nM range have been isolated and have been studied both for diagnostic and therapeutic applications, for example, for selective gene inactivation.

Cell-free protein expression. Techniques have been developed for protein synthesis *in vitro*, starting from a DNA template. Using an optimized *E. coli* lysate which contains an RNA polymerase, tRNAs, ribosomes, amino acids, and ATP, up to a few mg of protein can be synthesized within 24 h. The method has been advantageously used to explore bottlenecks in transcription and to express proteins such as proteases or antibacterial peptides which are toxic to a host organism. Equipment for the use of this technology is commercially available (ProteoMaster).

Knock out of gene functions. RNA molecules are involved in crucial steps of genetic information processing such as the splicing of exons and the synthesis of proteins. RNA interference, also termed post-translational gene silencing, has been recognized as a mechanism for regulating gene expression and mediating resistance to endogenous and exogenous pathogenic RNAs such as RNA viruses ("immune system of the genome"). In some of the above mechanisms, RNA can also be catalytically active (ribozymes), for example, splitting phosphodiester bonds in the absence of any protein. Many of these mechanisms are being explored for use in biotechnology. Antisense RNA – a technology that was discussed under the topic of gene silencing - has been successfully applied to eliminate polygalacturonase activity in ripening tomatoes (FlavrSavr®), thus leading to better aroma without wrinkling of the skin. *Trans*-cleaving ribozymes (i. e., ribozymes that cleave a foreign strand of RNA) have been explored in the therapy of HIV and breast and colorectal cancer up to the clinical phase II level. They can be applied, for example, by infusing transformed CD4+ lymphocytes or CD34+ hematopoietic precursor cells from the infected patient, which have been expanded *ex vivo*. In many eukaryotic cells, mRNA can be destroyed by a process termed RNA interference (RNAi): the presence of double-stranded RNA activates an RNAse able to recognise and digest matching endogenous mRNA, possibly by using the double-stranded RNA as a template. It is based on the random cutting of the double-stranded RNA by a RNase (DICER); after ATP-dependant enzymatic activation of the single-stranded fragments generated in this process, they can hybridize specifically with the mRNA and be recognized by an RNase, which then degrades the mRNA. For example, when a suitable synthetic double-stranded RNA is expressed behind an RNA polymerase III (Pol III) promoter, HIV gene expression in cotransfected cells is largely inhibited.

Gene therapy. The use of viral RNA vectors for human gene therapy is described elsewhere. mRNA extracts of human tumors were successfully used to transform monocytes of the same patient, resulting in mature dendritic cells loaded with tumor-specific RNA which, upon infusion, stimulated the immune system of the patient to form anti-tumor cytotoxic T-lymphocytes.

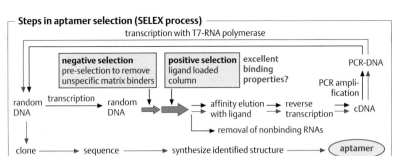

transcription with T7-RNA polymerase

negative selection
pre-selection to remove
unspecific matrix binders

positive selection
ligand loaded
column

excellent
binding
properties?

PCR-DNA

PCR ampli-
fication

random
DNA

transcription

random
DNA

affinity elution → reverse → cDNA
with ligand   transcription

removal of nonbinding RNAs

clone → sequence → synthesize identified structure → aptamer

Some RNA-derived aptamers binding to extracellular proteins

target	Kd(nM)	nucleic acid
acetylcholine receptor	2.0	RNA
basic fibroblast growth factor (PDGF)	0.35	RNA
interferon-γ	6.8	2'-modified RNA
keratinocyte growth factor (KGF)	0.0003	2'-modified RNA

Aptamer complex

with vitamin B12
(1DDY_5)

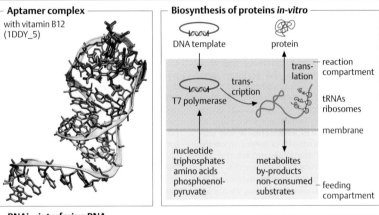

Biosynthesis of proteins *in-vitro*

DNA template

protein

T7 polymerase

trans-
cription

trans-
lation

reaction
compartment

tRNAs
ribosomes

membrane

nucleotide
triphosphates
amino acids
phosphoenol-
pyruvate

metabolites
by-products
non-consumed
substrates

feeding
compartment

RNAi – interfering RNA

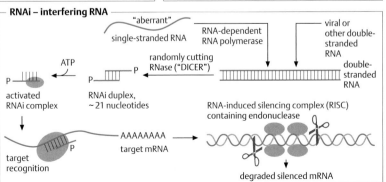

"aberrant"
single-stranded RNA

RNA-dependent
RNA polymerase

viral or
other double-
stranded
RNA

ATP

randomly cutting
RNase ("DICER")

double-
stranded
RNA

P

P — P

activated
RNAi complex

RNAi duplex,
~ 21 nucleotides

RNA-induced silencing complex (RISC)
containing endonuclease

P

AAAAAAAA
target mRNA

target
recognition

degraded silenced mRNA

243

Gene libraries and gene mapping

General. Even the small genomes of phages and viruses are much too large for direct sequencing of their DNA or RNA. Thus, genomic DNA is cut into fragments, cloned into vectors, stepwise approaching the size of fragments that can be sequenced. After fragments have been sequenced, the sequence information is analyzed by computer and made into a complete DNA- or RNA-sequence map of the genome (physical map). Due to the redundancy of identical base sequences, larger genome sequences can only be considered correct if enough markers have been identified (gene mapping). Of particular practical importance as markers are sequence-tagged sites (STS).

Gene libraries comprise a collection of DNA fragments that, together, constitute the complete genome. They are prepared by cutting genomic DNA into smaller DNA pieces, using ultrasound or restriction enzymes, and inserting the fragments into vectors. To prepare large fragments, restriction enzymes that cut at rare sequences are preferred, e. g., *Not*I, which recognizes 5′-GCGGCCGC-3′, occurring statistically only every $4^8 = 65\,536$ bp. The occurrence of a sequence in any genome depends strongly, however, upon the GC content of the DNA and on the presence of repeating DNA sequences.

Vectors. For preparing gene maps and for sequencing, genomic DNA fragments are cloned into vectors, by which they can be easily amplified through the transformation of host cells and from which they can be reisolated. The size of the genomic DNA determines the number of clones that are needed to create a complete gene library. For the larger eukaryotic genomes, the vectors of choice are artificial chromosomes (yeast/bacterial artificial chromosomes, YACs/BACs); they allow DNA inserts up to 2 Mbp to be packaged. For sequencing, these fragments are still much too large. Subcloning is usually carried out with λ-derived vectors, e. g., cosmids, which allow 30–45 kbp of foreign DNA to be packaged.

Gene mapping. The classical method of gene mapping relies on observing coupled phenotypes. For example, gene maps of Ascomycetes such as *Neurospora crassa* and *Saccharomyces cerevisiae* were obtained long ago by tetrad analysis. Because all the daughter cells are found in one ascus in the same sequence in which they were formed during meiosis, the spores can be easily isolated and analyzed for their phenotypic properties. Molecular genetics has strongly expanded these methods. For smaller genomes, restriction fragments can be generated and sequenced, often leading to direct physical mapping of a gene's position. By using well-chosen probes for PCR or DNA hybridization, one can also obtain genetic markers for larger genomes. An important procedure in this context is fluorescence in-situ hybridization (FISH) of a large DNA sequence. High resolution (within 10 kbp) can be obtained if this method is combined with "DNA combing". For this, a polylysine-covered slide is dipped into a solution containing large DNA fragments. If the slide is slowly ($0.3\,\mathrm{mm\,s^{-1}}$) pulled out of the solution, the DNA fragments align in parallel, offering perfect conditions for hybridization with a marker.

STS (sequence tagged sites) are DNA sequences 100–500 bp long, which occur just once in the whole genome. This implies that STSs do not include sequences found in repetitive DNA. STSs are often obtained from clone libraries containing large genome fragments, e. g. from a YAC or BAC library. If large eukaryotic genomes must be analyzed, chromosome-specific libraries are often used. Individual chromosomes can be isolated after staining with a fluorescent dye and sorting by flow cytometry (fluorescence-activated cell sorting, FACS), since the amount of dye that a chromosome binds depends on its size. After a collection of STS has been found for a given genome, it is simple to find out, with appropriate PCR primers, if they have a neighboring or far-distant position in the genome: if they are close together, a collection of overlapping gene fragments from a gene library should yield additional hybridizing gene fragments carrying the same STS. STS markers thus are excellent "mapping reagents" for molecular-coupling analysis of gene segments.

Gene libraries

organism	(haploid genome) b	number of required clones (P=95%)			
		λ-vector (EMBL4) a ≈ 17 kbp	cosmid a ≈ 35 kbp	BAC a ≈ 250 kbp	YAC a ≈ 1000 kbp
E. coli	4 800 000	850	410	56	13
S. cerevisiae	14 000 000	2500	1 200	167	41
Drosophila melanogaster	170 000 000	30 000	14 500	2036	508
tomato	700 000 000	123 500	59 000	8387	2096
man	3 000 000 000	529 000	257 000	35 948	8986
frog	23 000 000 000	4 053 000	1 969 000	275 602	68 901

$$N = \frac{\ln(1-P)}{\ln(1-a/b)}$$

P = probability

a = average length of DNA inserts in the vector (bp)
b = total size of the genome (bp)

Gene mapping

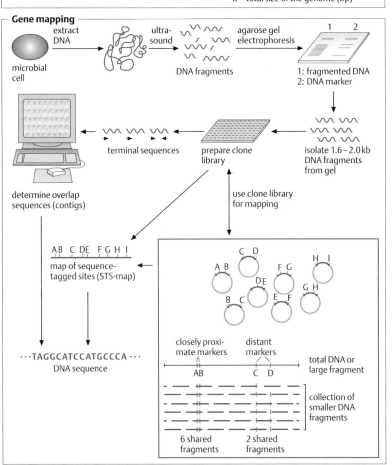

microbial cell

extract DNA

ultra-sound

DNA fragments

agarose gel electrophoresis

1: fragmented DNA
2: DNA marker

isolate 1.6 – 2.0 kb DNA fragments from gel

prepare clone library

terminal sequences

determine overlap sequences (contigs)

AB C DE F G H I
map of sequence-tagged sites (STS-map)

use clone library for mapping

···TAGGCATCCATGCCCA···
DNA sequence

closely proxi-mate markers
AB

distant markers
C D

total DNA or large fragment

collection of smaller DNA fragments

6 shared fragments

2 shared fragments

Genetic maps of prokaryotes

General. Genetic maps of microorganisms have been prepared by observing changes in the phenotype after conjugation (transfer of DNA from a donor to a recipient cell), after transduction (transfer of DNA pieces among bacteria by a phage), and after transformation (uptake of naked or plasmid DNA). Physical genome maps (the complete DNA sequence of a genome) have existed since 1995. They have been prepared by clone contig mapping or the shotgun procedure.

Genetic maps. Many changes in phenotype can be observed easily and rapidly in bacteria. Thus, loss of the capacity to form spores or flagella, or introduction of antibiotic resistance, can be used as a phenotypic marker. For the elucidation of metabolic pathways, blocked mutants, which have lost the ability to carry out one or more steps in a pathway have been studied for a long time; by adding the precursor molecule, this loss can be overcome. The short generation time of many bacteria (< 1 h) is a bonus for the microbial geneticist. When observation of two phenotype modifications in a recipient cell is combined with measurement of the time period required for conjugation or transduction, the distance of the genes coding for the two phenotypes can be estimated (linkage analysis). Prokaryotic gene maps are therefore dimensioned in minutes or centisomes. The time required for complete transfer of the *E. coli* genome into a recipient cell is 100 min at 37 °C.

Physical genome maps: clone contig maps. An important task in genome sequencing is to identify those clones in a genome library that contain neighboring DNA sequences (contiguous = neighboring). Clone "contigs" for a whole genome can be combined by clone fingerprinting techniques. Thus, the occurrence of overlapping restriction maps or sequence-tagged sites (STS) in two clones indicates that they contain overlapping parts of the genome sequence and can be used to deduce its overall sequence with the aid of computer programs. Wherever possible, phenotype-related markers from genetic maps are also used. For positional cloning of a gene from a nearby marker, chromosome walking is often used. In this approach, labeled RNA from the starting clone can be prepared for hybridization experiments or primers are constructed from the terminal sequence of a clone and used to prepare a PCR product from the contiguous clone, which is suspected to contain the desired gene. The DNA sequence of the complete genome is identified step by step using the incomplete genetic map as a template.

Physical genome mapping: shotgun method. This time-saving method is based on the concept that any DNA sequence of 600 b can be directly and rapidly determined. Thus, the genomic DNA is cut by restriction enzymes or ultrasound into small overlapping fragments, whose end-sequences are determined. A supercomputer is then used to compile the genome sequence from the sequenced overlaps. Due to the efficiency of modern teraflop computers, this procedure allows a typical microbial genome (1–5 Mbp) to be mapped within weeks or even days, depending on the number of sequencers available. Because genetic maps containing marker sequences are often available (when sequencing of the *E. coli* genome began in 1990, there were 1400 markers, corresponding to a mean distance between markers of 3300 bp on the 4.64 Mbp genome), the computed results obtained by shotgun sequencing can continuously be validated. By the end of 2002, > 200 microbial genomes had been completely or largely sequenced, and sequence inspection had led to abundant information about gene location. By comparing genomes and their functional units, a unified view about the metabolic modules involved in life is emerging.

Bioinformatics. The computer programs used in genome projects are aimed mainly at reliable determination of sequence homologies. Even rare sequencing errors (99 % precision) lead to errors in practically all recorded gene sequences and render it necessary to compare results from multiple sequencing experiments.

Genetic maps from conjugation experiments

sequential transfer of marker genes A, B, C during conjugation

cotransfer of closely proximate marker genes A, B during transduction or transformation

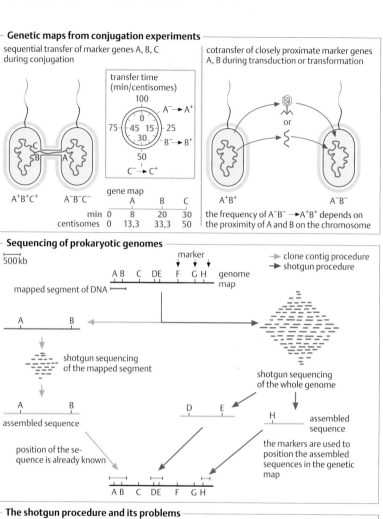

transfer time (**min**/centisomes)

$A^+B^+C^+$ $A^-B^-C^-$

gene map

	A	B	C
min	0	8	20	30
centisomes	0	13,3	33,3	50

A^+B^+ A^-B^-

the frequency of $A^-B^- \rightarrow A^+B^+$ depends on the proximity of A and B on the chromosome

Sequencing of prokaryotic genomes

500 kb

marker

A B C DE F G H   genome map

mapped segment of DNA

➤ clone contig procedure
➤ shotgun procedure

A   B

shotgun sequencing of the mapped segment

shotgun sequencing of the whole genome

A   B
assembled sequence

D   E

H   assembled sequence

position of the sequence is already known

the markers are used to position the assembled sequences in the genetic map

A B   C   DE   F   G H

The shotgun procedure and its problems

repetitive DNA sequences

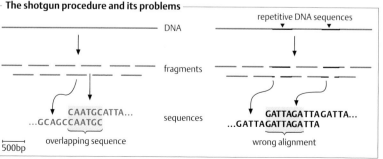

DNA

fragments

CAATGCATTA...
...GCAGCCAATGC

sequences

GATTAGATTAGATTA...
...GATTAGATTAGATTA

overlapping sequence

wrong alignment

500bp

247

Genetic maps of eukaryotes

General. Genetic maps of eukaryotes are produced, similar to prokaryotic maps, by coupling analysis of genetic phenotypes. Due to the diploid or polyploid set of chromosomes in eukaryotes, however, the gene sequence responsible for the phenotype may originate from different genotypes (heterozygous), which segregate during meiosis. Another feature of eukaryotic genomes is their much larger size, usually several billion base pairs. In addition, they contain introns and a significant number of repetitive DNA sequences, which hinder the search for unique sequences. In spite of these difficulties, the genomes of several eukaryotic organisms have already been completed (yeasts, fungi, worm, fly, human, mouse, rat, Arabidopsis, rice) and others approach completion (e.g., zebra fish).

Genetic maps. In experimentally accessible animals, genetic maps are based on pedigrees combined with linkage analysis: on observations of how phenotypic properties are linked during sexual reproduction, i.e., by meiotic crossing over (see genetics textbooks). Two phenotypes corresponding to genes that are close together on a chromosome are coinherited more frequently than two phenotypes due to genes that are farther apart. In consequence, the recombination frequency of coupled phenotypes leads to a virtual genetic map, whose dimension is the percentage frequency of recombination. This classical method is complemented today by many molecular genetics methods. The fingerprints of different relevant DNA fragments can be determined and compared by restriction mapping, and primers tagged with fluorescent markers can be used to locate genes within large DNA fragments or chromosomes by fluorescence in-situ hybridization (FISH).

Genome sequencing. The genomes of higher organisms contain repetitive DNA sequences (satellite DNA, Alu sequences, retrotransposons, etc.), which make unambiguous localization of a sequence within the overall genome difficult. Thus, SINE (short interspersed nuclear elements) ranging from 100 to 500 bp each, e.g., the Alu sequences,

which constitute up to 20% of a mammalian genome, and LINE (long interspersed nuclear elements), 6000–7000 bp long account for up to 10% of a mammalian genome. In addition, mini- and micro-satellite DNAs contribute another 5%. In humans, microsatellite DNA consists of 10–50 copies of a very short-sequence repetition such as AC or ACCC, which occurs >10 000 times and is distributed over the whole genome. Since each individual has a unique distribution of microsatellites, they are excellent genetic markers, e.g., in forensic investigations, and also in breeding domestic animals. Due to the high redundancy of repetitive sequences, genome sequencing by the shotgun approach, which is so useful for prokaryotic genomes, meets with considerable difficulty. In consequence, contig sequencing of overlapping clones is widely used, combined with genome walking, the use of sequence-tagged sites (STSs) and of expressed sequences tags (ESTs) as markers. STSs and ESTs provide complementary information. STSs span the whole genome, but do not discriminate between coding and noncoding regions and may include repetitive sequences. ESTs are short sequences derived from cDNA clones. Because cDNA is prepared by converting spliced mRNA into double-stranded DNA with reverse transcriptase, the sequences of ESTs contain no repetitive sequences and each EST has a unique sequence. If primers derived from EST sequences are hybridized with a genomic library, all introns and repetitive sequences are undetected, but clones that contain complete or fragments of expressed genes can be identified. As a consequence, STSs and ESTs complement each other very well as mapping reagents. Once the clones of a genome library have been correlated with a physical map of the genome by one or more of the above methods, subcloning of the cosmid, YAC, or BAC clones in λ phage libraries ensues, followed by DNA sequencing. Overlapping sequences are then analyzed with sequence-contig software to generate a complete sequence of the DNA of single chromosomes and, finally, of the whole genome. This computed map is validated with information from genetic maps.

Genetic maps

genes

m small wings

v shimmering eyes

w wide eyes

y yellow body

resulting position on a genetic map:

y	w		v	m
0	1,3		30,7	33,7

recombination frequency:

between m and v	3,0%
between m and y	33,7%
between v and w	29,4%
between w and y	1,3%

Chromosome nomenclature

chromosomes in the metaphase after Giemsa staining

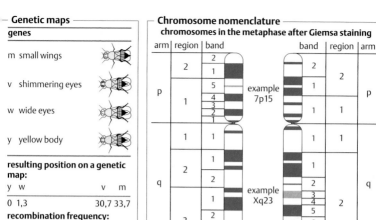

example 7p15

example Xq23

human chromosome 7   human X chromosome

Organization of eukaryotic genes

satellite DNA:
telomere and centromere region, minisatellites (16–64bp), microsatellites (2–4bp, up to 50 copies)

- ■ SINE (short interspersed repetitive elements), 100–500bp
- ■ LINE (long interspersed repetitive elements), up to 7000bp
- ■ gene
- □ centromere

telomere (5'-TTAGGG-3')

Gene structures of different organisms

a man: discontinuous exons, many introns, numerous repetitive elements, no operons

b yeast (*Saccharomyces cerevisiae*): tightly packed exons, rare introns, few repetitive elements

c maize (*Zea mays*): repetitive elements dominate, no operons

d bacteria (*Escherichia coli*): no introns, operons are frequent, rare repetitive sequences

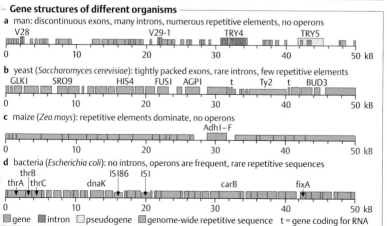

■ gene ■ intron □ pseudogene ■ genome-wide repetitive sequence t = gene coding for RNA

249

The human genome

General. Except for genome sequencing, genetic experiments with humans are neither legal nor desirable. However, because the pattern of population growth means that 6% of all people who ever lived on earth are living now, an extensive gene pool of humans is available. Genetic analyses of inheritance within families have led, over many decades of research, to a chromosomal map in which, even before the advent of human genomics, many hundreds of genetic diseases were roughly localized. Unambiguous localization of a disease in terms of the DNA sequence is, however, still rare. Within the human genome, which is ca. 3 billion (10^9) bp in size and distributed over 23 chromosomes (haploid set), only a few percent of the entire amount of DNA is sequence information coding for proteins. Most of the genome sequence consists of repetitive elements whose function is still unknown.

Genetic mapping. Key phenotypic markers in humans are inherited diseases, which in a few cases have been mapped by methods of human genetics (family analyses, chromosome analyses, functional and positional cloning of single genes). The Centre d'Etude du Polymorphisme Humain (CEPH), founded in 1984 in Paris, maintains tissue cultures of ca. 100 families from 3 generations, which on average comprise 4 grandparents, 2 parents, and an average of 8 children. In addition, populations with little mobility (e.g., Iceland, Tasmania) have been investigated in the hope of correlating genetic polymorphisms with the phenotypes of inherited diseases, leading to a functional gene map. Useful markers for such endeavors are microsatellites: they occur in high frequency and at the same time are highly variable, resulting in a probability of 70% that each individual is heterozygous for any given microsatellite. As a result, gene loci coupled to a microsatellite can be individually traced. In 1994, it became possible for the first time to generate a human genetic map with one marker per ca. 600 000 bp, combining the analysis of 291 meioses (304 individuals from 20 families of the CEPH collection) with the positions of 2335 microsatellites. This map has in the meantime been further resolved to one marker per ca. 100 000 bp.

Human genome sequencing. This large-scale project was begun in 1990 by an international, publicly funded consortium. Its strategy was based on contig sequencing of overlapping clones. Individual chromosomes were tagged with a fluorescent marker and separated by fluorescence-activated cell sorting (FACS). Their DNA was cut and transferred into ca. 300 000 BAC clones; clones with overlapping DNA sequences were identified by restriction mapping, chromosome walking, and sequence-tagged site (STS); their DNA was end-sequenced after subcloning, in both directions (to eliminate sequencing errors); and the overall sequence was computed using genetic maps for validation. Since 1996 this procedure was complemented by the sequencing of ca. 50 000 nonredundant expressed sequence tags (EST). Since 1998, a private company, Celera, started to compete with the public project, using a shotgun approach. To this end, human DNA was cut into 60 million fragments of ca. 2000 bp and 10 million sequences of ca. 10 000 bp, which were sequenced from both ends in lengths of 500 bp at a time (the 10 000-bp fragments were sequenced to ensure that repetitive sequences up to 5 kbp long were not wrongly assigned; it was recently shown, however, that this method is not free from error). The total length of DNA sequenced by this approach was ca. 35 billion bp, corresponding to a roughly 12-fold redundancy. Both projects have been completed, and the human genome sequence was published early in 2001. Surprisingly, only 30 000 to 40 000 open reading frames coding for primary gene products were found. As a result, the focus has now shifted to understanding the high diversity of primary gene products in different cell types – an objective of proteomics research. In addition, individual variation in genome sequence (e.g., SNPs = single nucleotide polymorphisms) and their phenotypic significance have become a major area of research.

The human chromosomes

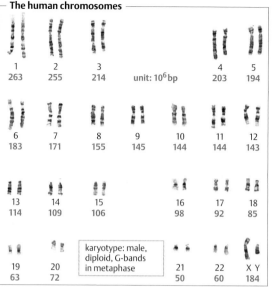

1 — 263
2 — 255
3 — 214
unit: 10⁶ bp
4 — 203
5 — 194

6 — 183
7 — 171
8 — 155
9 — 145
10 — 144
11 — 144
12 — 143

13 — 114
14 — 109
15 — 106
16 — 98
17 — 92
18 — 85

19 — 63
20 — 72

karyotype: male, diploid, G-bands in metaphase

21 — 50
22 — 60
X Y — 184

Mapping

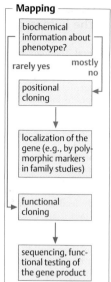

biochemical information about phenotype?

rarely yes   mostly no

positional cloning

localization of the gene (e.g., by polymorphic markers in family studies)

functional cloning

sequencing, functional testing of the gene product

Physical gene map with markers (chromosomes 1–5)

micro-satellites*   EST**
micro-satellites   EST
micro-satellites   EST
micro-satellites   EST
micro-satellites   EST

■ G bands

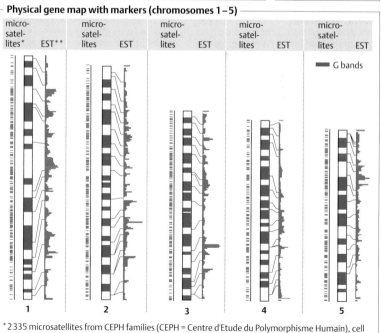

1   2   3   4   5

*2335 microsatellites from CEPH families (CEPH = Centre d'Etude du Polymorphisme Humain), cell bank from individuals of one family over 3 generations, **20 104 expressed sequence tags (EST)

Functional analysis of the human genome

General. The main reason for a functional analysis of the human genome is to understand the genetic basis of mono- and polygenic diseases. To this end, one attempts to localize and identify the gene or genes that are involved in or responsible for a diseased state and to compare deviations (polymorphisms) of the sequence(s) in sick and healthy people. Some diseases such as cystic fibrosis are already known to involve a single nucleotide difference (SNP, single nucleotide polymorphism). This approach may eventually lead to 1) individual genetic risk analyses, 2) individualized therapies (pharmacogenomics), 3) in the longer term, replacement of deficient genes (gene therapy), 4) identification of human target proteins that will be useful in the development of new drugs, and finally 5) a molecular physiology and pathophysiology of humans.

Genetic diagnostics. Since the completion of the human genome sequence, the number of human polymorphisms that have been deposited in gene databanks is rising exponentially (end of 2002: > 3 million SNPs). Each individual possesses many SNPs, distributed at a ratio of ca. 1 per 1000 bp or a total of ca. 3 million over the whole genome. If a SNP is located within the recognition sequence of a restriction enzyme, restriction analysis of this DNA section leads to an altered fragment pattern (RFLP, restriction fragment length polymorphism). Most SNPs and many RFLPs are silent mutations. Those located in microsatellite regions have proven very useful for the genotyping of inidivduals, e. g. for parental testing and in criminal investigations. The correlation of functional gene sequences with monogenic diseases is very difficult, however, and still impossible for polygenic diseases. As of now, human genetic investigations thus have remained in basic science, especially the coupling of phenotypes during inheritance. Many useful discoveries have also made by comparison with other genomes. Thus, even phylogenetically remote living beings such as the fruitfly, a worm, a fish, and yeast (Drosophila, Caenorhabditis, zebra fish, Saccharomyces) exhibit gene functions that are related to those in humans and, unlike humans, can be subjected to genetic experiments. In this context, the mouse genome, which has also been sequenced (ca. 3.3 billion bp) is of particular value, because the genetic and physiological relationship between mice and men is high. Knockout mice, where a specific gene or regulatory element has been impaired ("Alzheimer's" mouse, SCID-mouse), thus play a key role in fundamental research and in developing treatments. The results of such studies increase the chance of predicting, in the future, from a few individual gene sequences, the risk that an individual or his/her offspring will develop a certain disease. Of particular value to this end are transcription analysis of large gene clusters by DNA arrays and investigations of the proteome. No doubt these developments also raise many new ethical questions.

Targets for drug development. Once the genetic basis of a disease and the proteins involved in its phenotype have been elucidated, these proteins can be produced by genetic engineering and used as targets in a high-throughput screening for drug development. Due to the very complex structure and regulation of cells and the pharmacodynamic boundary conditions (resorption, organ distribution, metabolism of a drug), such monocausal approaches are, however, still far from perfect.

Pharmacogenomics. The vision of pharmacogenomics is to discover variations in an individual patient's genome by SNP analysis, translating the information into his or her individual metabolism and selecting drugs that are perfectly adapted to that person, in other words, to apply patient-specific therapy with greatly reduced side reactions. Individual genetic makeups, as determined by transcription analysis, may in the future be considered during the registration of new drugs.

Gene therapy. A rather complete understanding of the structure and function of the human genome is a prerequisite to gene therapy, i.e., the selective replacement of gene sequences in vivo.

Polymorphisms and SNP analysis

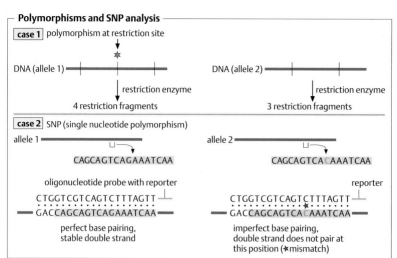

case 1 polymorphism at restriction site

DNA (allele 1)

restriction enzyme

4 restriction fragments

DNA (allele 2)

restriction enzyme

3 restriction fragments

case 2 SNP (single nucleotide polymorphism)

allele 1

CAGCAGTCAGAAATCAA

oligonucleotide probe with reporter

CTGGTCGTCAGTCTTTAGTT

GACCAGCAGTCAGAAATCAA

perfect base pairing,
stable double strand

allele 2

CAGCAGTCACAAATCAA

reporter

CTGGTCGTCAGTCTTTAGTT

GACCAGCAGTCACAAATCAA

imperfect base pairing,
double strand does not pair at
this position (∗mismatch)

Pharmacogenomics

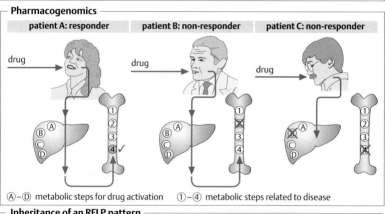

patient A: responder	patient B: non-responder	patient C: non-responder

drug

drug

drug

Ⓐ–Ⓓ metabolic steps for drug activation   ①–④ metabolic steps related to disease

Inheritance of an RFLP pattern

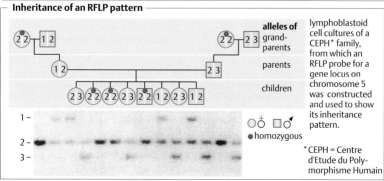

alleles of
grand-parents

parents

children

1 –
2 –
3 –

○○ ♀ □○ ♂
● homozygous

lymphoblastoid
cell cultures of a
CEPH* family,
from which an
RFLP probe for a
gene locus on
chromosome 5
was constructed
and used to show
its inheritance
pattern.

*CEPH = Centre
d'Etude du Poly-
morphisme Humain

253

DNA assays

General. Hybridization events that occur between two polynucleotide strands (DNA-DNA, DNA-RNA, or RNA-RNA) through hydrogen bonding can be visualized by tagging one of the strands with a radioactive, fluorescent, or otherwise detectable marker. Usually, the DNA or RNA to be analyzed is first amplified by PCR. To minimize false-positive or false-negative hybridization events, a thorough protocol for sample preparation (samples may be as diverse as urine, blood, tissue, plant materials, or fossils), the choice of the single-stranded probe, and the hybridization conditions are of utmost importance. DNA assays are now used in many areas, e. g., for genotyping pathogenic microorganisms, for genetic analysis of diseases, for monitoring genetically modified plants or foods derived from them, for parental testing, and in criminal investigations. The market volume for DNA assays already exceeds 1 billion US$ and is expected to grow more with the introduction of personal DNA-based diagnostic tests.

Equipment. Standard procedures for DNA assays are 1) comparative gel electrophoresis, 2) hybridization assays based on optical or electrochemical reporter groups, and 3) DNA arrays. In electrophoresis, the sample is analyzed in comparison to DNA of standard sizes. For assays based on hybridization, reporter groups can be linked to a single-stranded polynucleotide by standard chemical procedures, e. g., by biotin-avidin/streptavidin linkers. Alternatively, as in the LightCycler™ assay, an intercalating dye (SYBR Green™) is inserted into amplified DNA, allowing for real-time quantitative measurement of hybridization events. Another useful procedure is the TaqMan™ Assay, which is based on a "molecular beacon" (a fluorophore) attached to the end of the DNA probe. It interacts with a quencher at the other end of the probe. When a matching sequence is present in a sample that is subjected to PCR, the probe is amplified, the fluorophore is liberated, and the sample fluoresces.

Genotyping is widely used in forensic and paternal tests (genetic fingerprinting) and is emerging as a powerful method for the analysis of human single-nucleotide polymorphisms (SNPs). In paternal testing, a small number of microsatellite sequences of a male are compared with the microsatellite sequences of his putative offspring or, in criminal investigations, a suspect's sequences are compared with a sample containing DNA traces left at the crime scene. The probability that two individuals share the same microsatellite pattern is extremely low (except for identical twins). SNP analysis of a patient is an emerging technique to determine, e. g., the individual safety and efficacy of drugs (pharmacogenomics). Over 3 million human SNPs had been entered into the SNP database of the US-based National Center for Biotechnology Information (NCBI) as of the end of 2002. SNP analysis is also used in plant and animal breeding to secure pedigrees. In medical diagnosis, DNA assays are used to identify gene sequences and their aberrations and also the presence of infectious organisms in a blood, liquor or urine sample. A 1-μl blood sample is enough to demonstrate the presence of *Plasmodium falciparum*, the causative agent of malaria, in a patient. With DNA assays, food samples can be analyzed for the presence of extremely low levels of pathogenic microorganisms or genetically modified raw materials. Since genotyping assays generally start by amplifying the target DNA from the sample, the method implies that enough information is available about the sequence of the target DNA to prepare a synthetic probe that can hybridize with the target.

Prenatal genetic screening. is already sufficiently precise to detect monogenic diseases in a human embryo in weeks 9–12 of pregnancy. DNA is isolated from an embryonic cell, which can be obtained from a chorionic villus by a catheter. The extension of this method to risk analysis for mono- and polygenic diseases in the preimplantation stage (preimplantation diagnostics, PID) has already been established in principle, but is hotly debated with regard to ethics.

Forensic DNA analysis

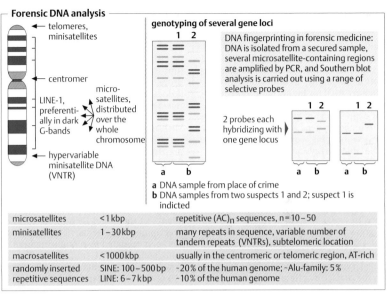

- telomeres, minisatellites
- centromer
- LINE-1, preferentially in dark G-bands → micro-satellites, distributed over the whole chromosome
- hypervariable minisatellite DNA (VNTR)

genotyping of several gene loci

DNA fingerprinting in forensic medicine: DNA is isolated from a secured sample, several microsatellite-containing regions are amplified by PCR, and Southern blot analysis is carried out using a range of selective probes

2 probes each hybridizing with one gene locus

a DNA sample from place of crime
b DNA samples from two suspects 1 and 2; suspect 1 is indicted

microsatellites	<1 kbp	repetitive (AC)$_n$ sequences, n = 10–50
minisatellites	1–30 kbp	many repeats in sequence, variable number of tandem repeats (VNTRs), subtelomeric location
macrosatellites	<1000 kbp	usually in the centromeric or telomeric region, AT-rich
randomly inserted repetitive sequences	SINE: 100–500 bp LINE: 6–7 kbp	~20% of the human genome; ~Alu-family: 5% ~10% of the human genome

DNA analysis of infectious diseases (examples)

- human papilloma virus (HPV)
- herpes virus, e.g., cytomegalovirus (CMS)
- HIV virus
- *Mycobacterium tuberculosis*
- *Borrelia burgdorferi*
- *Helicobacter pylori*
- *Neisseria meningitidis*

Genetic screening

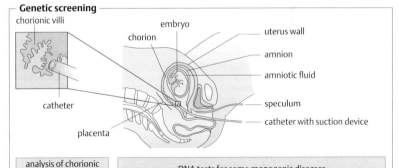

chorionic villi

- embryo
- chorion
- uterus wall
- amnion
- amniotic fluid
- speculum
- catheter with suction device

catheter
placenta

analysis of chorionic villi (weeks 9–12)	DNA tests for some monogenic diseases	
methods of analysis	**disease (frequency)**	**gene (position)**
chromosomal anomalies	sickle cell anemia (1:500; blacks)	(11 short)
biochemical tests	cystic fibrosis (1:2500)	CFTR (7q31)
FISH-based methods	hemophilia A and B (1:25000)	(Xq27)
DNA tests	Huntingdon's disease (1:20000)	IT15 (4p 16.3)
	phenylketonuria (1:10000)	PAH (12 long)
	inherited form of breast cancer (1:200, females)	BRCA1 (17q21)

DNA and protein arrays

General. DNA microarrays on solid surfaces (DNA chips, gene chips), or "liquid arrays", allow for simultaneous determination of many hybridization events. They are used for highly parallel genotyping, e.g., for detecting polymorphisms; for expression analysis, e.g., for studying differences in gene transcription; and for DNA sequencing. Commercially available DNA microarrays include "gene filters" made from nylon or nitrocellulose, which contain sets of cDNA fragments of yeast, mouse, man, or another organism. They offer densities of several hundred oligonucleotides cm^{-2}. Much higher densities of more than $10\,000-100\,000$ oligonucleotides cm^{-2} are obtained on rigid surfaces such as glass or plastic by two alternative methods: 1) stepwise chemical synthesis of the oligonucleotides on the surface by photolithographic techniques ("*in-situ* synthesis"); or by 2) microdeposition of DNA, RNA, or oligonucleotides with spotter equipment. "Liquid arrays", based on DNA-tagged microspheres, are a promising alternative for highly parallel DNA assays.

In-situ synthesizing is mostly done by coupling phosphoamidites, using nucleotide-specific protection groups that are removed by photochemical reactions. The distribution of each single step of synthesis over the whole array is controlled by photolithographic techniques (masks). This procedure enables the preparation of arrays with as many as $250\,000$ oligonucleotides cm^{-2}; recently, an array was presented containing 60 million DNA probes for human SNP analysis on a single wafer (20 cm diameter). The length of the oligonucleotides is usually limited to 25 bases or less because errors, due to incomplete coupling and deprotection, multiply during each step of synthesis. The manufacture of the many photolithographic masks required in this process results in high production costs for the prototype chip, but low costs for mass production.

256 **Microdeposition.** Instead of synthesizing DNA *in situ*, single-stranded DNA or cDNA of any length, or synthetic oligonucleotides, can also be microdeposited on a surface such as glass. For the coupling step, standard surface-chemistry procedures can be used. The polynucleotide is deposited either by contact spotting with pins or by non-contact spotting with drops, based on piezoelectric methods (inkjet printer technology). Commercial microspotters can reach speeds of $>10\,000$ DNA or cDNA fragments h^{-1}. Such methods have allowed arrays with ca. $10\,000$ oligonucleotides cm^{-2} to be prepared. Oligonucleotide arrays are useful for detecting incomplete hybridization due to mismatches (e.g., SNP analysis). Another application is the resequencing of DNA. To this end, the array must contain a complete single-stranded sequence of the target DNA in the form of overlapping fragments. This approach has been validated by sequencing human mitochondrial DNA (mtDNA), using an array of $16\,000$ 20mer oligonucleotides: 179 of the 180 known polymorphisms of mtDNA could be demonstrated in a single hybridization experiment.

Detection. Hybridization events are detected by labeling with radioactive or fluorogenic reporter groups, usually incorporated into the DNA probe during amplification. Detection is carried out by autoradiography, laser scanners, or CCD image analyzers. For label-free detection of hybridization events, the masses of primer-specific extension products of the probe can be separated and analyzed using matrix-assisted laser desorption/ionization time-of-flight mass spectrometry (MALDI-TOF)

"Liquid arrays". Single-stranded DNA-tagged polystyrene microspheres have been differentially labeled with two fluorophores, allowing the rapid sorting of up to 100 different DNA sequences by their spectral addresses in a fluorescence-activated cell sorter (FACS).

Protein arrays. Instead of DNA, a large number of proteins can be immobilized on a glass surface. For example, target proteins of kinases can be tested if all putative targets are immobilized on one array and kinase-catalyzed phosphorylation is carried out in the presence of ^{32}P-labeled ATP, followed by autoradiography.

Oligonucleotide arrays

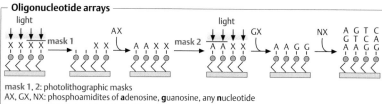

mask 1, 2: photolithographic masks
AX, GX, NX: phosphoamidites of **a**denosine, **g**uanosine, any **n**ucleotide

DNA array (expression profiling)

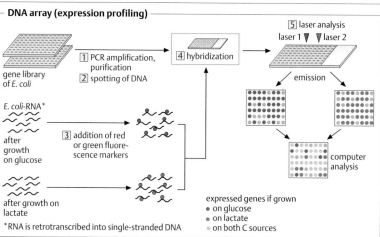

gene library
of *E. coli*

[1] PCR amplification, purification
[2] spotting of DNA

[4] hybridization

[5] laser analysis
laser 1 ▼ ▼ laser 2

emission

computer analysis

E. coli-RNA*
after growth on glucose

[3] addition of red or green fluorescence markers

after growth on lactate

expressed genes if grown
- on glucose
- on lactate
- on both C sources

*RNA is retrotranscribed into single-stranded DNA

Applications and detection methods

array density	applications
high, > 10 000 DNA sequences	identification of genes, transcription (expression) profiling
medium, 1 000 – 10 000 DNA sequences	analysis of mutations, alleles, polymorphisms
low, < 1 000 DNA sequences	genetic predispositions, acute diseases, infections

DNA microarray

hybridization with fluorescence- or gold-/silver-labeled DNA

detection by CCD, laser fluorescence scanning or absorption (silver)

DNA filter assay

hybridization with radioactively labelled DNA

detection by autoradiography

mass-spectrometric DNA assay

single-stranded DNA is probed by primer-specific extension

detection by sequence-specific molecular mass using MALDI-TOF mass spectrometry

spectral DNA assay ("liquid array")

microspheres are labelled with single-stranded DNA probes and coded with 2 fluorophores

detection by interrogating spectral addresses in a FACS system

Reporter groups

General. Reporter molecules play an important role in both basic and applied research. They are used 1) in cytochemical analysis of cells and histochemical analysis of tissues, 2) for sorting cells in a cell sorter, 3) for visualization of binding events, e. g., of antibodies, receptors, or DNA, and 4) in many procedures used in genetic engineering (e. g., for cloning or the investigation of promoters). Frequently used reporter atoms and molecules include radioactive isotopes, fluorophores, and enzymes. For linking reporters to proteins or DNA, the biotin-streptavidin and digoxigenin systems are often used. Frequently used reporters in genetic engineering are the genes coding for coding forc β-galactosidase, luciferase, and GFP (green fluorescent protein).

Radioactive markers. In radioimmunoassays (RIA), reactants are labeled with ^{131}I or ^{35}S, and radioactivity is determined in a scintillation counter. In molecular genetics experimentation, ^{32}P-labeled phosphate is often introduced into DNA or RNA, using DNA polymerase I, the Klenow fragment, or RNA polymerase. Radioactively labeled DNA or RNA can then be used to detect hybridization events by autoradiography or, somewhat faster, in a phosphoimager. These methods are still being used in many laboratories, in spite of strict radiation-safety regulations and competition from faster protocols, because they are highly sensitive.

Fluorophores. Fluorophores such as fluoresceine or rhodamine are highly sensitive reporter molecules, down to the picomolar range. Fluorophore-labeled antibodies are used in histochemical and cytochemical investigations. In combination with a cell sorter (FACS = fluorescence-activated cell sorter), they permit labeled cells to be separated from unlabeled cells at speeds of 1000 cells min^{-1} or more. This procedure is used in cell biology, e. g., for separating B cells from T cells after labeling specific antigens displayed on their surface, and for separation of mixed cultures of bacteria after labeling taxon-specific sequences of their 16S rRNA by hybridization (FISH = fluorescence in-situ hybridization). Fluorescence reporters for DNA are SYBR-Green and ethidium bromide.

Enzymes. Compared to the reporter molecules mentioned so far, enzymes have the advantage of further enhancing an assay's sensitivity by signal amplification. This property is especially useful for quantitative analytical determinations. Reporter enzymes that are used often are alkaline phosphatase and horseradish peroxidase. Their reactions can be assayed in biosensors, by electrochemistry, or photometrically. If fluorometric or chemiluminometric enzyme assay protocols are used, the sensitivity may reach pico- or even attomolar detection limits.

Digoxigenin and biotin-streptavidin. These chemicals are often used to couple a reporter molecule to the biomolecule to be analyzed. Digoxigenin (M_R = 390.52) is a steroid that can be coupled to a nucleotide via its hydroxy group without interfering with hybridization events. It binds with high affinity (10^{-9} M) to a specific antibody that can be labeled with various reporter molecules. A similar system is biotin-streptavidin (affinity 10^{-10} M).

Genetic markers. Introducing genes that code for reporter proteins enables cloning events to be rapidly analyzed. "Blue-white" screening, for instance, is based on a plasmid-borne DNA sequence coding for a fragment of β-galactosidase. If foreign DNA is inserted within this fragment, it can no longer complement the chromosomal DNA that codes for the remaining part of this enzyme, and externally added chromogenic substrate X-gal (5-bromo-4-chloro-3-indolyl-β-D-galactopyranoside) can no longer be hydrolyzed to a blue dye. Another very useful genetic marker is GFP (green fluorescent protein), whose gene has been cloned from a jellyfish. Without any additional substrates, GFP shows strong fluorescence. The GFP gene can be used as a reporter to demonstrate promoter function, gene expression, or gene regulation. Luciferase genes cloned from fireflies or photobacteria are used for the same purpose, but visualization of the expressed reporter proteins requires the addition of an enzyme substrate (luciferin or decanal, respectively).

Radioactive labeling

by filling up with ^{32}P-dATP

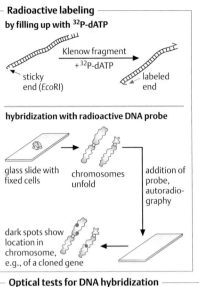

sticky end (EcoRI) → Klenow fragment + ^{32}P-dATP → labeled end

hybridization with radioactive DNA probe

glass slide with fixed cells

chromosomes unfold

addition of probe, autoradio-graphy

dark spots show location in chromosome, e.g., of a cloned gene

Cell sorter (FACS*)

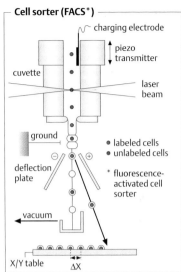

charging electrode

piezo transmitter

cuvette

laser beam

ground

● labeled cells
● unlabeled cells

* fluorescence-activated cell sorter

deflection plate

⊖ ⊕

vacuum

X/Y table   ΔX

Optical tests for DNA hybridization

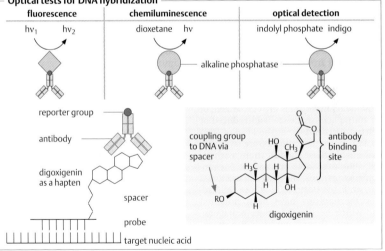

fluorescence	chemiluminescence	optical detection
hv_1 hv_2	dioxetane hv	indolyl phosphate indigo

alkaline phosphatase

reporter group

antibody

digoxigenin as a hapten

spacer

probe

target nucleic acid

coupling group to DNA via spacer

HO CH_3

H_3C

H   H

RO   H   OH

digoxigenin

antibody binding site

Frequently used reporter genes

gene	externally added substrate	detection
lacZ': gene fragment of β-galactosidase	5-bromo-4-chloro-3-indolyl-β-D-galactopyranoside (X-Gal)	visual (blue-white)
lux: luciferase from firefly or photobacteria	luciferin or decanal	luminescence
GFP: green fluorescent protein from jellyfish	–	fluorescence

259

Protein design

General. Protein design or protein engineering implies the modification of protein sequences by genetic methods. Protein engineering techniques are used to 1) understand enzyme mechanisms, 2) modify enzyme or antibody binding sites at will, and 3) alter global properties of an enzyme, such as its stability to high temperature, extreme pH, proteases, its solubility, or its antigenicity. If a known protein structure is used as the starting point, and individual amino acids or sequences are replaced by site-directed mutagenesis, the protocol is termed rational protein design. The genetic exchange of amino acids at random and selection of hits by their improved properties is called directed evolution.

General methods. Both rational protein design and directed evolution require that the gene coding for the protein be available. For rational protein design, structural information about the protein is required; this can be derived from the x-ray structure, from NMR structural data, and from a structural model that was derived from the tertiary structures of closely related proteins by homology modeling.

Mutagenesis. In rational protein design, individual amino acids are exchanged, or an amino acid sequence is introduced or deleted. This is done at the level of DNA by PCR. Several protocols are available that enable these modifications with rapid, simple, reliable procedures. For random mutagenesis, the gene can be cloned into an E. coli host strain whose DNA repair mechanisms have been impaired and which is cultured under mutagenic conditions. Alternatively, a PCR protocol is used which, by addition of Mn^{2+} ions or other means, leads to an artificially high number of errors during DNA amplification (1–3%). Gene shuffling, still another method, is based on the concept of creating a library of DNA fragments of related genes (their sequence identity must be ca. 80%) and recombining the fragments by PCR methods, followed by high-throughput screening for the desired properties.

Rational protein design. The tertiary structure of a protein is usually obtained by x-ray crystallography, sometimes also by 3D NMR techniques from ^{13}C- and ^{15}N-labeled proteins. Coordinates of > 18 000 protein structures are available from protein databases, which are internationally accessible through the Internet. If a protein of interest shows sequence homology of > 30% to a protein whose coordinates are available, homology modeling of the unknown structure based on the known coordinates usually provides a structural model of the unknown protein that is sufficiently precise for mutagenesis experiments. Until recently, such simulations were only possible in vacuum, due to limited computer power. With the advent of supercomputers and highly parallel computers, the modeling of protein structures, of mutant proteins, and of their binding to substrates or antigens can be done in solvent (this may require molecular mechanics calculations (force-field calculations) of the interactions of several tens of thousands of atoms!). Although improving, the predictions derived from such in-silico methods must usually be optimized by several cycles of simulations and genetic experiments (protein engineering cycles).

Directed evolution. In contrast to rational protein design, structural models are not necessary for this technique. For an enzyme, the encoding gene is subjected to random mutagenesis, the many mutant genes are expressed in a mutant library, and the mutants are assayed for the desired properties. For optimizing the binding selectivities of antibodies, the phage display technique has been effectively used: it allows large libraries of mutant antibodies (up to 10^{10}) to be screened. For enzymes, the quality of the enzyme assay is of crucial importance and determines both the speed of obtaining and the quality of the mutants. In recent years, this method has yielded excellent results for the improvement of industrial enzymes, e. g., in altering their substrate specificity or their thermo- and alkali stability. Robotic systems or FACS equipment (flow-activated cell sorters) are generally used for these methods.

260

Concepts

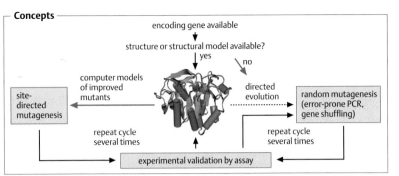

encoding gene available

structure or structural model available?

computer models of improved mutants

site-directed mutagenesis

directed evolution

random mutagenesis (error-prone PCR, gene shuffling)

repeat cycle several times

repeat cycle several times

experimental validation by assay

Examples

enzyme	method	mutation	effect
detergent protease	rational design	^{222}met→ala	oxidation stable
human insulin	rational design	^{22}lys→pro	slower degradation
tissue plasminogen activator (hTPA)	rational design	deletion of kringle 2	slower degradation
penicillin acylase	directed evolution	5 positions	better solvent stability
P450 fatty acid hydroxylase	directed evolution	4 positions	greatly modified substrate specificity

Error-prone PCR

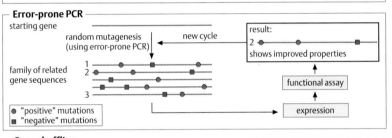

starting gene

random mutagenesis (using error-prone PCR)

new cycle

result: 2 shows improved properties

family of related gene sequences

"positive" mutations

"negative" mutations

functional assay

expression

Gene shuffling

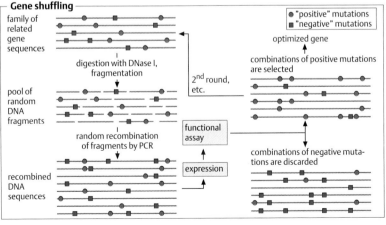

family of related gene sequences

"positive" mutations

"negative" mutations

optimized gene

digestion with DNase I, fragmentation

2nd round, etc.

combinations of positive mutations are selected

pool of random DNA fragments

random recombination of fragments by PCR

functional assay

expression

combinations of negative mutations are discarded

recombined DNA sequences

Gene therapy

General. Among the ca. 15 000 human diseases that have been reported, over 90 % cannot be cured by rational therapy. Most of them originate in inherited or acquired genetic defects. Gene therapy is the attempt to replace genes having impaired function by healthy genes. As of the end of 2002, over 600 gene therapeutic protocols have been used for the treatment of > 3500 patients, mostly in the USA. Gene therapy directed towards somatic human cells is generally accepted, but gene transfer into human sperm or egg cells (germline), which would lead to inheritable new properties of the recipient's offspring, is not and is subject to a moratorium. It is practical to distinguish gene therapy ex vivo, in which human cells are multiplied (expanded) and transformed outside the human body before they are re-infused, and in-vivo gene therapy, which is direct therapy of patients with genetic material.

General concepts. For most diseases, the genetic basis is still largely unknown. But even gene therapy of monogenic diseases having a known genetic cause meets with great difficulty: the immunological barriers of the body and the cellular control mechanisms directed towards foreign nucleic acids must be overcome. Presently, experimentation is focused on 1) recombination of defective gene sequences with added cDNA of correct sequence, 2) silencing of genes via antisense or interfering RNA, and 3) repair of defective DNA sequences with RNA-DNA chimeras. Preferred vectors include retroviruses (ca. 35 % of all protocols), adenoviruses (ca. 27 %), or cationic liposomes for lipofection at a level of ca. 12 %. Liposomes permit larger cDNA fragments to be transferred; however, the space available for foreign DNA inserts in a viral capsid is small – between 4 kbp (adenovirus) and 30 kbp (Herpes virus). Liposomes can be administered as an aerosol through the respiratory tract, reaching cells via endocytosis. Viral vectors are usually applied subcutaneously, into the muscle, or directly into a tumor. The transfer of the patient's own bone marrow after modification with DNA (autologous cell therapy) and the direct application of DNA have also been described.

Individual protocols. About ⅔ of all gene therapeutic protocols are focused on tumor treatments. To this end, a tumor-suppressor gene such as BRCA1 or p53, a cytokine gene such as IL-2, a histocompatibility antigen such as HLA-B7, or a so-called suicide gene is administered. Protocols for the treatment of monogenic diseases often concern human severe combined immunodeficiency (SCID), which is caused by defects in the gene coding for adenosine deaminase (ADA). Gene therapy of infectious diseases (vaccination with naked DNA) is another major field of research.

Gene transfer ex vivo uses protocols successfully established for bone marrow transplantation. A preferred cell type is the hemapoietic stem cell, the precursor cell for all cells of the immune and blood systems. If it should become possible to replace a genetic deficiency in these yet undifferentiated cells outside the body by gene transfer, these transgenic cells should, after transfusion into the donor, differentiate into "healthy" immune and blood cells. At present, however, cultivation of undifferentiated adult stem cells is very difficult, and using embryonic stem cells for this purpose is a matter of controversial debate.

Gene transfer in vivo. In some of the many attempts to replace defective genes by transfecting a patient with vectors, liposomes, or DNA, at least part of the target cells were transformed. Sometimes, the clinical condition of the patient improved significantly; for example, 4 of the 5 adolescent SCID-X1 patients who received gene therapy live at home again with their parents. The selective transfection of a predetermined cell type, however, is still largely an unsolved problem.

Conclusion. Although gene therapy of monocausal diseases in humans is possible in principle, many unsolved problems remain. The death of an adolescent gene therapy patient in the USA, due to uncontrolled propagation of the adenovirus used as the vector, has led to further increases in safety measures during therapy.

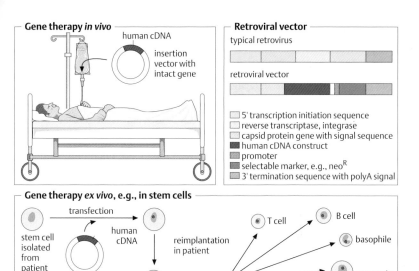

Gene therapy *in vivo*

human cDNA

insertion vector with intact gene

Retroviral vector

typical retrovirus

retroviral vector

☐ 5' transcription initiation sequence
☐ reverse transcriptase, integrase
☐ capsid protein gene with signal sequence
■ human cDNA construct
▨ promoter
▨ selectable marker, e.g., neo[R]
▨ 3' termination sequence with polyA signal

Gene therapy *ex vivo*, e.g., in stem cells

stem cell isolated from patient

transfection

human cDNA

vector with intact gene

reimplantation in patient

cell differentiation

all mature cells contain the new gene

T cell

B cell

basophile

monocyte

neutrophile

eosinophile

Vectors for gene therapy

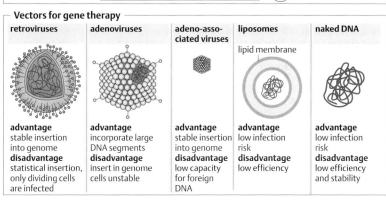

retroviruses	adenoviruses	adeno-associated viruses	liposomes	naked DNA
			lipid membrane	
advantage stable insertion into genome **disadvantage** statistical insertion, only dividing cells are infected	**advantage** incorporate large DNA segments **disadvantage** insert in genome cells unstable	**advantage** stable insertion into genome **disadvantage** low capacity for foreign DNA	**advantage** low infection risk **disadvantage** low efficiency	**advantage** low infection risk **disadvantage** low efficiency and stability

Experiments on gene therapy (end of 2002)

disease	examples/transferred genes
cancer (> 2 400 patients, > 400 protocols)	histocompatibility antigens, tumor-suppressor genes, suicide genes, IL-2, IL-7 and IL-12
monogenic diseases (> 300 patients, > 80 protocols)	SCID ADA gene, cystic fibrosis, factor IX, chronic granulomatosis
infectious diseases, mostly AIDS (> 400 patients, > 40 protocols)	transgenic T-lymphocytes, DNA vaccines
other diseases (> 100 patients, > 60 protocols)	VEGF121 (atheriosclerosis), rheumatoid arthritis

Proteomics

General. The term proteome was coined in 1995 and describes the total of all proteins encoded by a genome. In higher organisms, alternative splicing and post-translational modifications may lead, on average, to the formation of 10 proteins per gene. Proteomics is focussed on understanding how proteins are differentially expressed and modified by the genome and how this determines their interaction and function (functional genomics).

Methods. The core procedure of proteomics research is the separation and identification of a large number of proteins. The *Escherichia coli* proteome, for example, contains ca. 4000 proteins. Sample preparation is critical and requires different protocol for membrane proteins than for cytoplasmic proteins. In eukaryotes, protein extracts of individual cell types are used preferentially, providing information on differential expression. The most important method for separating proteins is 2-dimensional polyacrylamide gel electrophoresis (2D PAGE), in which separation in the first dimension is based on isoelectric point, and in the second, on molar mass. Resolution can be improved by controlling the pH gradient. Semi-quantitative analysis of 2D PAGE gels is achieved by staining (silver), scanning, and computer analysis. High-throughput systems allow about 100 gels to be analyzed per week. Identification of rarely expressed proteins (10–1000 copies per cell), correlation of post-translationally modified proteins with the correct precursor protein, and quantification are major bottlenecks. One method of achieving quantitative analysis is to use recombinant antibody libraries, but the proteins' identities must be known. If quantities > 1 µg of protein are expressed, N-terminal sequencing may be possible if the N-terminus is not acylated; then the sequence can be matched with a computer database. For identification of less frequently expressed proteins, mass spectrometry is the method of choice. Using MALDI-TOF mass spectrometry (matrix-assisted laser-desorption-ionization time-of-flight), approximate molecular masses are obtained. Alternatively, proteins can be digested in the gel by trypsin, resulting in a peptide mixture whose mass again is analyzed by MALDI-TOF; comparison with a tryptic digest *in silico* provides information about the identity of the protein. This procedure can be used only if the genome sequence is available and all proteins have been assigned. If this is not true, electrospray mass spectrometry (ESI-TOF) can help to identify fragmentation patterns of the unknown protein, leading to protein sequence tags (PST), which can often be assigned to entries in a protein database. The sensitivity of both methods is several femtomoles per protein spot. To achieve this sensitivity, however, several hundred thousand copies of this particular protein must be expressed in the cell.

Applications. Using proteomics, cell-specific or induced variations in the expressed protein pattern of a cell can be analyzed. Examples are 1) differences in the expression pattern of the proteome of *E. coli* after growth on glucose or lactate as a C source, 2) comparison of the protein patterns of pancreatic cells from a healthy and a diabetic person, 3) toxicological investigations on altered protein patterns in the liver after medication. Often, marker proteins for specific diseases have already been identified. Proteomics can also help to identify protein functions in a cell. To this end, scientists try to build a proteome map for the cell type under investigation and to understand protein interactions in this "cell factory". A useful method is the genetic introduction of his tags, which can be used to purify the expressed protein and associated proteins by affinity purification and establish their identity by mass spectrometry.

Human proteomics initiative. Currently (mid-2002), the SWISS-PROT database contains 8300 annotated human protein sequences. These entries are associated with about 21 900 literature references, 21 200 experimental or predicted post-translational modifications, 2340 splice variants, and 13 450 polymorphisms (most of which are linked with disease states).

One genome – various proteomes

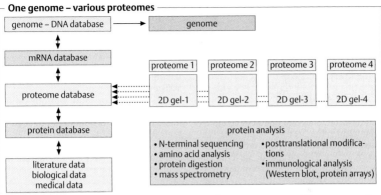

| genome – DNA database | → | genome |

mRNA database ↕

proteome database ↕ ← proteome 1 (2D gel-1) ← proteome 2 (2D gel-2) proteome 3 (2D gel-3) proteome 4 (2D gel-4)

protein database ↕

literature data
biological data
medical data

protein analysis
- N-terminal sequencing
- amino acid analysis
- protein digestion
- mass spectrometry
- posttranslational modifications
- immunological analysis (Western blot, protein arrays)

a higher organism with 100 000 genes may contain > 1 million protein species, which are expressed in modified forms by different cell types (> 50 000 protein species per cell type)

Methods of proteome analysis

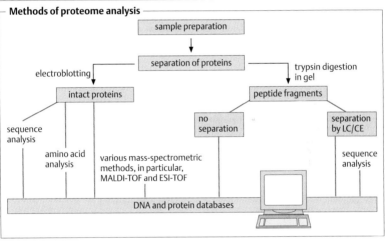

sample preparation
↓
separation of proteins

electroblotting → intact proteins

trypsin digestion in gel → peptide fragments

- sequence analysis
- amino acid analysis
- various mass-spectrometric methods, in particular, MALDI-TOF and ESI-TOF

no separation

separation by LC/CE → sequence analysis

DNA and protein databases

2D gel of yeast proteins

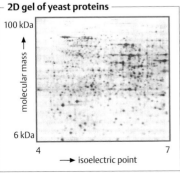

100 kDa

molecular mass →

6 kDa

4 → isoelectric point 7

Applications

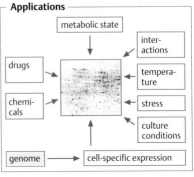

metabolic state

drugs

chemicals

genome → cell-specific expression

inter-actions

temperature

stress

culture conditions

Drug screening

General. New drugs and agrochemicals have traditionally been searched for by trial-and-error methods. Recent insights in the life sciences and the technical concepts of genetic engineering have replaced these methods by more rational approaches. They are based on the premise that drugs act on one or several targets within the human organism. Frequent targets are enzymes, receptors, or ion channels. The targets of agricultural agents may also be plant proteins that are involved in photosynthesis. In consequence of developments in genomics and proteomics, the functions of these targets are now much better understood. Targets can be prepared in sufficient quantities by recombinant technology, and their interaction with natural or synthetic compounds can be experimentally investigated. Lead structures identified by this approach can be modified by chemical synthesis, using combinatorial-chemistry approaches, resulting in novel, tailor-made, highly efficient drugs, or agrochemicals.

Identification and preparation of targets. The identification of targets is still difficult, especially for diseases of multigenic origin (the usual situation). By combining studies of inheritance patterns, for example, in genetically isolated populations such as those in Iceland or Tasmania, with SNP, allele, or proteome analysis of patients suffering from a disease, valuable progress has been made recently. Once an enzyme, an ion channel, or a receptor has been identified as the putative cause of a disease, animal experiments are used to validate the hypothesis by knockout or RNA$_i$ experiments. If this approach leads to the validation of a target, it can be used in a high-throughput assay for screening new drugs. If, e.g., a G-coupled receptor has been identified as a target, it is expressed in the membranes of mouse fibroblast cells, while firefly luciferase is coexpressed as a reporter enzyme in the cells' cytoplasm. If a ligand binds to the receptor, signal transduction via cAMP-responsive elements raises the level of intracellular cAMP. Once luciferin, the luciferase substrate, is added to such cells, ligand binding can now be quantified by luciferase activity, which depends on cAMP.

High-throughput screening. A widely distributed system in industry entails robot-assisted screening of large chemical libraries (100 000 or more substances) for interaction with a drug target. Combinatorial chemistry can increase the size of such libraries nearly without limit. A decisive factor for success is the availability of a rapid, valid assay for interaction between drug and target. Assays are usually carried out in 96-well, sometimes 384-well microtiterplates. Silicon-wafer technology in combination with confocal laser spectroscopy has further reduced the scale to nanoliter volumes, allowing the assay of > 10 000 chemicals per day.

Rational drug design. Although the x-ray structures of most drug targets are still unknown, scientists have long attempted to postulate their drug-binding properties by systematic measurement of their binding of chemical analogues, leading to quantitative structure–activity relationships (QSAR). Based on a model of the binding-site structure derived from such analyses, they attempt to optimize the structure of a new drug by using molecular modeling and other computational approaches.

Future aspects. In the pharmaceutical industry, only one new drug reaches the market for each ca. 50 000 chemical substances investigated. Worldwide, only ca. 30 new chemical entities (NCEs) are introduced each year as new active agents in drugs. In view of the many incurable diseases, and also for economic reasons, the global pharmaceutical industry attempts to increase its success rate by using target-based screening. Another concept (pharmacogenomics) is based on analysis of individual disease-linked polymorphisms. If individual targets can be identified by, for example, DNA assays, drugs could be tailored to fit each individual protein involved in regulation and metabolism, leading to personalized medications with fewer side effects.

Drug screening

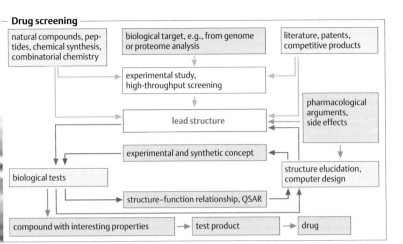

Human drug targets

	number[*]	example	application
enzymes	8000	acetylcholinesterase	Alzheimer drug
receptors	15000	serotonin receptor	schizophrenia
ion channels	3000	Ca ion channel	old-age diseases

[*] man, estimates

High-throughput screening

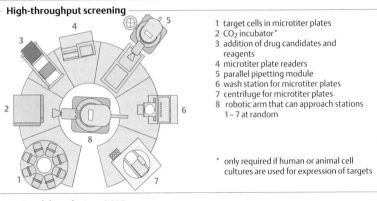

1 target cells in microtiter plates
2 CO_2 incubator[*]
3 addition of drug candidates and reagents
4 microtiter plate readers
5 parallel pipetting module
6 wash station for microtiter plates
7 centrifuge for microtiter plates
8 robotic arm that can approach stations 1–7 at random

[*] only required if human or animal cell cultures are used for expression of targets

Rational drug design, QSAR

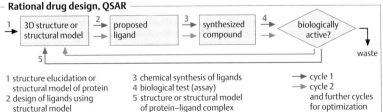

1 structure elucidation or structural model of protein
2 design of ligands using structural model
3 chemical synthesis of ligands
4 biological test (assay)
5 structure or structural model of protein–ligand complex

→ cycle 1
→ cycle 2 and further cycles for optimization

Bioinformatics

General. The rapid progress of molecular biology would be unthinkable without the breathtaking developments in computing and telecommunications – two technologies that ensure storage, rapid sorting, and worldwide retrieval of large amounts of biological data. The communication platform for data exchange is the World Wide Web, which is part of the Internet, founded as early as 1970 for scientific telecommunications within the USA. Today, the Web spans the globe with > 35 million Web servers and > 100 million users, a figure that is increasing every year. Apart from its dominant commercial use, the Internet enables the global exchange of scientific data and their collective or individual retrieval and organization. The terms bioinformatics and biocomputing encompass the handling and processing of information on DNA and protein sequences, as well as protein structures, which are increasingly directed toward understanding the genomes and proteomes of whole organisms, including humans.

Sequence information. The broad implementation and high speed of DNA sequencing has led to a steep increase in the number of stored DNA sequences available to the public. By the end of 2002, the total number of DNA nucleotide bases that have been sequenced exceeded 20 trillion (2×10^{10}). The storage of these data in a single global databank, with 3 mirror sites in the USA, Japan, and Europe (UK), and their access through the Web, permits scientists to compare new DNA sequence data in real time with known sequences and to use this information to reach conclusions regarding any identity or homology between their DNA or protein sequences and previously defined genes and proteins. The most important databases for protein sequence information are the PIR (Protein Information Resource) in Washington, D.C., with ca. 280 000 entries, and the SwissProt database in Geneva with ca. 120 000 entries.

Structure information. Relative to sequence information, data on tertiary structures of proteins have grown less quickly, because they rely essentially on x-ray analysis, which includes time-consuming preparation of crystals and their heavy-metal derivatives. This experimental bottleneck is a particular drawback for the analysis of large protein complexes and for membrane proteins, which constitute about 30 % of all known proteins. Protein structure analysis by multidimensional NMR, an alternative procedure applicable to proteins in solution, is presently limited to proteins with molecular mass < 30 kDa. In spite of these experimental difficulties, the number of protein structures available from the Protein Data Bank (PDB) by the end of 2002 exceeded 18 000 structures and structure variants, with an increase of about 3000 per year. If the homology of a protein sequence (for a protein of unknown tertiary structure) to the sequence of a protein of experimentally determined structure exceeds 30 %, a structural model of the unknown protein can usually be built on the computer, using homology modeling. In the structure databases one can find, at present, about 100 to 150 different protein architectures. Although the number of theoretically possible protein sequences is astronomically high (for a protein made up of 300 amino acids, the possible number of different sequences is 20^{300}), two secondary structural motifs (the α helix and the β sheet) are relatively stable. This greatly reduces the number of probable protein architectures. In addition, larger structural motifs seem to be constituted in a modular manner. The number of different basic architectures is believed not to exceed 1500. *Structural genomics* attempts to predict the structure and function of all proteins in the proteome and to use this information to elucidate overall protein architectures.

Functional analysis. Many specialized websites focus on individual enzymes such as acetylcholinesterases, ion channels, receptors, cytokines, or genomic information. In addition, there are attempts to analyze and alter the metabolic fluxes in whole organisms (metabolic engineering) and their complex signaling networks (system dynamics).

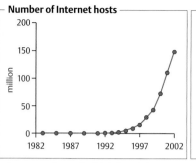

Number of Internet hosts

million

200
150
100
50
0

1982   1987   1992   1997   2002

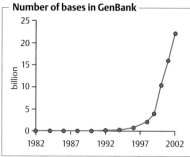

Number of bases in GenBank

billion

25
20
15
10
5
0

1982   1987   1992   1997   2002

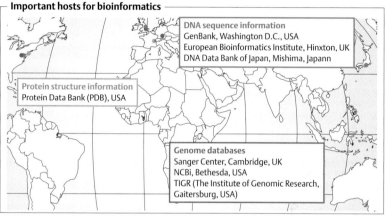

Important hosts for bioinformatics

DNA sequence information
GenBank, Washington D.C., USA
European Bioinformatics Institute, Hinxton, UK
DNA Data Bank of Japan, Mishima, Japann

Protein structure information
Protein Data Bank (PDB), USA

Genome databases
Sanger Center, Cambridge, UK
NCBi, Bethesda, USA
TIGR (The Institute of Genomic Research,
Gaitersburg, USA)

Important websites (a small selection)

DNA sequences		
GenBank	>22 billion base pairs	www.ncbi.nlm.nih.gov/Genbank/index.html
EMBL Database	(December 2002)	www.ebi.ac.uk/embl/index.html
DNA-Database of Japan		www.ddbj.nig.ac.jp
(3 mirror sites)		

genome informations		
Sanger Center	human genome, mouse genome, many other genomes	www.sanger.ac.uk
National Center for Biotechnology Information (NCBI)	human genome, mouse genome, many other genomes	www.ncbi.nlm.nih.gov
TIGR data base	hundreds of completed and unfinished genomes	www.tigr.org

protein structures		
Protein Data Bank (PDB)	>18 000 protein structures	www.pdb.bnl.gov

proteome informations		
Swiss Institute of Bioinformatics (ExPASy)	>8 000 annotated human protein sequences, and more	us.expasy.org

Metabolism

General. In spite of > 4 billion years of life on earth, which have led to a great variety of living organisms, their basic building blocks and their patterns of metabolism and replication are based on variations of just a few basic principles. The number of species by far exceeds 1 million, but only a few thousand different enzyme functions have been classified, and several hundred thousand proteins (many of which share high homology with simple unicellular eukaryotes such as *Saccharomyces cerevisiae*) are enough to build and maintain a higher organism such as humans. All living beings on earth form a sensitive ecological network, which includes many thousands of specialized species that survive under nearly all imaginable environmental conditions (ecological niches). A major distinction is made between autotrophic organisms, which can use CO_2 as the main carbon source, and heterotrophic organisms, which need organic compounds for growth. Another distinction is made between aerobic organisms, growing in air, and anaerobic organisms. With respect to the details of their metabolisms, organisms may differ, for example, by how they assimilate glucose: through fructose-1,6-bisphosphate (glycolysis) or by a pentose phosphate or a 2-keto-3-deoxy-6-phosphogluconate pathway. With the accomplishment of total genome sequencing, a breathtaking view into even finer structural and metabolic variations has become possible, which helps us to better understand how organisms are adapted to their specific environments. We are also making great strides in understanding regulatory networks in organisms and their environmental interactions (system dynamics, biocybernetics) and are learning to simulate complex living systems *in silico*. In biotechnology, a key interest is to modulate metabolism, e. g., for increased yields of a product, for elimination of a by-product, or, in breeding, for introduction or elimination of a phenotypic trait. Traditional methods of crossing or mutation, followed by selection, are now backed up more and more by genetic engineering.

Autotrophic metabolism. Autotrophic organisms reduce CO_2 to carbon sources such as glucose. Phototrophic organisms such as plants, algae, and cyanobacteria derive the energy required for this process from light, which they convert, in their photoreaction centers, into chemical energy stored as adenosine triphosphate (ATP, "the universal energy currency"). Lithotrophic organisms derive energy from the oxidization of inorganic compounds, e. g., S, N, metal ions. Autotrophic organisms of importance in biotechnology are transgenic plants, nitrifying bacteria, and Thiobacilli used in mineral leaching.

Heterotrophic anaerobic metabolism. Heterotrophic anaerobic organisms are used in biotechnology in the production of ethanol, acetone, 1-butanol, and lactic acid. They generate ATP by catabolism of sugars. Methane-forming archaea, which develop in anaerobic sludge treatment, also belong in this group; they exhibit some unusual metabolic steps. Energy generation in anaerobic metabolism proceeds with low yield efficiency.

Heterotrophic aerobic metabolism. Most microorganisms used in biotechnology exhibit heterotrophic aerobic metabolism. The bulk of their energy is generated through the respiratory chain, which feeds off the citric acid cycle. With ca. 36 moles of ATP generated per mole of glucose, the energy yield is very high. Several biotechnological products, such as citric acid or glutamic acid, are derived from the citric acid cycle, which as a result must be replenished by anaplerotic pathways in overproducing organisms.

Secondary metabolism. Many organisms form metabolites that are not involved in primary cell functions (secondary metabolites). In plants, secondary metabolites are of key importance in defense mechanisms against pathogens and predators and in attracting insects for fertilization and dissemination. In microorganisms, the physiological function of secondary metabolites is less clear. Often, these are important biotechnological products such as antibiotics, alkaloids, colorants, or aroma compounds.

Metabolic pathways – a network

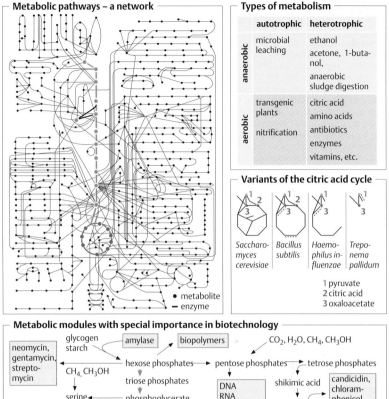

- metabolite
- — enzyme

Types of metabolism

		autotrophic	heterotrophic
anaerobic		microbial leaching	ethanol
			acetone, 1-butanol,
			anaerobic sludge digestion
aerobic		transgenic plants	citric acid
			amino acids
		nitrification	antibiotics
			enzymes
			vitamins, etc.

Variants of the citric acid cycle

Saccharomyces cerevisiae | *Bacillus subtilis* | *Haemophilus influenzae* | *Treponema pallidum*

1 pyruvate
2 citric acid
3 oxaloacetate

Metabolic modules with special importance in biotechnology

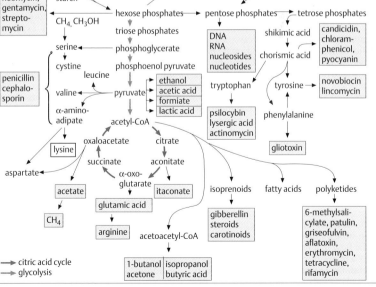

⟶ citric acid cycle
⟶ glycolysis

neomycin, gentamycin, streptomycin

glycogen starch — amylase — biopolymers

CO_2, H_2O, CH_4, CH_3OH

CH_4, CH_3OH

hexose phosphates → pentose phosphates → tetrose phosphates

triose phosphates

serine

phosphoglycerate

DNA RNA nucleosides nucleotides

shikimic acid

candicidin, chloramphenicol, pyocyanin

cystine

phosphoenol pyruvate

chorismic acid

penicillin cephalosporin

leucine

valine

pyruvate

ethanol
acetic acid
formiate
lactic acid

tryptophan

tyrosine

novobiocin lincomycin

α-amino-adipate

acetyl-CoA

psilocybin lysergic acid actinomycin

phenylalanine

gliotoxin

lysine

oxaloacetate

citrate

aspartate

succinate

aconitate

α-oxo-glutarate

itaconate

isoprenoids

fatty acids

polyketides

acetate

glutamic acid

CH_4

arginine

acetoacetyl-CoA

gibberelin steroids carotinoids

6-methylsalicylate, patulin, griseofulvin, aflatoxin, erythromycin, tetracycline, rifamycin

1-butanol acetone | isopropanol butyric acid

Metabolic engineering

General. Increasingly in-depth knowledge of the metabolism, its pathways, and their regulation has led to attempts to describe the entire metabolic flux of whole cells, using mathematical modeling, computer simulations, and fast, precise measurement techniques. Based on these analyses, investigators are attempting to modulate metabolism, regulatory functions, and signal networks, to direct the metabolic flow towards a desired product or to adapt it to a desired carbon source (metabolic engineering). The experimental setup usually includes genetic engineering of a key step in synthesis or catabolism (metabolic design).

Metabolic flux analysis. Once the biochemistry of a metabolic pathway is understood, a system of balance equations can be formulated for each intermediate and added up in a stoichiometric matrix. Using these equations under a given set of conditions, the metabolic flux within the pathway can be calculated. A necessary condition is that mass transfer to and from the environment through the cell membrane (substrate uptake and product release) must be quantified. This may not be sufficient, however, to solve the system of balance equations unambiguously. In the standard case of metabolic fluxes through additional pathways, it is necessary to determine enzyme activities, to measure expression profiles of genes using DNA arrays, or to carry out marker experiments using probes that are labeled with radioactive (^{14}C, ^{3}H, ^{32}P, or ^{35}S) or stable isotopes (^{13}C or N^{15}). To this end, cells are grown on well defined, isotope-labeled substrate mixtures, and the metabolites, occasionally also the macromolecules, are analyzed for their isotope distribution, using nuclear magnetic resonance (NMR) or mass spectrometry (MS). In favorable cases, the use of such isotopomers (isotope-labeled isomers) may allow generation of a mathematical model for metabolic flow in a branched pathway or even in a whole cell.

Metabolic control analysis. If one wants to understand which enzyme in a pathway is limiting to the metabolic flux, the hierarchy of flux control must be analyzed. In other words, we need to understand which fractional change in the activities of all the enzymes involved has the largest impact on the overall pathway. Usually, this type of control is distributed among several enzymes. To determine each enzyme's control coefficient, its expression can be modified by genetic engineering, and the resulting change in metabolic flux can be quantitatively analyzed. Alternatively, the flux control coefficients can be calculated on the basis of mathematical models that adequately describe the kinetic parameters of the pathway network under investigation. For experimental validation of such models, intracellular metabolites are measured under transient process conditions. This is done, for instance, by stimulating a stationary cell culture by adding a substrate (e. g., a glucose spike) and measuring the response of the cell by analyzing all relevant internal pool concentrations.

Applications. In microorganisms, these methods are mostly used to 1) enhance the spectrum of substrates, 2) enlarge their biodegradation potential, and 3) increase yields for the production of metabolites. For instance, the lactose operon was successfully cloned and expressed in *Zymomonas mobilis*, allowing for the use of whey for ethanol production, and in *Corynebacterium glutamicum*, for glutamic acid and lysine production from whey. By cloning genes for the catabolism of aromatic compounds into *Pseudomonas* sp. B13, engineered mutants were obtained which, unlike the wild strain, could grow on chlorinated or methylated aromatic compounds. The increase in product formation by metabolic engineering is studied for many purposes, e. g., for the production of amino acids, ethanol, biopolymers, and vitamins, and also for producing secondary metabolites such as antibiotics. For example, a thorough analysis of competitive metabolic fluxes during the biosynthesis and secretion of L-lysine in *Corynebacterium glutamicum* led to a strategy that increased lysine yields by 50 %, by using molecular genetics for redirection of flux.

Block diagram of the metabolism of an aerobic, heterotrophic organism

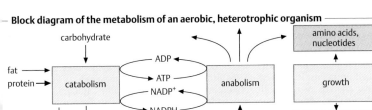

Metabolic flux during lysine production

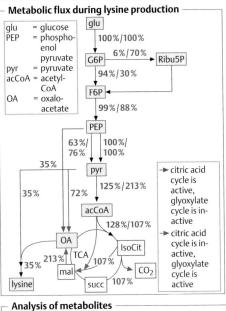

glu = glucose
PEP = phosphoenol pyruvate
pyr = pyruvate
acCoA = acetyl-CoA
OA = oxaloacetate

→ citric acid cycle is active, glyoxylate cycle is inactive

→ citric acid cycle is inactive, glyoxylate cycle is active

Metabolite analysis using NMR

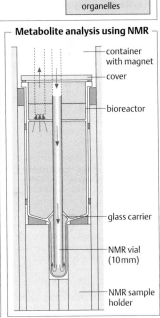

- container with magnet
- cover
- bioreactor
- glass carrier
- NMR vial (10 mm)
- NMR sample holder

Analysis of metabolites

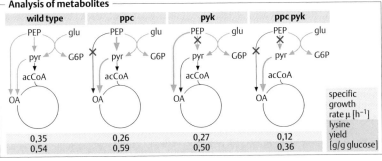

	wild type	ppc	pyk	ppc pyk	
specific growth rate μ [h⁻¹]	0,35	0,26	0,27	0,12	specific growth rate μ [h^{-1}]
lysine yield [g/g glucose]	0,54	0,59	0,50	0,36	lysine yield [g/g glucose]

273

Systems biology

General. Systems biology is a new field of research aiming at a holistic description of cell functions. It is based on the functional analysis of metabolic, regulatory, and signaling compounds, which are aggregated to functional modules. These functional modules are combined to provide a cell model that is based on experiment and allows interactive predictions of a cell's behavior. A long-term objective of systems biology is, for example, the support of clinical trials by simulation studies. Using an analogy, systems biology is a dynamic roadmap of the cell, which includes traffic patterns, why such patterns emerge, and how we can control them.

Key components. Before compiling information on a biological system, detailed data must become available on the *system's structure*. This includes the functions and interactions of genes, protein structures, an biochemical pathways and the mechanisms by which the intracellular and multicellular structures are modulated. A second key property of living cells is their capacity to undergo dynamic changes, in response to internal or external factors. *Simulation studies* describe these changes and allow the identification of essential mechanisms underlying specific behaviors. *Control factors* which minimize malfunctions must be elucidated. Finally, *design principles* and simulations help to cluster the above modules into a functional system.

Measurement. Any analysis of the biological regulation present in a complex living system requires extensive databases, for example, the sequencing and expression profiling of genes (genome and transcriptome), proteome analysis, and the simultaneous measurement of enzyme activities and metabolites (metabolome) within milliseconds, to follow metabolic events under defined conditions in a highly parallel format with high throughput. Single-molecule measurements, robotic nano-devices, and femtolasers that permit visualization of molecular interactions are typical examples ("interactome"). Microfluidic systems such as the micro total analysis system (μ-TAS) allow picoliters of samples to be measured rapidly and precisely, and many established techniques such as PCR and protein separation by capillary electrophoresis are being scaled-down to follow, for example, mRNA levels and protein modifications.

Robustness. Biological systems exhibit a considerable robustness, which protects the cell if a subsystem runs out of control. Phenomenological parameters related to robustness are: 1) the ability to cope with environmental changes; 2) a system's relative insensitivity to kinetic changes; and 3) a graceful degradation of the system's functions after damage, rather than catastrophic failure. This behavior is attained by various control systems such as negative feedback and feedforward control, by redundancy and structural stability of vital functions, and by modularity, that is, physical or functional insulation of subsystems so that failure of one module does not spread to others.

Computational tools originally designed for general engineering purposes are frequently used in systems biology. Because data integration, management, visualization, and analysis of large-scale cellular systems is a challenging task, integrated modeling and simulation software has been developed. More recently, Systems Biology Mark-Up Language (SBML), along with Cell Mark-Up Language (CellML), have evolved as promising standards for XML-based computer-readable model definitions. These standards enable models to be exchanged irrespective of the computational tool used. Systems Biology Workbench (SBW) is built on SBML and provides a framework of modular open-source software that can be shared among research institutions for collective projects.

Consortia and applications. Major projects are under way in the U.S.A. (e.g., the Alliance for Cellular Signaling, AfCS), in Japan (e.g., the Kitano Project), and in Germany (systems biology of the liver cell).

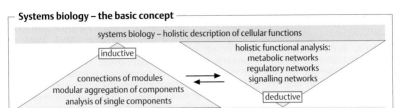

Systems biology – the basic concept

systems biology – holistic description of cellular functions

inductive

holistic functional analysis:
metabolic networks
regulatory networks
signalling networks

deductive

connections of modules
modular aggregation of components
analysis of single components

biological information/knowledge

Terms and experimental methods

term	quantitative data on	methods
metabolome	metabolites	robot-assisted rapid LC – MS, enzyme assays ^{15}N ^{13}C tracer analysis based on NMR
genome	gene function	knock-out experiments
transcriptome	differential mRNA formation	expression profiles using cDNA arrays
proteome	differential protein synthesis and modification	MALDI-TOF or electron spray mass spectro- metry, 2D gel electrophoresis
"interactome"	differential protein-protein interactions	two hybrid system, atom force microscopy, fluorescence resonance electron transfer (FRET)

Signalling network

transforming growth factor beta (TGF-β)
signal transductionnetwork in mammalian
cells, based on 553 reactions and 561 pools.
Most of the pools are proteins or
enzymes activated by phosphorylation

red dots: reaction steps
blue dots: pools

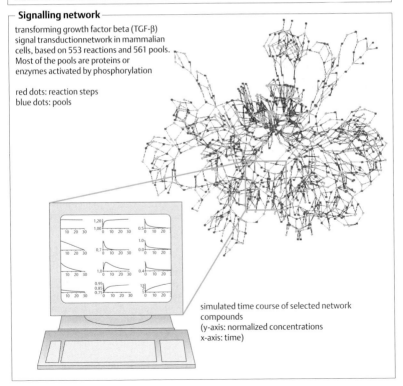

simulated time course of selected network
compounds
(y-axis: normalized concentrations
x-axis: time)

Safety in genetic engineering

General. Early in the development of rDNA technology, scientists expressed concerns about the safety of the techniques employed to transfer genes from one organism to another. After a short moratorium, self-imposed by scientists in the USA and the UK after the Asilomar conference in 1975, safety in genetic engineering is now controlled in all industrialized countries by regulations, in some by laws. There are, however, significant differences in the details, e. g., in containment requirements.

United States NIH guidelines. Risk assessment is ultimately a subjective process. The investigator must make an initial risk assessment based on the risk group (RG) of an agent. Agents are classified into four risk groups according to their relative pathogenicity for healthy adult humans by the following criteria: 1) risk group 1 (RG1) agents are not associated with disease in healthy adult humans; 2) risk group 2 (RG2) agents are associated with human disease that is rarely serious and for which preventive or therapeutic interventions are *often* available; 3) risk group 3 (RG3) agents are associated with serious or lethal human disease for which preventive or therapeutic interventions *may be* available; 4) risk group 4 (RG4) agents are likely to cause serious or lethal human disease for which preventive or therapeutic interventions are *not usually* available.

European guidelines. In Europe, a statutory regulatory system is in force. It requires the assessment of the risks associated with the use of genetically modified organisms (GMOs). All work with living GMOs carried out within the European Union is regulated tightly by EU directives (European Directive on the contained use of genetically modified microorganisms (GMMs)). Each country implements the EU directives into its own legal system. In the UK the legislation covers all organisms, not only microorganisms. The Health and Safety Executive (HSE) operates and enforces legislation in Great Britain that controls the safety of activities involving genetically modified organisms in containment (Guidance on the genetically modified organisms (contained use) regulations 2000). GMM are classified into 4 classes, depending on the risk assessment. When the assessed control measures fall between two containment levels, the activity must be classified at the higher level. The HSE must be notified of some activities involving modified organisms.

Trained personnel. All staff who carry out genetic experimentation must have adequate professional training and must periodically receive additional formal instruction. A project leader is responsible for the professional conduct of all experiments and for their documentation. In some countries, a biological safety officer must be appointed for one or several laboratories who, in the name of the central executive officer of the institution, supervises the technical status of the laboratories and consults with the project leaders.

Laboratory equipment. Laboratories used for genetic engineering must first be cleared by the regulatory office in charge. They must be signposted, and admission is regulated. Depending on the experiments that are to be done, they must comply with different biosafety or containment levels (BSL), and construction features must meet these standards (e. g., clean benches with negative pressure, locks). The requirements for the discharge of wastewater, exhaust air, and waste increase with increasing safety levels. This also relates to health checks. In Germany, e. g., all employees working with RG2 agents or higher must undergo regular health checkups. Even stricter regulations are in place for recombinant product production facilities. In some countries, the public is invited to participate in the registration of production plants starting at level RG2.

Documentation. Generally speaking, a regulatory office must be notified about the initiation of work at a genetic engineering laboratory, and the lab must be registered. All genetic experiments must be documented. In Germany, documents concerning RG1 agents must be kept for 10 years; those for RG2 agents and higher, for 30 years.

Genetic engineering laboratory RG1 and RG2 (according to German legislation)

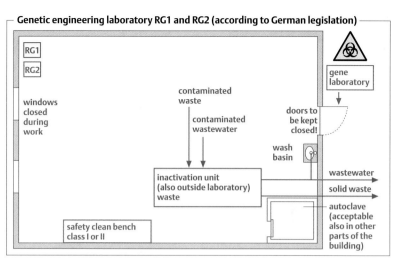

Genetic engineering production facilities

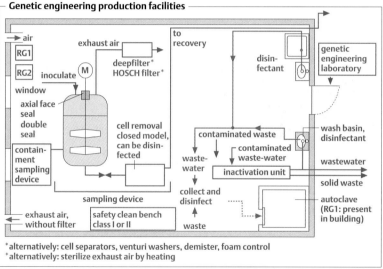

*alternatively: cell separators, venturi washers, demister, foam control
*alternatively: sterilize exhaust air by heating

Risk assessment of genetically engineered microorganisms

	risk for man and environment	examples
RG1	no risk	lab strains of *E. coli*, bakers' yeast, transgenic plants and animals
RG2	low	some Pseudomonas strains, Xanthomonas
RG3	moderate	*Mycobacterium tuberculosis*, plant viruses
RG4	high	microorganisms strongly pathogenic to humans

Regulation of products derived from biotechnology

General. The manufacturing and sale of genetically engineered products is regulated by a complex set of rules intended to serve the safety of the consumer and the environment. In spite of differences in national regulations, these rules are similar in most countries. Because the USA pioneered genetic engineering, this survey is focused on US regulations. Most regulations are issued by the Department of Agriculture (USDA) (planting, field tests), the Environmental Protection Agency (EPA) (pesticides, genetically modified microorganisms), and by the Food and Drug Administration (FDA) (food products and additives, biopharmaceuticals)

Registration of a (bio)pharmaceutical. Pharmaceutical products are registered by the higher health authorities of a nation, in the USA by the Food and Drug Administration (FDA). For registration of a drug, the following documents must be provided: 1) extensive data on its efficacy and safety, 2) precise documentation of the manufacturing process at certified sites (ISO 9001), and 3) documents concerning continuing quality control. The registration procedure usually starts with a preclinical phase, where efficacy and safety are investigated in the research laboratory and in animal experiments. If promising results are obtained, a process for manufacturing the drug is established and it receives the status of an experimental drug, entering a series of clinical experiments on humans. First, a small group of healthy volunteers is used to test drug safety (clinical phase I). This is followed by determination of efficacy and safety on groups of patients (clinical phase II). If all results are promising, testing of efficacy, safety, and side reactions is done with large groups of patients (clinical phase III). All data must then be submitted to the FDA, which can decide to register the drug or ask for further experiments. This process takes 11 years for chemical drugs, but just 9 years for biopharmaceuticals, at an average price exceeding 100 million US$ (only 1 among 50 000 preclinical candidates reaches registration). Parallel to the registration steps, the manufacturing process is standardized and documented. It must be proven that the product is free of chemical, microbiological, and genetic contaminants. This implies scrupulous control of the quality of air, water, starting materials, and storage conditions during manufacturing. In fermentation processes, the absence of biological contaminants (mycotoxins, retroviruses, etc.) must be ensured.

Agricultural products. Since 1990, > 25 agricultural biotechnology products have been cleared by the US regulatory system to commercialization in the marketplace; e.g., in 1996, the EPA approved the use of a genetically engineered *Bacillus thuringiensis* as a pesticide. In 1990 and 1994, respectively, the FDA approved the commercial use of recombinant chymosin (rennet) for cheese making and marketing of the FlavrSavr tomato. In the past few years, the USDA and FDA have approved the commercial use of insect-resistant recombinant corn, cotton and potato; of herbicide-tolerant rapeseed, cotton, soybean, and corn: of slow-ripening tomatoes; and of rapeseed with an altered oil composition.

Food products and additives may be obtained by fermentation steps or by genetic modification. They include transgenic fruits and vegetables, meat and fish, and staple products such as flour, sugar, or milk from transgenic organisms. Food additives that are prepared by fermentation include enzymes, thickeners such as dextran or xanthan gum, aroma compounds, and colorants. In the USA, since 1996 none of the above products has been subject to labeling. The 1997 European Union legislation on novel foods and novel food ingredients however, requires labeling on food packages containing genetically modified organisms or their proteins or DNA. For products that have been manufactured by a genetically modified organism but which do not remain in the end product, labeling is not required.

Biopharmaceuticals

	under develop-ment (2000)	registered (2000)	%	average time to registration
traditional drugs	1 000	27	2,7	11,3 years
biopharmaceuticals	369	32	8,7	7,8 years

in the USA, ~ 50 therapeutic biopharmaceuticals were registered by the end of 2001

Registration of a pharmaceutical

1 discovery, efficacy
2 animal tests
3 clinical testing phase I: safety for healthy humans (20–30 volunteers)
4 clinical testing phase II: efficacy and safety (100–300 patients)
5 clinical testing phase III: large-scale experiments in hospitals on efficacy, safety and side effects (1000–5000 patients)
6 engineering of plant
7 construction
8 manufacturing procedure under GMP
9 inspection of plant
MS1 registration as experimental drug
MS2 submission of all documents on testing and manufacturing to registration office

patent protection ceases after 20 years

Criteria for registration

efficacy	effects are proven and of national economic relevance*
safety	based on animal and clinical experiments, side effects can be estimated and are low compared to efficacy
manufacturing safety	the manufacturing process is standardized and follows GMP rules. The preparation is free of side products, pyrogens, viruses, bacteria, and infectious DNA
other	price relates to benefit

*except for orphan drugs (drugs active against rare diseases)

Enzymes as food additives: conditions for registration

recommendations of the JECFA*	
enzymes obtained from animals or plants	toxicological studies not required
enzymes from GRAS microorganisms	limited toxicological tests (acute toxicity)
enzymes from other microorganisms	chronic toxicology tests
enzymes from pathogenic microorganisms	not permitted

GRAS = generally recognized as safe   JECFA = joint FAO/WHO Expert Committee on Food Additives

279

Ethical considerations and acceptance

General. Various aspects of genetic engineering and cell biology have raised science-philosophical and ethical questions. These include 1) how to protect a person's genetic information, 2) how to justify gene therapy, in particular of haploid (germline) cells, leading to inheritable traits, 3) the cloning of human embryos, 4) stem-cell therapy, and 5) animal welfare, e. g., knockout animals used in drug development and biopharmaceutical production in transgenic animals, and 6) the military use of biotechnology and genetic engineering. Although modern biotechnology is socially accepted in principle, it is widely held that benefits must be balanced against risks.

Individual genetic information. The advantages of genetic screening, in particular after having completed the sequencing of the human genome, make it reasonable to believe that individual dispositions toward genetic diseases will become predictable in the future, e. g., by prenatal diagnostics. This raises the question of how doctors and society should handle this information (e. g., should abortion be accepted if an embryo displays a genetic disposition, but not a deterministic risk, of developing an incurable disease?). We also need to clarify to what extent governmental bodies may store individual genetic data for, e. g., the purpose of faster screening in criminal investigations, or if insurance companies and employers may access these data to balance insurance or employment vs. health risks.

Gene therapy. In a democratic society, somatic gene therapy depends on the informed consent of the patient, and risks are low. However, if gene therapy is applied to haploid eggs or sperm, modified genetic patterns may be inherited by the offspring without their consent. The necessary techniques are rapidly developing in animal experiments, but their acceptance and success in human therapy is far from clear.

Genetic manipulation. The use of microorganisms in the production or transformation of fine chemicals is hardly debated, even if genetic manipulations are involved. Concerns about genetically engineered plants are usually confined to questions of whether the food products obtained from such plants are safe, and if the long-term ecological consequences of their cultivation can be controlled. In comparison, genetic manipulation of animals raises more public concern. The goals of transgenic animal experiments (production of biopharmaceuticals, use of knockout animals in drug research) play a minor role in these concerns.

Animal and human clones. Following the production of identical clones in sheep, mice, pigs, cattle, and goats, the reproductive cloning of humans has met with major ethical discussions. These include the population genetic consequences of techniques such as the sexing of progeny or in-vitro fertilization and embryo transfer.

Stem cell therapy. A main discussion point is at which stage of development human life has started and must be protected and if therapeutic alternatives can be developed, e. g., by the expansion of adult stem cells.

Military or terrorist use. Standard biotechnological procedures can be used to produce agents of warfare, such as *Bacillus anthracis* spores or smallpox virus. Genetic engineering techniques could be used to develop bioweapons that overrun the defensive mechanisms of the immune system. Although bioweapons have been outlawed by the Geneva Convention, several nations are suspected of producing them.

Public acceptance. In the USA, which is at the forefront of most developments in genetic research and gene technology (2/3 of all gene therapy experiments are done in the USA), the public generally hass fewer reservations than in Europe and Japan. In most nations, the medical use of biotechnology and genetic engineering meets with public support. The production of transgenic animals meets with considerable public criticism in Europe, although in practice transgenic mice have become a standard tool for biomedical research. Transgenic plants meet little enthusiasm in European nations, but are standard crops in the USA and Canada and of increasing importance in nations like China. Genetically modified enzymes are widely used in technology and food technology, without major public opposition.

Science-philosophy arguments concerning genetic engineering

categorical argument

Some human activities such as genetic engineering are fundamentally reprehensible. Developing this technology, "man plays God" and claims competencies beyond his capacities, degrading nature to the course of his technical manipulations.

pragmatic argument

The key objective of genetic engineering is to reduce the suffering of diseased individuals. The procedures which are applied must, however, be safe, and the patient must be able to decide if he or she wishes to apply genetic diagnosis or therapy.

social policy argument

The social effects of genetic engineering cannot be estimated. In genetic therapy, wrong priorities are chosen, better prophylaxis would be more desirable. We start down a slippery slope that will lead us involuntarily to inhumane practices towards the next generations ("eugenics bottom up")

Problematic areas of genetic research

topic	state of the art	regulation or trend
cloning of humans	cloning of animals possible	not permitted
use of embryonic stem cells	growing expertise	permitted, but regulated
artificial insemination, sexing, surrogate mothers	state of the art in animals	artificial insemination permitted, sexing and surrogate mothers forbidden
prenatal diagnosis	cytological methods established, DNA-based diagnosis partially established	permitted, abortion permitted after medical indication
identifying genetic risks by genetic screening	possible for some monogenic diseases	under debate if one gene defect is predictive and if diagnosis is acceptable for incurable diseases; strict data protection required towards employers, insurance companies
knockout animals for drug research	widely established	generally accepted, but hotly debated by animal protection groups
food and biopharmaceutical production using transgenic animals or plants	many techniques established	debated in view of consumer protection, animal protection, ecological consequences
transgenic microorganisms or cell lines for production of biopharmaceuticals	established	widely accepted

Public acceptance of genetic engineering (survey 2001)

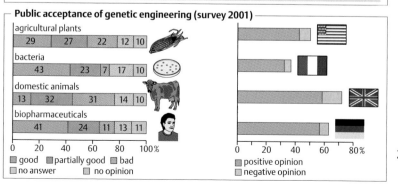

agricultural plants: 29 | 27 | 22 | 12 | 10

bacteria: 43 | 23 | 7 | 17 | 10

domestic animals: 13 | 32 | 31 | 14 | 10

biopharmaceuticals: 41 | 24 | 11 | 13 | 11

0 20 40 60 80 100%

■ good ■ partially good ■ bad
■ no answer ■ no opinion

0 20 40 60 80%

■ positive opinion
■ negative opinion

Patents in biotechnology

General. The governments of all industrialized countries can, through their patent offices, grant inventors a monopoly for a limited period of time. Generally such a monopoly is restricted to a period of 20 years. There are different kinds of intellectual property protection, such as patents or utility models on the commercial use of the manufacture of materials, processes for their manufacture, and the uses thereof. For a patent to be granted the invention must be novel, inventive, and applicable. Therefore, in *patents on products or compounds* the material described in the invention, such as a chemical compound, mixture, or a piece of equipment must be novel, inventive, and of economic value. The same applies to *process patents* In addition to these three prerequisites the invention has to be disclosed in such a manner that a person skilled in the art can practice it without undue burden. Discoveries, scientific theories, mathematical methods, and aesthetic creations are not patentable. Neither are software or doing business as such, nor therapeutic, diagnostic, or surgical methods for humans and animals patentable. However, pharma-ceutical compounds or mixtures are patentable. Intellectual property rights may not be granted on inventions that would be contrary to "ordre publique" or morality.

Patenting process. A patent application, which in many cases has been written by a patent attorney, can be filed with the patent office by a natural or legal person. Within 6 months after publication of the European patent application the examination request must be filed. It is then analyzed by an examiner for patentability. The text of the application, is published by the patent office 18 months after priority deposition. In general the patent is issued between 4 to 5 years after deposition. After the patent has been granted, third parties can oppose within a 9 month period. The priority date in most countries is the date of submission of the application to the patent office ("first to file"), but in the USA it is the date of documentation of the invention ("first to invent"), for example, evidence of the invention in a laboratory journal. In the USA and Japan, patents can be submitted up to 12 and 6 months, respectively, after an invention has been published ("period of grace"); under the European Patent Covention, any kind of prior publication prevents patenting. Patents, once granted, remain valid for 20 years after the filing date if the maintenance fees are paid. The costs for a patent application range between € 1000 and 10 000 or more, and its prosecution, maintenance and enforcement are very expensive. To prevent high translation costs for registration with foreign patent offices, a patentee can file an application according to the PCT treaty in one of seven official languages. After the international preliminary examination report on the patentability of the application has been issued, the patentee can decide whether to persue the application in the more than 110 PCT contract states, which requires translating the application text into the languages of the states the patentee has chosen.

Patents in biotechnology. Materials isolated from living beings or manufactured by biotechnological processes are patentable under the rules mentioned above. Living beings especially bred for economic purposes are also patentable, e. g., microbial production strains or cell lines, transgenic plants and animals, but not germplasms. The requirement for "enablement" necessary for a person skilled in the art to carry out the invention may often require deposition of a sample of the organism necessary for the manufacture of a product with a depositary institution (Budapest treaty, 1965). For claims concerning proteins or DNA-based products, however, indication of the amino acid or nucleic acid sequence is usually enough. Mutagenized genes or proteins or polymorphic DNA may be patented. Gene sequences of expressed sequence tags (ESTs) are not patentable as long as there is no revelation of their specific properties.

Patent categories and their relation to biotechnology

examples from biotechnology

	substance patents		process patents
substan-ces	cloned genes, recombinant proteins, monoclonal antibodies, plasmids, promoters, vectors, cDNA sequences, monovalent vaccines	processes	DNA isolation, DNA synthesis, preparation of vectors, purification processes for proteins
substance mixtures	polyvalent vaccines, bioinsecticides, pharmaceutical preparations, microorganisms, transgenic plants and animals	process protocols	hybridization assays, diagnostic procedures, PCR methods, analysis of mutants
equip-ment	pulse-field gel electrophoresis, DNA sequencer, biolistics gene cannon	appli-cations	applications of bioinsecticides, fermentation protocols for genetically modified microorganisms

Steps toward a patent

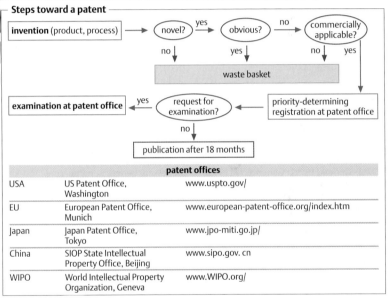

	patent offices	
USA	US Patent Office, Washington	www.uspto.gov/
EU	European Patent Office, Munich	www.european-patent-office.org/index.htm
Japan	Japan Patent Office, Tokyo	www.jpo-miti.go.jp/
China	SIOP State Intellectual Property Office, Beijing	www.sipo.gov. cn
WIPO	World Intellectual Property Organization, Geneva	www.WIPO.org/

Biotechnology patents

	world		USA		Germany		UK		France		Japan		China	
	1990	2000	1990	2000	1990	2000	1990	2000	1990	2000	1990	2000	1990	2000
molecular genetics	2830	15117	1275	8530	225	947	164	847	128	571	636	1634	3	362
fermentation	3581	3288	705	746	408	312	130	116	115	100	1632	1219	36	162
pharmaceuticals	11297	23533	3341	10026	1210	1926	671	1244	703	1060	3216	4309	56	692

data from Chemical Abstracts CAS online, sections 3 "Biochemical Genetics", 16 "Fermentation and Bioindustrial Engineering", and 63 "Pharmaceuticals". The country chosen by a transnational enterprise for primary deposit of patent applications does not necessarily indicate where the invention was made

283

International aspects of bio-technology

General. Like most technical and scientific breakthroughs in human history, bio-technology and genetic engineering have resulted from a mosaic of many national contributions. Often, quantum jumps in basic research or an economic crisis have helped to move a step ahead; this is certainly true for the development of fermentation industries in Europe and Japan. Genetic engineering, however, is clearly a brainchild of the USA. Since ca. 1971, a growing number of academic entrepreneurs in this country have started to found companies, which in many cases attracted private venture capital, to be finally sold out to larger pharmaceutical companies or to grow into an independent larger company. In the meantime, there are > 1400 venture-capital based biocompanies in the USA and in Europe, and ca. 100 in Japan, focused on biotechnology, genetic engineering, and bioinformatics.

USA. The first companies to be founded around molecular genetics were Cetus (1971), Genex (1972), Genentech (1976), Biogen (1978), and Amgen (1980). The initial incentive was usually a new technology that had been secured by patents (e. g., for Genentech, the heterologous expression of genes according to the Cohen-Boyer patent). Some of these companies succeeded, while still funded by venture capital, in inventing and patenting a large number of innovations within just a few years. Genentech obtained patents on the production of human insulin in *E. coli* (1976) and of human factor VIII (1978) and human TPA (1980) in animal cells. License fees from these patents and, to some extent, their own marketing efforts led to a significant increase in turnover and stock value. In 1992, Roche, Switzerland, acquired 70% of the Genentech shares for 3 billion US$; shortly after, it also acquired Cetus with its global patent on PCR. However, as of today, the number of successfully operating startup companies is still small, compared to the invested capital. Large pharmaceutical companies such as Novartis, Roche, Aventis, GlaxoSmithKline, Merck, Bayer, Pfizer,

Bristol-Myers Squibb, etc., all operate transnational research and development centers, to use the large potential of genomic and post-genomic research for the development of novel drugs fitting into their marketing strategies. Many of these centers are located in the USA.

Europe. In the beginning of genetic engineering, most companies in continental Europe were slow to set up their own research laboratories in this field, and venture companies also developed slowly. But now Europe is about to catch up with the USA, and a growing number of successful small and medium enterprises have been established in most European countries, with the UK and Germany sharing the top position. Although the total number of biotech companies in Europe now exceeds that of the USA (at the end of 2001, there were 1879 biotech companies in Europe and 1457 in the USA, together > 60% of the global number of 4284), US companies employ roughly twice the workforce, have about 3-fold more sales and 10-fold more pharmaceutical products in phase III clinical tests.

Japan. Venture companies are rare, but nearly all large companies have strong stakes in genetic engineering and cell biology. However, only a few companies that have diversified into this area from other business have been successful. Examples are Suntory and Kirin Beer.

Innovation and performance. Most of the innovations that help to advance biotechnology originate in the industrialized nations of Europe, Japan, increasingly China, but above all, the USA. If scientific publications in key areas such as molecular genetics and cell biology are taken as an indicator, the US contributes about 40% of all global communications.

Governmental programs. As one of the megatechnologies of the 21st century, the development of biotechnology is supported in all industrialized nations, and also in many developing nations such as China. The development of core competencies and the training of personnel both play important roles. A key task is to train innovative young minds for the challenges of new multidisciplinary technologies that involve biology, chemistry, ecology, engineering, and informatics, and last but not least, economics.

International comparison of performance - publications in chemical abstracts, 2000

	world	USA	Germany	UK	France	Japan	China
molecular genetics	55834	22140	3706	3716	2646	7480	2911
fermentation	7499	1202	528	308	220	1710	998
pharmaceuticals	35469	12706	2618	1890	1471	6042	2105

Performance of biotech companies

	USA		Europe	
	2000	2001	2000	2001
total number of companies	1379	1457	1734	1879
of which are public companies	344	342	107	104
employees (× 1000)	130	107	67	87
R&D expenditure (billion US $ or €)	14.7	15.7	5.5	7.5
% of which are public companies	10.2	11.5	2.9	4.2
sales (billion US $ or €)	26.6	28.5	9.9	13.7
of which public companies	16.5	18.3	5.2	7.5
net loss (billion US $ or €)	6.2	6.9	1.5	1.5
of which public companies	4.1	4.8	0.4	0.6

European biotech companies (2001)

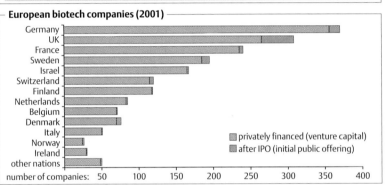

Biotech developmental products (2001)

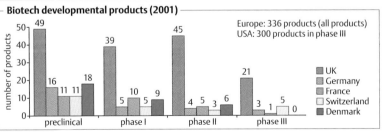

Financing (million US $ or €)

	USA		Europe	
	2000	2001	2000	2001
IPO	4997	208	3138	175
additional seed capital	24951	5330	2603	723
venture capital	2778	2392	1228	1374
total	32726	7930	6969	2272

Index

A

aberrant RNA 243
abscisic acid 166, 167
absorption chromatography 26, 30
acarbose **124, 125**
acceptance of genetic engineering **280, 281**
acetate 271
acetic acid **18, 19**, 271
acetic acid bacteria s. Acetobacter
acetic acid, glacial 18
Acetobacter 6, **18**, 22, 191
Acetobacter aceticum **18**, 19
Acetobacter suboxydans 54, 55, 65
acetogenesis 16
acetoin 8, 9
acetolactate 8
acetolactate decarboxylase 8
acetone **16, 17**, 271
acetoxymethoxyphenylethylamine 65
acetylcholine receptor 243
acetylcholinesterase 74, 75
Achromobacter obae 28, 29
Acinetobacter calcoaceticum 58, 59
Acremonium chrysogenum 38, 41, 53, 186
acridine orange, as mutagen 193
acrylamide **70, 71**
acrylonitril **70, 71**
Actinomycetes 35, 36
actinomycin 35, 42, 271
Actinoplanes utahensis 124, 125
activated sludge 104
active transport 27
acyclovir **44, 45**
acylamino acids 32, 33, 71
ADA s. adenosine deaminase
adaptation 98
adapter 224, 225
additives 78, 79, 80
adenine 57, 216, **217**
adenosine 57
adenosine deaminase (ADA) 241, 262
adenosine triphosphate (ATP) 56, 57, 68, 69, 270
adenosine-5'-monophosphate (AMP) 56
adenovirus 120, 176
adenovirus, for gene therapy 262, **263**
adenylate kinase 73
adhesin, for tooth decay prophylaxis 174
adipositas 124
adriamycin 34, 48
adsorbable organic halogens (AOX) 88

adsorption chromatography **212**
adult respiratory stress 124
aeration 40
aeration basin **104, 105**, 106
aeration rate **201**
aeration system 100
aerobic 10, 36
aerobic fermentation 100
aerobic metabolism 270, 271
affinity chromatography 122, 123, **212, 213**
agar 190
agarose gel 212
agarose gel electrophoresis **230, 231**
agricultural products, registration 278
Agrobacter rhizogenes 170
Agrobacterium tumefaciens **170, 171**, 234
AI s. artificial insemination
AICA-riboside 56
AIDS 134, 135, 140, 143, 176
airlift bioreactor 18, 20, 56, **102**, 103, 168, 169, **202, 203**
airlift cultivation, of human cells 131
airlift reactor, for mammalian cells 146, 147
Ajinomoto 26
Äkta system 212
alanine 24, 25
alanine aminotransferase, ALAT 75
alcohol dehydrogenase 18, 19, 73
alcohol determination s. ethanol
alcohol oxidase promoter 238, **239**
alcohol-free beer **8**
aldehyde dehydrogenase 18, 19
aldolases 70, 98
algae 35
alginate 208
alginates, microbial 60, 61
alkaline phosphatase 75, 150, 258, 259
alkanes 20, 58
alkylating agents, as mutagen 193
allele 158
allele, analysis 257
allergens 143
allergy 78
allografting, of tissues 130
alpha1-antitrypsin **118, 119, 124, 125**, 162, 163
Alu sequences, of DNA 248
Alzheimer disease, animal model 160, 252
Amgen 284
amidases **32, 33**
amikacin 46, 47
amines 70, 71

289

291

293

301

recrystallization *s.* crystallization
red rot fungi 186
red wine 6
redox potential 112, 152
redundancy, in biological systems 274
regioselective 64, 66, 70, 71, 76
registration of enzymes **68, 69,** 76, **278, 279**
registration of pharmaceuticals **278, 279**
regulation-defective mutants 28, 192
Reichstein S → cortexolon
Reichstein-Gruessner process 54, 55
release test 140
Remicade antibody 147
remission 74
renewable resources 80, 164, **174**
rennet 11, 69, 76, 77, **92,** 93
rennin **92**
rennin, microbial 92
repeated fed-batch 18
Repilysin 122
replacement vectors 160
replica plating **192, 193**
replication, of DNA 35, 216
reporter enzyme 72, 150
reporter gene 170, 234, 236, 238, 239
reporter group 144, 150, 254, 256, 258, **259**
reporter reaction 152
repression 27, 29, 56
repressor protein 238
residence time 107, 108
residual sugar 6
resistance *s.* antibiotic resistance
respiration **101**
respiration quotient **202**
restorer gene 164
restriction analysis, of DNA 222, 230, 250
restriction digest 99
restriction endonucleases *s.* restriction enzymes
restriction enzyme 159, **222, 223,** 224, 228
restriction enzyme, frequency of restriction sites **223**
restriction fragment length polymorphism, RFLP 158, **159,** 252, 253
restriction fragments 244
restriction mapping of DNA 222, 230, 246, 248
restriction sites, of DNA 220, 234, 235
reteplase 122, 123
retinal 62, 63
retrogradation 80
retrotransposon 248

retrovirus 176, 177, 206, 224
retrovirus, for gene therapy 262, **263**
retrovirus, propagation cycle **177**
retrovirus, recombinant 162
reverse genetics **52,** 226
reverse osmosis 16, 211
reverse transcriptase 176, **177,** 219,220, 221, **224,** 225, 226, 228
reverse transcriptase, PCR 220, 226, **228**
Reynolds number 200, 201
RFLP *s.* restriction fragment length polymorphism
rhamnose 90
rhamnose lipid 58, 59
rheumatoid arthritis 133
Rhizomania virus 172, 173
Rhizomucor miehei 71, 77, 99
Rhizopus oligosporus 10, 11
Rhizopus oryzae **186,** 187
Rhizopus sp. 69, 80, 90
Rhodococcus erythropolis 58, 59
Rhodotorula glutinis 30, 31, 102
RIA *s.* radioimmunoassay
riboflavin **54, 55**
ribonuclease *s.* RNase
ribonucleic acid *s.* RNA
ribosomal RNA **182**
ribosome 35, 44, 46, 48, 50, 181, 218, 219, 238
ribosome-inactivating protein, RIP 172
ribozyme, hammerhead 172
ribozymes 242
rice (*Oryza sativa*), genome 170
rice blight 47
rice genome 170
rice starch → starch
rice wine 6
rifampicin 34, 35, **50, 51**
rifamycin 35, **50, 51,** 271
Ri-plasmid **170**
RISC RNA-induced silencing complex 243
risk groups (RG), in genetic engineering **276, 277**
Rituxan antibody 147
RNA 218, 271
RNA genome 176
RNA matrix 220
RNA polymerase 219, 238
RNA promoter, tobacco mosaic virus 240
RNA world, primordial 242
RNA, autolysis **102**
RNA-DNA chimera 262
RNase 220
RNase, in RNAi complexes **243**

313

T

Literature

For this book, a large number of informations from books, book sections, reviews, publications and personal communications were used. It is impossible to cite all these sources.

In view of the presumed interests of the reader, I have chosen to provide a limited selection of about 600 literature citations, which are ordered according to the entries of this book.

In addition, the following handbooks, series and review journals were used:
- Kirk-Othmer Encyclopedia of Chemical Technology, 4th Edition, Interscience-Wiley
- Ullmann's Encyclopedia of Industrial Chemistry, 5th Edition, Wiley-VCH
- Advances in Biochemical Engineering, Springer-Verlag
- Biotechnology – A Comprehensive Handbook, 2nd Edition, Wiley-VCH
- Current Opinions in Biotechnology, Current Biology Publications
- Current Opinions in Microbiology, Current Biology Publications
- Current Opinions in Structural Biology, Current Biology Publications
- Current Opinions in Genetics and Cell Biology, Current Biology Publications
- Trends in Biotechnology, Elsevier Publishers

The internet has become a most valuable resource for scientific information. Since there is no guarantee for continuous access to a website, however, they are not cited in this book.

Textbooks

HJ Rehm, G Reed, A Pühler, P Stadler, eds., Biotechnology – A Multi-Volume Comprehensive Treatise, 2nd Edition. Wiley-VCH, ISBN 1-56081-602-3

B Atkinson, F Maviura (1991) Biochemical Engineering and Biotechnology Handbook, 2nd Edition, Mamillan Publishers, ISBN 1-56159-012-6

J Black (1999) Microbiology – Principles and Explorations, 4th Edition, Prentice Hall, ISBN 0-13-920711-2

T Brown (2001) Gene Cloning and DNA Analysis: An Introduction, 4th edition, Blackwell Science Inc, ISBN 0-6320-5901-X

T Brown (1999) Genomes John Wiley & Sons, ISBN 1-85996-201-7

B Dixon (1996) Power unseen: how Microbes rule the World, W. H. Freeman, ISBN 0-7167-4550-X

A Demain, J Davies (1999) Industrial Microbiology and Biotechnology, 2nd Edition, ASM Press, ISBN 1-55581-128-0

BR Glick, JJ Pasternak (1998) Molecular Biotechnology: Principles and Applications of Recombinat DNA, 2nd Edition, Amer. Society for Microbiology, ISBN 1-5558-1136-1

AL Lehninger, DL Nelson, MM Cox (2000), Principles of Biochemistry, 3rd Edition, Worth Publishing, ISBN 1-5725-9931-6

J Lengeler, G Drews, H Schlegel (1999) Biology of the Prokaryotes Thieme, 3-13-108411-1

P Präve, U Faust, W Sittig, D Sukatsch (1987) Fundametals of Biotechnology, VCH Publishing, ISBN 0-8957-3224-6

L Stryer (1995), Biochemistry, 4th Edition, W H Freeman and Co, ISBN 0-7167-3687-X

G Walsh, D Headon (1994) Protein Biotechnology, John Wiley & Sons, ISBN 0-471-944396-

OP Ward (1989) Fermentation Biotechnology: Principles, Processes, and Products, Prentice Hall, ASIN 0-1331-5052-6

J Watson, M Gilman, J Witkowski, M Zoller (1992) Recombinant DNA, 2nd Edition, Scientific American Books, ISBN 0-7167-2282-8

Alcoholic beverages 6

I Pretorius, F Bauer (2002), Meeting the consumer challenge through genetically customized wine-yeast strains. Trends Biotechnol 20, 426.

V Singleton, C Butzke (1998) Wine in Kirk-Othmer Encyclopedia of Chemical Technology 4th Edition Supplement, 733. Interscience-Wiley, ISBN 0-471-52696-7

R Kunkee, H Eschnauer (1996) Wine in Ullmann's Encyclopedia of Industrial Chemistry 5th edition A28, 269. Wiley-VCH, ISBN 3-527-20128-9

H Dittrich (1995) Wine and Brandy in Biotechnology 2nd Edition, 463. Vol. 9, G Reed, T Nagodawithana (ed.). VCH-Wiley, ISBN 3-527-28319-6

Beer 8

M Linko, A Haikara, A Ritala, M Penttilä (1998), Recent advances in the malting and brewing industry *J. Biotechnol.* 65, 85

J Russell, G Stewart (1995) *Brewing* in Biotechnology 2nd Edition, 419. Vol. 9, G Reed, T Nagodawithana (ed.). VCH-Wiley, ISBN 3-527-28319-6

J Nissen (1992) *Beer* in Kirk-Othmer Encyclopedia of Chemical Technology 4th Edition 4, 22. Interscience-Wiley, ISBN 0-471-52672-X

L Narziss, H Esslinger (1985) *Beer* in Ullmann's Encyclopedia of Industrial Chemistry 5th edition A3, 421. Wiley-VCH, ISBN 3-527-20103-3

Fermented food 10

M Saarela, G Mogensen, R Fonden, J Matto, T Mattila-Sandholm (2000), Probiotic bacteria: safety, functional and technological properties *J Biotechnol* 84, 197.

A Forde, GF Fitzgerald (2000), Biotechnological approaches to the understanding and improvement of mature cheese flavour *Curr Opin Biotechnol* 11, 484.

DB Archer (2000), Filamentous fungi as microbial cell factories for food use. *Curr Opin Biotechnol* 11, 478

H Fleming, K Kyung, F Breidt (1995) *Vegetable Fermentations* in Biotechnology 2nd Edition, 629. Vol. 9, G Reed, T Nagodawithana (ed.). VCH-Wiley, ISBN 3-527-28319-6

R Shaver, K Batajoo (1995) *Fermented Feeds and Feed Products* in Biotechnology 2nd Edition, 769. Vol. 9, G Reed, T Nagodawithana (ed.). VCH-Wiley, ISBN 3-527-28319-6

L Beuchat (1995) *Indigenous Fermented Foods* in Biotechnology 2nd Edition, 505. Vol. 9, G Reed, T Nagodawithana (ed.). VCH-Wiley, ISBN 3-527-28319-6

N Olson (1995) *Cheese* in Biotechnology 2nd Edition, 353. Vol. 9, G Reed, T Nagodawithana (ed.). VCH-Wiley, ISBN 3-527-28319-6

G Burkhalter, C Steffen, Z Puhan (1986) *Cheese, Processed Cheese, and Whey* in Ullmann's Encyclopedia of Industrial Chemistry 5th edition A6, 163. Wiley-VCH, ISBN 3-527-20106-8

Food and lactic acid fermentation 12

RP Ross, S Morgan, C Hill (2002) Preservation and fermentation: past, present and futureInt J Food Microbiol 79, 3

M Teuber (1993) *Lactic Acid Bacteria* in Biotechnology Second, Completely Revised Edition, 325. Vol. 1, H Sahm (ed.). VCH Verlagsgesellschaft, ISBN 3-527-28337-4

Ethanol 14

M Galbe, G Zacchi (2002), A review of the production of ethanol from softwood. *Appl Microbiol Biotechnol* 59, 618.

MJ Davies (2001), 'Corn-to-car' ethanol production *Trends Biotechnol* 19, 380.

AE Wheals, LC Basso, DM Alves, HV Amorim (1999), Fuel ethanol after 25 years *Trends Biotechnol* 17, 482.

CS Gong, NJ Cao, J Du, GT Tsao (1999), Ethanol production from renewable resources *Adv Biochem Eng Biotechnol* 65, 207

N Kosaric (1996) *Ethanol – Potential Source of Energy and Chemical Products* in Biotechnology 2nd Edition, 121. Vol. 6, M Roehr (ed.). VCH-Wiley, ISBN 3-527-28316-1

T Senn, H Pieper (1996) *Ethanol – Classical Methods* in Biotechnology 2nd Edition, 5. Vol. 6, M Roehr (ed.). VCH-Wiley, ISBN 3-527-28316-1

J Logsdon (1994) *Ethanol* in Kirk-Othmer Encyclopedia of Chemical Technology 4th Edition 9, 812. Interscience-Wiley, ISBN 0-471-52677-0

B Maiorella, H Blanch, C Wilke (1984), Biotechnology Report: Economic Evaluation of Alternative Ethanol Fermentation Processes *Biotechnol. Bioeng.* 26, 1003

1-Butanol, acetone 16

P Durre, M Bohringe, S Nakotte, S Schaffer, K Thormann, B Zickner (2002) Transcriptional regulation of solventogenesis in Clostridium acetobutylicum. J Mol Microbiol Biotechnol.4, 295

P Dürre, H Bahl (1996) *Microbial Production of Acetone/Isopropanol* in Biotechnology

2nd Edition, 229. Vol. 6, M Roehr (ed.). VCH-Wiley, ISBN 3-527-28316-1

P Durre, R Fischer, A Kuhn, K Lorenz, W Schreiber, B Sturzenhofecker, S Ullmann, K Winzer, U Sauer (1995), Solventogenic enzymes of Clostridium acetobutylicum: catalytic properties, genetic organization, and transcriptional regulation FEMS Micrbiol. Rev 17, 251

Acetic acid/vinegar 18

S Arnold, T Becker, A Delgado, F Emde, A Enenkel (2002), Optimizing high strength acetic acid bioprocess by cognitive methods in an unsteady state cultivation. J Biotechnol. 97, 133

H Ebner, S Sellmer, H Follmann (1996) Acetic Acid in Biotechnology 2nd Edition, 381. Vol. 6, M Roehr (ed.). VCH-Wiley, ISBN 3-527-28316-1

H Ebner, H Follmann, S Sellmer (1996) Vinegar in Ullmann's Encyclopedia of Industrial Chemistry 5th edition A27, 403. Wiley-VCH, ISBN 3-527-20127-0

H Ebner, H Follmann, S Sellmer (1995) Vinegar in Biotechnology 2nd Edition, 579. Vol. 9, G Reed, T Nagodawithana (ed.). VCH-Wiley, ISBN 3-527-28319-6

Citric acid 20

L Karaffa, E Sandor, Fekete, A Szentirmai (2001) The biochemistry of citric acid accumulation by Aspergillus niger. Acta Microbiol Immunol Hung. 48, 429

M Roehr, C Kubicek, J Kominec (1996) Citric Acid in Biotechnology 2nd Edition, 307. Vol. 6, M Roehr (ed.). VCH-Wiley, ISBN 3-527-28316-1

G Blair, P Staal (1993) Citric Acid in Kirk-Othmer Encyclopedia of Chemical Technology 4th Edition 6, 354. Interscience-Wiley, ISBN 0-471-52674-6

F Verhoff (1986) Citric Acid in Ullmann's Encyclopedia of Industrial Chemistry 5th edition A7, 103. Wiley-VCH, ISBN 3-527-20107-6

Lactic acid, gluconic acid 22

J Kascak, J Kominek, M Roehr (1996) Lactic Acid in Biotechnology 2nd Edition, 293. Vol. 6, M Roehr (ed.). VCH-Wiley, ISBN 3-527-28316-1

M Roehr, C Kubicek, J Kominek (1996) Gluconic Acid in Biotechnology 2nd Edition, 347. Vol. 6, M Roehr (ed.). VCH-Wiley, ISBN 3-527-28316-1

M Roehr, C Kubicek, J Kominek (1996) Further Organic Acids in Biotechnology 2nd Edition, 363. Vol. 6, M Roehr (ed.). VCH-Wiley, ISBN 3-527-28316-1

R Datta (1995) Hydroxycarboxylic Acids in Kirk-Othmer Encyclopedia of Chemical Technology 4th Edition 13, 1042. Interscience-Wiley, ISBN 0-471-52682-7

S Chahal (1990) Lactic Acid in Ullmann's Encyclopedia of Industrial Chemistry 5th edition A15, 97. Wiley-VCH, ISBN 3-527-20115-7

H Hustede, H-J Haberstroh, E Schinzing (1989) Gluconic Acid in Ullmann's Encyclopedia of Industrial Chemistry 5th edition A12, 449. Wiley-VCH, ISBN 3-527-20112-2

Amino acids 24

W Leuchtenberger (1996) Amino Acids – Technical Production and Use in Biotechnology 2nd Edition, 465. Vol. 6, M Roehr (ed.). VCH-Wiley, ISBN 3-527-28316-1

N Esaki, S Nakamori, T Kurhara, S Furuyoshi, K Soda (1996) Enzymology of Amino Acid Production in Biotechnology 2nd Edition, 503. Vol. 6, M Roehr (ed.). VCH-Wiley, ISBN 3-527-28316-1

K Araki (1992) Amino Acids, Survey in Kirk-Othmer Encyclopedia of Chemical Technology 4th Edition 2, 504. Interscience-Wiley, ISBN 0-471-52669-X

A Kleemann, W Leuchtenberger, B Hoppe, H Tanner (1985) Amino Acids in Ullmann's Encyclopedia of Industrial Chemistry 5th edition A2, 57. Wiley-VCH, ISBN 3-527-20102-5

L-Glutamic acid 26

PG Peters-Wendisch, B Schiel, VF Wendisch, E Katsoulidis, B Mockel, H Sahm, BJ Eikmanns (2001) Pyruvate carboxylase is a major bottleneck for glutamate and lysine production by Corynebacterium glutamicum. J Mol Microbiol Biotechnol. 3, 295

T Kawakita (1992) L-Monosodium Glutamate in Kirk-Othmer Encyclopedia of Chemical Technology 4th Edition 2, 571.

Interscience-Wiley, ISBN 0-471-52669-X

T Kawakita, C Sano, S Shioya, M Takehara, S Yamaguchi (1990) *Monosodium Glutamate* in Ullmann's Encyclopedia of Industrial Chemistry 5th edition A16, 711. Wiley-VCH, ISBN 3-527-20116-5

Methionine, L-lysine, L-threonine 28

J Ohnishi, S Mitsuhashi, M Hayashi, S Ando, H Yokoi, K Ochiai, M Ikeda (2002) A novel methodology employing Corynebacterium glutamicum genome information to generate a new L-lysine-producing mutant. Appl Microbiol Biotechnol. 58, 217

AA de Graaf, L Eggelin, H Sahm (2001) Metabolic engineering for L-lysine production by Corynebacterium glutamicum. Adv Biochem Eng Biotechnol. 73, 9

Aspartame, L-phenylalanine, and L-aspartic acid 30

D Kuhn, P Durrschmidt, J Mansfeld, R Ulbrich-Hofmann (2002) Boilysin and thermolysin in dipeptide synthesis: a comparative study. Biotechnol Appl Biochem. 36, 71

J Fry, The world market for intense sweeteners (1999) World Rev Nutr Diet. 85:201

T Sato, T Tosa T. Production of L-aspartic acid (1993) Bioprocess Technol. 16, 15

Amino acids via enzymatic transformation 32

A Bommarius, M Schwarm, K Drauz (2001), Comparison of Different Chemoenzymatic Process Routes to Enantiomerically Pure Amino Acids *Chimia* 55, 50

H Griengl, H Schwab, M Fechter (2000), The synthesis of chiral cyanohydrins by oxynitrilases *Trends Biotechnol* 18, 252.

Antibiotics: occurence, applications, mechanism of action 34

D Borders (1992) *Antibiotics, Survey* in Kirk-Othmer Encyclopedia of Chemical Technology 4th Edition 2, 893. Interscience-Wiley, ISBN 0-471-52669-X

G Cauwenbergh (1992) *Antiparasitic Agents, Antimycotics* in Kirk-Othmer Encyclopedia of Chemical Technology 4th Edition 3, 473. Interscience-Wiley, ISBN 0-471-52671-1

M Plempel, H Böshagen, J McGuire (1985) *Antimycotics* in Ullmann's Encyclopedia of Industrial Chemistry 5th edition A3, 77. Wiley-VCH, ISBN 3-527-20103-3

M Ohno, M Otsuka, M Yagisawa, S Kondo, H Öppinger, H Hoffmann, D Sukatsch, L Hepner, C Male (1985) *Antibiotics* in Ullmann's Encyclopedia of Industrial Chemistry 5th edition A2, 467. Wiley-VCH, ISBN 3-527-20102-5

Antibiotics: industrial production, resistance 36

A Dessen, AM Di Guilmi, T Vernet, O Dideberg (2001) Molecular mechanisms of antibiotic resistance in gram-positive pathogens Curr Drug Targets Infect Disord. 1, 63

K Chater, M Bibb (1997) *Regulation of Bacterial Antibiotic Production* in Biotechnology 2nd Edition, 57. Vol. 7, Hv Döhren, H Kleinkauf (ed.). VCH-Wiley, ISBN 3-527-28310-2

Hv Döhren, U Gräfe (1997) *General Aspects of Secondary Metabolism* in Biotechnology 2nd Edition, 1. Vol. 7, Hv Döhren, H Kleinkauf (ed.). VCH-Wiley, ISBN 3-527-28310-2

β-Lactam antibiotics: structure, biosynthesis, mechanism of action 38

P Moreillon, JM Entenza(2001) Antibiotic resistance: learning from animal feeds and animal experimentation. Clin Microbiol Infect. 7 Suppl 5, 13

JF Martin (1998) New aspects of genes and enzymes for beta-lactam antibiotic biosynthesis. Appl Microbiol Biotechnol. 50, 1

K Lindner, D Bonner, W Koster (1992) *Monobactams* in Kirk-Othmer Encyclopedia of Chemical Technology 4th Edition 3, 107. Interscience-Wiley, ISBN 0-471-52671-1

J Roberts (1992) *Cephalosporins* in Kirk-Othmer Encyclopedia of Chemical Technology 4th Edition 3, 28. Interscience-Wiley, ISBN 0-471-52671-1

R Southgate, N Osborne (1992) *Carbapenems and Penems* in Kirk-Othmer Encyclopedia of Chemical Technology 4th Edition 3, 1. Interscience-Wiley, ISBN 0-471-52671-1

β-Lactam antibiotics: manufacture 40

MA Wegman, MHA Janssen, F van Rantwijk, RA Sheldon (2001), Towards Biocatalytic Synthesis of β-Lactam Antibiotics. *Adv Synth Catal* 343, 559.

E van de Sandt, E de Vroom (2000), Innovations in cepahalosporin and penicillin production: painting the antibiotics industry green. *Chimica OGGI*, 72.

MA Penalva, RT Rowlands, G Turner (1998), The optimization of penicillin biosynthesis in fungi *Trends Biotechnol* 16, 483.

P Skatrud, T Schwecke, Hv Liempt, M Tobin (1997) *Advances in the Molecular Genetics of β-Lactam Antibiotic Biosynthesis* in Biotechnology 2nd Edition, 247. Vol. 7, Hv Döhren, H Kleinkauf (ed.). VCH-Wiley, ISBN 3-527-28310-2

Amino acid and peptide antibiotics 42

H Kleinkauf, Hv Döhren (1997) *Peptide Antibiotics* in Biotechnology 2nd Edition, 277. Vol. 7, Hv Döhren, H Kleinkauf (ed.). VCH-Wiley, ISBN 3-527-28310-2

Glycopeptide, polyether and nucleoside antibiotics 44

A Srinivasan, JD Dick, TM Perl (2002) Vancomycin resistance in staphylococci, Clin Microbiol Rev.15, 430

J Kallen, V Mikol, V Quesniaux, M Walkinshaw, E Schneider-Scherzer, K Schörgendorfer, G Weber, H Fliri (1997) *Cyclosporins: Recent Developments in Biosynthesis, Pharmacology and Biology, and Clinicla Applications* in Biotechnology 2nd Edition, 535. Vol. 7, Hv Döhren, H Kleinkauf (ed.). VCH-Wiley, ISBN 3-527-28310-2

Aminoglycoside antibiotics 46

W Piepersberg, J Distler (1997) *Aminoglycosides and Sugar Components in Other Secondary Metabolites* in Biotechnology 2nd Edition, 397. Vol. 7, Hv Döhren, H Kleinkauf (ed.). VCH-Wiley, ISBN 3-527-28310-2

Tetracyclines, chinones, chinolones, and other aromatic antibiotics 48

U Gräfe, K Dornberger, H Salz (1997) *Biotechnical Drugs as Antitumor Agents* in Biotechnology 2nd Edition, 641. Vol. 7, Hv Döhren, H Kleinkauf (ed.). VCH-Wiley, ISBN 3-527-28310-2

J Hlavka, G Ellestad, I Chopra (1992) *Tetracyclines* in Kirk-Othmer Encyclopedia of Chemical Technology 4th Edition 3, 331. Interscience-Wiley, ISBN 0-471-52671-1

T Nagabhushan, G Miller, K Varma (1992) *Chloramphenicol and Analogues* in Kirk-Othmer Encyclopedia of Chemical Technology 4th Edition 2, 961. Interscience-Wiley, ISBN 0-471-52669-X

Macrolide antibiotics 50

H Kirst (2001) *Antibiotics, Macrolides* in Kirk-Othmer Encyclopedia of Chemical Technology 4th Edition 3, 169. Interscience-Wiley, ISBN 0-471-52671-1

P Zhong, V Shortridge (2001) The emerging new generation of antibiotic: ketolides, Curr Drug Targets Infect Disord. 1,125

New pathways to antibiotics 52

R Gokhale, D Tuteja (2001) *Biochemistry of Polyketide Synthases* in Biotechnology 2nd Edition, 341. Vol. 10, H Rehm (ed.). VCH-Wiley, ISBN 3-527-28320-X

Y Xue, DH Sherman (2001) Biosynthesis and combinatorial biosynthesis of pikromycin-related macrolides in Streptomyces venezuelae. Metab Eng. 3, 15

M Chartrain, PM Salmon, DK Robinson, BC Buckland (2000), Metabolic engineering and directed evolution for the production of pharmaceuticals *Curr Opin Biotechnol* 11, 209.

Vitamins 54

RD Hancock, R Viola (2002), Biotechnological approaches for L-ascorbic acid production. T*rends Biotechnol* 20, 299

S Shimizu (2001) *Vitamins and Related Compounds: Microbial Production* in Biotechnology 2nd Edition, 319. Vol. 10, H Rehm (ed.). VCH-Wiley, ISBN 3-527-28320-X

V Kuellmer (2001) *Ascorbic Acid* in Kirk-Othmer Encyclopedia of Chemical

Technology 4th Edition 25, 17. Interscience-Wiley, ISBN 0-471-52694-0

J Scott (1998) *Vitamins, Survey* in Kirk-Othmer Encyclopedia of Chemical Technology 4th Edition 25, 1. Interscience-Wiley, ISBN 0-471-52694-0

F Yoneda (1998) *Riboflavin (B2)* in Kirk-Othmer Encyclopedia of Chemical Technology 4th Edition 25, 132. Interscience-Wiley, ISBN 0-471-52694-0

J Scott (1998) *Vitamin B12* in Kirk-Othmer Encyclopedia of Chemical Technology 4th Edition 25, 193. Interscience-Wiley, ISBN 0-471-52694-0

M Eggersdorfer, G Adam, M John, W Hähnlein, L Labler, K-U Baldenius, L Bussche-Hünnefeld, E Hilgemann, P Hoppe, R Stürmer, F Weber, A Rüttimann, G Moine, H-P Hohmann, R Kurth, J Paust, H Pauling, B-J Weimann, B Kaesler, B Oster, U Fechtel, K Kaiser, B Potzolli, M Casutt, T Koppe, M Schwarz, U Hengartner, A Saizieu, C Wehrli, R Blum (1996) *Vitamins* in Ullmann's Encyclopedia of Industrial Chemistry 5th edition A27, 443. Wiley-VCH, ISBN 3-527-20127-0

Nucleosides und nucleotides 56

A Kuninaka (1996) *Nucleotides and Related Compounds* in Biotechnology 2nd Edition, 561. Vol. 6, M Roehr (ed.). VCH-Wiley, ISBN 3-527-28316-1

R Suhadolnik, N Reichenbach (1992) *Nucleosides and Nucleotides* in Kirk-Othmer Encyclopedia of Chemical Technology 4th Edition 3, 214. Interscience-Wiley, ISBN 0-471-52671-1

Biosurfactants und biocosmetics 58

RS Makkar, SS Cameotra (2002), An update on the use of unconventional substrates for biosurfactant production and their new applications. *Appl Microbiol Biotechnol* 58, 428.

EZ Ron, E Rosenberg (2001) Natural roles of biosurfactants Environ Microbiol. 3, 229

IM Banat, RS Makkar, SS Cameotra (2000), Potential commercial applications of microbial surfactants. *Appl Microbiol Biotechnol* 53, 495.

N Kosaric (1996) *Biosurfactants* in Biotechnology 2nd Edition, 659. Vol. 6, M

Roehr (ed.). VCH-Wiley, ISBN 3-527-28316-1

UA Ochsner, T Hembach, A Fiechter (1996), Production of rhamnolipid biosurfactants *Adv Biochem Eng Biotechnol* 53, 89

Microbial polysaccharides 60

R van Kranenburg, IC Boels, M Kleerebezem, WM de Vos (1999), Genetics and engineering of microbial exopolysaccharides for food: approaches for the production of existing and novel polysaccharides *Curr Opin Biotechnol* 10, 498.

I Sutherland (1998), Novel and established applications of microbial polysaccharides *TIBTech* 16, 41

A Becker, F Katzen, A Puhler, L Ielpi (1998) Xanthan gum biosynthesis and application: a biochemical/genetic perspective Appl Microbiol Biotechnol. 50,145

I Sutherland (1996) *Extracellular Polysaccharides* in Biotechnology 2nd Edition, 613. Vol. 6, M Roehr (ed.). VCH-Wiley, ISBN 3-527-28316-1

G Cote, J Ahlgren (1995) *Microbial Polysaccharides* in Kirk-Othmer Encyclopedia of Chemical Technology 4th Edition 16, 578. Interscience-Wiley, ISBN 0-471-52685-1

Biomaterials 62

A Steinbuchel, S Hein (2001), Biochemical and molecular basis of microbial synthesis of polyhydroxyalkanoates in microorganisms *Adv Biochem Eng Biotechnol* 71, 81

MB Hinman, JA Jones, RV Lewis (2000), Synthetic spider silk: a modular fiber *Trends Biotechnol* 18, 374.

IY Galaev, B Mattiasson (1999), 'Smart' polymers and what they could do in biotechnology and medicine *Trends Biotechnol* 17, 335.

D Oesterhelt (1998), The structure and mechanism of the family of retinal proteins from halophilic archaea *Curr. Opin. in Struct. Biology* 8, 489

H Heslot (1998), Artificial fibrous proteins: a review *Biochimie* 80, 19

G Brauneg, G Lefebvre, KF Genser (1998), Polyhydroxyalkanoates, biopolyesters from renewable resources:

physiological and engineering aspects *J Biotechnol* 65, 127.

S Fahnestock, L Bedzyk (1997), Production of synthetic spider dragline silk protein in Pichia pastoris *Appl Microbiol Biot4echnol* 47, 33

A Steinbüchel (1996) *PHB and Other Polyhydroxyalkanoic Acids* in Biotechnology 2nd Edition, 403. Vol. 6, M Roehr (ed.). VCH-Wiley, ISBN 3-527-28316-1

A Salerno, I Goldberg (1993), Cloning, expression and characterization of a synthetic analog to the bioadhesive precorsor protein of the sea mussel Mytilus edulis *Appl Microbiol Biotechnol* 39, 221

Biotransformation 64

J Schrader, R Berger (2001) *Biotechnological Production of Terpenoid Flavor and Fragrance Compounds* in Biotechnology 2nd Edition, 373. Vol. 10, H Rehm (ed.). VCH-Wiley, ISBN 3-527-28320-X

U Bornscheuer (2000) *Industrial Biotranformations* in Biotechnology 2nd Edition, 277. Vol. 8b, D Kelly (ed.). VCH-Wiley, ISBN 3-527-28324-2

K Faber (2000) *Biotransformations in Organic Chemistry*, 4th Edition ed., Springer-Verlag, ISBN 3-540-66334-7

K Faber, R Patel (2000), Chemical biotechnology. A happy marriage between chemistry and biotechnology: asymmetric synthesis via green chemistry *Curr Opin Biotechnol* 11, 517.

A Liese, K Seelbach, C Wandrey (2000) *Industrial Biotransformations*, Wiley-VCH, ISBN 3-527-30094-5.

J Rabenhorst (2000) *Biotechnological Production of Natural Aroma Chemicals by Fermentation Processes* in Biotechnology 2nd Edition, 333. Vol. 8b, D Kelly (ed.). VCH-Wiley, ISBN 3-527-28324-2

B Schulze, MG Wubbolts (1999), Biocatalysis for industrial production of fine chemicals *Curr Opin Biotechnol* 10, 609.

D Kelly (1999) *Biotransformations - Practical Aspects* in Biotechnology 2nd Edition, 25. Vol. 8a, D Kelly (ed.). VCH-Wiley, ISBN 3-527-30104-6

M Turner (1997) *Perspectives in Biotransformation* in Biotechnology 2nd Edition, 5. Vol. 8a, D Kelly (ed.). VCH-Wiley, ISBN 3-527-28318-8

S Shimizu, J Ogawa, M Kataoka, M Kobayashi (1997), Screening of novel microbial enzymes for the production of biologically and chemically useful compounds *Adv Biochem Eng Biotechnol* 58, 45

M Turner (1999) *Biotransformations - Practical Aspects* in Biotechnology 2nd Edition, 25. Vol. 8a, D Kelly (ed.). VCH-Wiley, ISBN 3-527-30104-6

O Sebek, J Rosazza (1995) *Microbial Transformations* in Kirk-Othmer Encyclopedia of Chemical Technology 4th Edition 16, 611. Interscience-Wiley, ISBN 0-471-52685-1

J Prenosil, Ö Kut, I Dunn, E Heinzle (1989) *Immobilized Biocatalysts* in Ullmann's Encyclopedia of Industrial Chemistry 5th edition A14, 1. Wiley-VCH, ISBN 3-527-20114-9

Steroid biotransformations 66

C Duport, R Spagnoli, E Degryse, D Pompon (1998) Self-sufficient biosynthesis of pregnenolone and progesterone in engineered yeast Nat Biotechnol.16, 186

R Müller (1994) *Steroids* in Ullmann's Encyclopedia of Industrial Chemistry 5th edition A25, 309. Wiley-VCH, ISBN 3-527-20125-4

Enzymes 68

S Miot, J Boulay (2001), Protein technologies and commercial enzymes *Curr Opin Biotechnol* 12, 329.

A Curtis (2000) *Carbon-Carbon Bond Formation Using Enzymes* in Biotechnology 2nd Edition, 5. Vol. 8b, D Kelly (ed.). VCH-Wiley, ISBN 3-527-28324-2

Flintsch, G Watt (2000) *Enzymes in Carbohydrate Chemistry: Formation of Glycosidic Linkages* in Biotechnology 2nd Edition, 243. Vol. 8b, D Kelly (ed.). VCH-Wiley, ISBN 3-527-28324-2

D Kelly, J Mahdi (2000) *Lyases* in Biotechnology 2nd Edition, 41. Vol. 8b, D Kelly (ed.). VCH-Wiley, ISBN 3-527-28324-2

G Robinson, S Jackman, J Stratford (2000) *Halocompounds* in Biotechnology 2nd Edition, 173. Vol. 8b, D Kelly (ed.). VCH-Wiley, ISBN 3-527-28324-2

JM Woodley (2000), Advances in enzyme technology – UK contributions. *Adv Biochem Eng Biotechnol* 70, 93.

H Holland (1999) *Hydroxylation and Dihydroxylation* in Biotechnology 2nd Edition, 475. Vol. 8a, D Kelly (ed.). VCH-Wiley, ISBN 3-527-30104-6

D Hoople (1999) *Cleavage and Formation of Amide Bonds* in Biotechnology 2nd Edition, 243. Vol. 8a, D Kelly (ed.). VCH-Wiley, ISBN 3-527-30104-6

K Foster, S Frackman, J Jolly (1995) *Production of Enzymes as Fine Chemicals* in Biotechnology 2nd Edition, 73. Vol. 9, G Reed, T Nagodawithana (ed.). VCH-Wiley, ISBN 3-527-28319-6

R Scopes (1994) *Protein Purification - Principles and Practice*, 3d Edition ed., Springer-Verlag, ISBN 0-387-94072-3

R Perham, M Grassl, G Michal, B Rexer, A Scheltinga, C Gölker, S Fukui, A Tanaka, H Uhlig, W Goldstein, H Hagen, S Pedersen, B Poldermans, E Reimerdes, W Leuchtenberger, U Plöcker, H Waldmann, G Whitesides, G-B Kresse, K Wulff, G Henninger, L Flohé, W Günzler, C Kessler, K Aunstrup (1987) *Enzymes* in Ullmann's Encyclopedia of Industrial Chemistry 5th Edition A9, 341. Wiley-VCH, ISBN 3-527-20109-2

Enzyme catalysis 70

K Jaeger, T Eggert (2002), Lipases for biotechnology. *Curr Opin Biotechnol* 13, 390.

S Fetzner (2002) Oxygenases without requirement for cofactors or metal ions Appl Microbiol Biotechnol. 60, 243

S Panke, MG Wubbolts (2002), Enzyme technology and bioprocess engineering. *Curr Opin Biotechnol* 13, 111.

A Schmid, JS Dordick, B Hauer*, et al.* (2001), Industrial biocatalysis today and tomorrow. *Nature* 409, 258.

J McGregor-Jones (2000) *Synthetic Applications of Enzyme-Catalyzed Reactions* in Biotechnology 2nd Edition, 351. Vol. 8b, D Kelly (ed.). VCH-Wiley, ISBN 3-527-28324-2

UT Bornscheuer, RJ Kazlauskas (1999) *Hydrolases in Organic Synthesis*, Wiley-VCH, ISBN 3-527-30104-6.

A Bunch (1999) *Nitriles* in Biotechnology 2nd Edition, 277. Vol. 8a, D Kelly (ed.). VCH-Wiley, ISBN 3-527-30104-6

R Kazlauskas, U Bornscheuer (1999) *Biotransformations with Lipases* in Biotechnology 2nd Edition, 37. Vol. 8a, D Kelly (ed.). VCH-Wiley, ISBN 3-527-30104-6

D Witiak, A Hopper (1996) *Chiral Pharmaceuticals* in Kirk-Othmer Encyclopedia of Chemical Technology 4th Edition 18, 511. Interscience-Wiley, ISBN 0-471-52687-8

A Zaks (1994) *Enzymes in Organic Synthesis* in Kirk-Othmer Encyclopedia of Chemical Technology 4th Edition 9, 672. Interscience-Wiley, ISBN 0-471-52677-0

Analytical enzymes 72

A Usmani (1995) *Medical Diagnostic Reagents* in Kirk-Othmer Encyclopedia of Chemical Technology 4th Edition 16, 88. Interscience-Wiley, ISBN 0-471-52685-1

G Kresse (1995) *Analytical Use of Enzymes* in Biotechnology 2nd Edition, 137. Vol. 9, G Reed, T Nagodawithana (ed.). VCH-Wiley, ISBN 3-527-28319-6

E Kopetzki, K Lehnert, P Buckel (1994) Enzymes in diagnostics: achievements and possibilities of recombinant DNAS technology Clin Chem 40, 688

Enzyme tests 74

P Gherson, H Lanza, M elavin, D Vlastelica (1992) *Automated Instrumenation, Clincial Chemistry* in Kirk-Othmer Encyclopedia of Chemical Technology 4th Edition 3, 751. Interscience-Wiley, ISBN 0-471-52671-1

Enzymes as additives 76

O Kirk, TV Borchert, CC Fuglsang (2002), Industrial enzyme applications. *Curr Opin Biotechnol* 13, 345.

W Aehle, O Misset (1999) *Enzymes for Industrial Applications* in Biotechnology Second, Completely Revised Edition, 189. Vol. 5a, U Ney, D Schomburg (ed.). VCH Verlagsgesellschaft, ISBN 3-527-28315-3

H Uhlig (1998) *Industrial Enzymes and their Applications*, John Wiley & Sons, ISBN 0-471-19660-6.

H Olsen (1995) *Use of Enzymes in Food Processing* in Biotechnology 2nd Edition, 663. Vol. 9, G Reed, T Nagodawithana (ed.). VCH-Wiley, ISBN 3-527-28319-6

P Nielsen, H Malmos, T Damhus, B Diderichsen, H Nielsen, M Simonsen, H Schiff, A Oestergaard, H Olsen, P Eigtved, T Nielsen, J Xing (1994) *Industrial Enzyme Applications* in Kirk-Othmer

Encyclopedia of Chemical Technology 4th Edition 9, 567. Interscience-Wiley, ISBN 0-471-52677-0

Detergent enzymes 78

R Gupta, QK Beg, P Lorenz (2002) Bacterial alkaline proteases: molecular approaches and industrial applications. Appl Microbiol Biotechnol. 59, 15 0-8247-9995-X

JHv Ee, O Misset, EJ Baas (1997) *Enzymes in Detergency* in Surfactant Science Series *Vol. 69*, MJ Schick, FM Fowkes (ed.). Marcel Dekker, Inc., ISBN 0-8247-9995-X

J Lynn (1993) *Detergency* in Kirk-Othmer Encyclopedia of Chemical Technology 4th Edition 7, 1072. Interscience-Wiley, ISBN 0-471-52675-4

G Jakobi, A Löhr, M Schwuger, D Jung, W Fischer, P Gerike, K Künstler (1987) *Detergents* in Ullmann's Encyclopedia of Industrial Chemistry 5th edition A8, 315. Wiley-VCH, ISBN 3-527-20108-4

Enzymes for starch hydrolysis 80

C Bertoldo, G Antranikian (2002) Starch-hydrolyzing enzymes from thermophilic archaea and bacteria. Curr Opin Chem Biol. 6, 151

R Whistler, J Daniel (1997) *Starch* in Kirk-Othmer Encyclopedia of Chemical Technology 4th Edition 22, 699. Interscience-Wiley, ISBN 0-471-52691-6

R Daniel, R Whistler, A Voragen, W Pilnik (1994) *Starch and Other Polysaccharides* in Ullmann's Encyclopedia of Industrial Chemistry 5th edition A25, 1. Wiley-VCH, ISBN 3-527-20125-4

Enzymes and sweeteners 84

MM Silveira, R Jonas (2002), The biotechnological production of sorbitol. *Appl Microbiol Biotechnol* 59, 400.

R Hebeda (1997) *Syrups* in Kirk-Othmer Encyclopedia of Chemical Technology 4th Edition 23, 582. Interscience-Wiley, ISBN 0-471-52692-4

T Lee (1997) *Sweeteners* in Kirk-Othmer Encyclopedia of Chemical Technology 4th Edition 23, 556. Interscience-Wiley, ISBN 0-471-52692-4

R Hebeda (1995) *Carbohydrate-Based Sweeteners* in Biotechnology 2nd Edition,

737. Vol. 9, G Reed, T Nagodawithana (ed.). VCH-Wiley, ISBN 3-527-28319-6

G-WR Lipinski (1995) *Sweeteners* in Ullmann's Encyclopedia of Industrial Chemistry 5th edition A26, 23. Wiley-VCH, ISBN 3-527-20126-2

H Schiweck, M Clarke (1994) *Sugar* in Ullmann's Encyclopedia of Industrial Chemistry 5th edition A25, 345. Wiley-VCH, ISBN 3-527-20125-4

F Schenck (1989) *Glucose-Containing Syrups* in Ullmann's Encyclopedia of Industrial Chemistry 5th edition A12, 457. Wiley-VCH, ISBN 3-527-20112-2

Enzymes for the hydrolysis of cellulose and polyoses 86

N Thompson (1995) *Hemicellulose* in Kirk-Othmer Encyclopedia of Chemical Technology 4th Edition 13, 54. Interscience-Wiley, ISBN 0-471-52682-7

H Krässig, J Schurz, R Steadman, K Schliefer, W Albrecht (1986) *Cellulose* in Ullmann's Encyclopedia of Industrial Chemistry 5th edition A5, 375. Wiley-VCH, ISBN 3-527-20105-X

Enzymes in pulp and paper processing 88

L Viikari, M Tenkanen, A Suurnäkki (2001) *Biotechnology in the Pulp and Paper Industry* in Biotechnology 2nd Edition, 523. Vol. 10, H Rehm (ed.). VCH-Wiley, ISBN 3-527-28320-X

A Gutierrez, JC del Rio, MJ Martinez, AT Martinez (2001), The biotechnological control of pitch in paper pulp manufacturing *Trends Biotechnol* 19, 340.

M Schulein (2000), Protein engineering of cellulases. *Biochim Biophys Acta* 1543, 239.

A Breen, FL Singleton (1999), Fungi in lignocellulose breakdown and biopulping *Curr Opin Biotechnol* 10, 252.

JF Dean, PR LaFayette, KE Eriksson, SA Merkle (1997), Forest tree biotechnology *Adv Biochem Eng Biotechnol* 57, 1

A Suurnakki, M Tenkanen, J Buchert, L Viikari (1997), Hemicellulases in the bleaching of chemical pulps *Adv Biochem Eng Biotechnol* 57, 261

J Genco (1996) *Pulp* in Kirk-Othmer Encyclopedia of Chemical Technology 4th Edition 20, 493. Interscience-Wiley, ISBN 0-471-52689-4

M Lyne (1996) *Paper* in Kirk-Othmer Encyclopedia of Chemical Technology 4th Edition 18, 1. Interscience-Wiley, ISBN 0-471-52687-8

Pectinases 90

DR Kashyap, PK Vohra, S Chopra, R Tewari (2001) Applications of pectinases in the commercial sector: a review Bioresour Technol. 77, 215

J Baird (1994) *Gums* in Kirk-Othmer Encyclopedia of Chemical Technology 4th Edition 12, 842. Interscience-Wiley, ISBN 0-471-52681-9

Enzymes and milk products 92

C Hall (1995) *Milk and Milk Products* in Kirk-Othmer Encyclopedia of Chemical Technology 4th Edition 16, 700. Interscience-Wiley, ISBN 0-471-52685-1

BA Law, F Mulholland (1991) The influence of biotechnological developments on cheese manufacture. Biotechnol Genet Eng Rev. 9, 369

J Stein, K Imhof (1990) *Milk and Dairy Products* in Ullmann's Encyclopedia of Industrial Chemistry 5th edition A16, 589. Wiley-VCH, ISBN 3-527-20116-5

Enzymes in baking and meat processing 94

B Belderok (2000) Developments in bread-making processes Plant Foods Hum Nutr. 55, 1

G Spicher, J Brümmer (1995) *Baked Goods* in Biotechnology 2nd Edition, 241. Vol. 9, G Reed, T Nagodawithana (ed.). VCH-Wiley, ISBN 3-527-28319-6

G Spicher, Y Pomeranz (1985) *Bread and Other Baked Products* in Ullmann's Encyclopedia of Industrial Chemistry 5th edition A4, 331. Wiley-VCH, ISBN 3-527-20104-1

Enzymes in leather and textile treatment 96

P Hamlyn (1995), The Impact of Biotechnology on the Textile Industry *Textiles Magazine* 3, 6

E Heideman (1990) *Leather* in Ullmann's Encyclopedia of Industrial Chemistry 5th edition A15, 259. Wiley-VCH, ISBN 3-527-20115-7

Procedures for obtaining novel technical enzymes 98

J Beilen, Z Li (2002), Enzyme technology: an overview. *Curr Opin Biotechnol* 13, 338.

ET Farinas, T Bulter, FH Arnold (2001), Directed enzyme evolution. *Curr Opin Biotechnol* 12, 545.

KA Powell, SW Ramer, SB Del Cardayre, *et al.* (2001), Directed Evolution and Biocatalysis. *Angew Chem Int Ed Engl* 40, 3948.

MT Reetz (2001), Combinatorial and Evolution-Based Methods in the Creation of Enantioselective Catalysts. *Angew Chem Int Ed Engl* 40, 284.

FH Arnold, JC Moore (1997), Optimizing industrial enzymes by directed evolution *Adv Biochem Eng Biotechnol* 58, 1

Bakers' yeast and fodder yeasts 100

C Caron (1995) *Commercial Production of Baker's Yeast and Wine Yeast* in Biotechnology 2nd Edition, 321. Vol. 9, G Reed, T Nagodawithana (ed.). VCH-Wiley, ISBN 3-527-28319-6

Single cell protein, single cell oil 102

C Ratledge (1997) *Microbial Lipids* in Biotechnology 2nd Edition, 133. Vol. 7, Hv Döhren, H Kleinkauf (ed.). VCH-Wiley, ISBN 3-527-28310-2

N Scrimshaw, E Murray (1995) *Nutritional Value and Safety of "Single Cell Protein"* in Biotechnology 2nd Edition, 221. Vol. 9, G Reed, T Nagodawithana (ed.). VCH-Wiley, ISBN 3-527-28319-6

J Litchfield (1994) *Nonconventional Foods* in Kirk-Othmer Encyclopedia of Chemical Technology 4th Edition 11, 871. Interscience-Wiley, ISBN 0-471-52680-0

W Babel, H-D Pöhland, K Soyez (1993) *Single Cell Proteins* in Ullmann's Encyclopedia of Industrial Chemistry 5th edition A24, 165. Wiley-VCH, ISBN 3-527-20124-6

Biotechnology and environmental processes

S Kjelleberg (2002), Environmental biotechnology. *Curr Opin Biotechnol* 13, 199.

K Riedel, G Kunze, A Konig (2002), Microbial sensors on a respiratory basis for wastewater monitoring. *Adv Biochem Eng Biotechnol* 75, 81.

P Nisipeanu (1999) *Laws, Statutory Orders and Directives on Waste and Wastewater Treatment* in Biotechnology 2nd Edition, 141. Vol. 11a, J Winter (ed.). VCH-Wiley, ISBN 3-527-28321-8

Aerobic waste water treatment 104

C Gallert, J Winter (2000) *Perspectives of Waste, Wastewater, Off-Gas, and Drinking Water Management* in Biotechnology 2nd Edition, 479. Vol. 11c, J Winter (ed.). VCH-Wiley, ISBN 3-527-28336-6

W Fritsche, M Hofrichter (2000) *Aerobic Degradation by Microorganisms* in Biotechnology 2nd Edition, 145. Vol. 11b, J Klein (ed.). VCH-Wiley, ISBN 3-527-28323-4

L Hartmann (1999) *Historical Development of Wastewater Treatment Processes* in Biotechnology 2nd Edition, 5. Vol. 11a, J Winter (ed.). VCH-Wiley, ISBN 3-527-28321-8

C Gallert, J Winter (1999) *Bacterial Metabolism in Wastewater Treatment* in Biotechnology 2nd Edition, 17. Vol. 11a, J Winter (ed.). VCH-Wiley, ISBN 3-527-28321-8

R Kayser (1999) *Activated Sludge Process* in Biotechnology 2nd Edition, 253. Vol. 11a, J Winter (ed.). VCH-Wiley, ISBN 3-527-28321-8

P Baumann, B Dorias (1999) *Trickling Filter Systems* in Biotechnology 2nd Edition, 335. Vol. 11a, J Winter (ed.). VCH-Wiley, ISBN 3-527-28321-8

P Koppe, A Stozek, V Neitzel (1999) *Municipal Wastewater and Sewage Sludge* in Biotechnology 2nd Edition, 161. Vol. 11a, J Winter (ed.). VCH-Wiley, ISBN 3-527-28321-8

K Rosenwinkel, U Austermann-Haun, H Meyer (1999) *Industrial Wastewater Sources and Treatment Strategies* in Biotechnology 2nd Edition, 191. Vol. 11a, J Winter (ed.). VCH-Wiley, ISBN 3-527-28321-8

P Weiland (1999) *Agricultural Waste and Wastewater Sources and Management* in Biotechnology 2nd Edition, 217. Vol. 11a, J Winter (ed.). VCH-Wiley, ISBN 3-527-28321-8

H Kroiss, K Svardal (1999) *CSTR-Reactors and Contact Processes in Industrial Waste-water Treatment* in Biotechnology 2nd Edition, 479. Vol. 11a, J Winter (ed.). VCH-Wiley, ISBN 3-527-28321-8

H Jördening, K Buchholz (1999) *Fixed Film Stationary Bed and Fluidized Bed Reactors* in Biotechnology 2nd Edition, 493. Vol. 11a, J Winter (ed.). VCH-Wiley, ISBN 3-527-28321-8

A Schramm, R Amann (1999) *Nucleic Acid-Based Techniques for Analyzing the Diversity, Structure, and Dynamics of Microbial Communities in Wastewater Treatment* in Biotechnology 2nd Edition, 85. Vol. 11a, J Winter (ed.). VCH-Wiley, ISBN 3-527-28321-8

G Andrews (1993) *Aerobic Waste Water Process Models* in Biotechnology Second, Completely Revised Edition, 407. Vol. 4, K Schügerl (ed.). VCH Verlagsgesellschaft, ISBN ISBN 3-527-28314-5

Anaerobic waste water and sludge treatment 106

C Gallert, J Winter (2002) Solid and liquid residues as raw materials for biotechnology Naturwissenschaften. 89, 483

Y Sekiguchi, Y Kamagata, H Harada (2001), Recent advances in methane fermentation technology *Curr Opin Biotechnol* 12, 277.

B Schink (2000) *Principles of Anaerobic Degradation of Organic Compounds* in Biotechnology 2nd Edition, 169. Vol. 11b, J Klein (ed.). VCH-Wiley, ISBN 3-527-28323-4

M McInerney (1999) *Anaerobic Metabolism and its Regulation* in Biotechnology 2nd Edition, 455. Vol. 11a, J Winter (ed.). VCH-Wiley, ISBN 3-527-28321-8

S Phythian (1999) *Esterases* in Biotechnology 2nd Edition, 193. Vol. 8a, D Kelly (ed.). VCH-Wiley, ISBN 3-527-30104-6

H Märkl (1999) *Modeling of Biogas Reactors* in Biotechnology 2nd Edition, 527. Vol. 11a, J Winter (ed.). VCH-Wiley, ISBN 3-527-28321-8

D Schürbüscher, C Wandrey (1993) *Anaerobic Waste Water Process Models* in Biotechnology Second, Completely Revised Edition, 441. Vol. 4, K Schügerl (ed.). VCH Verlagsgesellschaft, ISBN 3-527-28314-5

Biological treatment of exhaust air 108

K Engesser, T Plaggemeier (2000) *Microbiological Aspects of Biological Waste Gas Purification* in Biotechnology 2nd Edition, 275. Vol. 11c, J Winter (ed.). VCH-Wiley, ISBN 3-527-28336-6

K Fischer (2000) *Biofilters* in Biotechnology 2nd Edition, 321. Vol. 11c, J Winter (ed.). VCH-Wiley, ISBN 3-527-28336-6

D Chitwood, J Devinny (2000) *Commercial Applications of Biological Waste Gas Purification* in Biotechnology 2nd Edition, 357. Vol. 11c, J Winter (ed.). VCH-Wiley, ISBN 3-527-28336-6

T Plaggemeier, O Lämmerzahl (2000) *Treatment of Waste Gas Pollutants in Trickling Filters* in Biotechnology 2nd Edition, 333. Vol. 11c, J Winter (ed.). VCH-Wiley, ISBN 3-527-28336-6

M Reiser (2000) *Waste Gas Treatment: Membrane Processes and Alternative Techniques* in Biotechnology 2nd Edition, 345. Vol. 11c, J Winter (ed.). VCH-Wiley, ISBN 3-527-28336-6

E Schippert, H Chmiel (2000) *Bioscrubbers* in Biotechnology 2nd Edition, 305. Vol. 11c, J Winter (ed.). VCH-Wiley, ISBN 3-527-28336-6

M Waweru, V Herrygers, HV Langenhove, W Verstraete (2000) *Process Engineering of Biological Waste Gas Purification* in Biotechnology 2nd Edition, 259. Vol. 11c, J Winter (ed.). VCH-Wiley, ISBN 3-527-28336-6

Biological soil treatment 110

J Widada, H Nojiri, T Omori (2002), Recent developments in molecular techniques for identification and monitoring of xenobiotic-degrading bacteria and their catabolic genes in bioremediation. *Appl Microbiol Biotechnol* 60, 45.

W Ulrici (2000) *Contaminated Soil Areas, Different Countries and Contaminants, Monitoring of Contaminants* in Biotechnology 2nd Edition, 5. Vol. 11b, J Klein (ed.). VCH-Wiley, ISBN 3-527-28323-4

J Klein (2000) *Possibilities, Limits, and Future Developments of Soil Bioremediation* in Biotechnology 2nd Edition, 465. Vol. 11b, J Klein (ed.). VCH-Wiley, ISBN 3-527-28323-4

M Koning, K Hupe, R Stegmann (2000) *Thermal Processes, Scrubbing/Extraction,*

Bioremediation and Disposal in Biotechnology 2nd Edition, 305. Vol. 11b, J Klein (ed.). VCH-Wiley, ISBN 3-527-28323-4

T Held, H Dörr (2000) *In situ Remediation* in Biotechnology 2nd Edition, 349. Vol. 11b, J Klein (ed.). VCH-Wiley, ISBN 3-527-28323-4

R Unterman, M DeFlaun, RJSteffan (2000) *Acvanced in situ Bioremediation - A Hierarchy of Technology Choices* in Biotechnology 2nd Edition, 399. Vol. 11b, J Klein (ed.). VCH-Wiley, ISBN 3-527-28323-4

C Wischnak, R Müller (2000) *Degradation of Chlorinated Compounds* in Biotechnology 2nd Edition, 241. Vol. 11b, J Klein (ed.). VCH-Wiley, ISBN 3-527-28323-4

V Schulz-Berendt (2000) *Bioremediation with Heap Technique* in Biotechnology 2nd Edition, 319. Vol. 11b, J Klein (ed.). VCH-Wiley, ISBN 3-527-28323-4

K Blotevogel, T Gorontzy (2000) *Microbial Degradation of Compounds with Nitro Functions* in Biotechnology 2nd Edition, 273. Vol. 11b, J Klein (ed.). VCH-Wiley, ISBN 3-527-28323-4

M Kästner (2000) *"Humification" Process or Formation of Refractory Soil Organic Matter* in Biotechnology 2nd Edition, 89. Vol. 11b, J Klein (ed.). VCH-Wiley, ISBN 3-527-28323-4

F-M Menn, J Easter, G Sayleri (2000) *Genetically Engineered Microorganisms and Bioremediation* in Biotechnology 2nd Edition, 5. Vol. 11b, J Klein (ed.). VCH-Wiley, ISBN 3-527-28323-4

W Chen, F Bruhlmann, RD Richins, A Mulchandani (1999), Engineering of improved microbes and enzymes for bioremediation *Curr Opin Biotechnol* 10, 137.

R Prince (1998) *Bioremediation* in Kirk-Othmer Encyclopedia of Chemical Technology 4th Edition Supplement, 48. Interscience-Wiley, ISBN 0-471-52696-7

M Dua, A Singh, N Sethunathan, AK Johri (2002), Biotechnology and bioremediation: successes and limitations. *Appl Microbiol Biotechnol* 59, 143.

Microbial leaching, biofilms, and biocorrosion 112

HC Flemming (2002) Biofouling in water systems – cases, causes and counter-measures Appl Microbiol Biotechnol. 59, 629

H Brandl (2001) *Microbial Leaching of Metals* in Biotechnology 2nd Edition, 191. Vol. 10, H Rehm (ed.). VCH-Wiley, ISBN 3-527-28320-X

G Gadd (2001) *Accumulation and Transformation of Metals by Microorganisms* in Biotechnology 2nd Edition, 225. Vol. 10, H Rehm (ed.). VCH-Wiley, ISBN 3-527-28320-X

W Sand (2001) *Microbial Corrosion and its Inhibition* in Biotechnology 2nd Edition, 265. Vol. 10, H Rehm (ed.). VCH-Wiley, ISBN 3-527-28320-X

H Ehrlich (1997), Microbes and metals *Appl Microbiol Biotechnol* 48, 687

Medical biotechnology

P Buckel (1998), Toward a new natural medicine. Naturwissenschaften 8, 155

P Buckel (1996), Recombinant proteins for therapy Trends Pharmacol. Sci 17, 450

Insulin 114

S Shoelson (1995) *Insulin and Other Antidiabetic Agents* in Kirk-Othmer Encyclopedia of Chemical Technology 4th Edition 14, 662. Interscience-Wiley, ISBN 0-471-52683-5

AF Bristow (1993) Recombinant-DNA-derived insulin analogues as potentially useful therapeutic agents. Trends Biotechnol. 11, 301

F Schmidt (1985) *Antidiabetic Drugs* in Ullmann's Encyclopedia of Industrial Chemistry 5th edition A3, 1. Wiley-VCH, ISBN 3-527-20103-3

BH Frank, RE Chance (1983), Two routes for producing human Insulin utilizing recombinant DNA technology. *Muench. med. Wschr.* 125, 14.

Growth hormone and other hormones 116

C Hew, G Fletcher (1997), Transgenic fish for aquaculture *Chemistry & Industry* 311

G Becker, W MacKellar, R Riggin, V Wroblewski (1995) *Hormones, Human Growth Hormone* in Kirk-Othmer Encyclopedia of Chemical Technology 4th Edition 13, 406. Interscience-Wiley, ISBN 0-471-52682-7

W Engeland (1995) *Hormones, Survey* in Kirk-Othmer Encyclopedia of Chemical Technology 4th Edition 13, 357. Interscience-Wiley, ISBN 0-471-52682-7

J Sandow, E Scheiffele, M Haring, G Neef, K Prezewowsky, U Stache (1989) *Hormones* in Ullmann's Encyclopedia of Industrial Chemistry 5th edition A13, 89. Wiley-VCH, ISBN 3-527-20113-0

Hemoglobin, serum albumen, and lactoferrin 118

TM Chang (1999), Future prospects for artificial blood *Trends Biotechnol* 17, 61.

W Bell (1992) *Blood, Coagulants and Anticoagulants* in Kirk-Othmer Encyclopedia of Chemical Technology 4th Edition 4, 333. Interscience-Wiley, ISBN 0-471-52672-X

S Yamashita (1985) *Blood* in Ullmann's Encyclopedia of Industrial Chemistry 5th edition A4, 201. Wiley-VCH, ISBN 3-527-20104-1

Blood clotting agents 120

R Jiang, T Monroe, R McRogers, PJ Larson (2002) Manufacturing challenges in the commercial production of recombinant coagulation factor VIII Haemophilia. 8 Suppl, 1

S Soukharev, D Hammond, NM Ananyeva, JA Anderson, CA Hauser, S Pipe, EL Saenko (2002) Expression of factor VIII in recombinant and transgenic systems Blood Cells Mol Dis. 28, 234

EGD Tuddenheim (1997) *Haemophilia: molecular biology at the centre of human disease* in Molecular Biology in Medicine, 1. TM Cox, J Sinclair (ed.). Blackwell Science Ltd., ISBN 0-632-02785-1

Anticoagulants und thrombolytic agents 122

F Markwardt (2002), Hirudin as alternative anticoagulant – a historical review. *Semin Thromb Hemost* 28, 405.

BH Bendixen, L Ocava (2002 Evaluation and management of acute ischemic stroke Curr Cardiol Rep. 4, 149

W Bode, H Renatur (1997) Tissue-type plasminogen activator: variants and crystal/solution structures demarcate structural determinants of function. *Curr Opin Struct Biol* 6, 865

H Jayaram, G Ahluwalia, D Cooney (1994) *Therapeutic Enzyme Applications* in Kirk-Othmer Encyclopedia of Chemical Technology 4th Edition 9, 621. Interscience-Wiley, ISBN 0-471-52677-0

Enzyme inhibitors 124

A Muscate, C Levinson, G Kenyon (1994) *Enzyme Inhibitors* in Kirk-Othmer Encyclopedia of Chemical Technology 4th Edition 9, 646. Interscience-Wiley, ISBN 0-471-52677-0

The immune system 126

B Corthesy (2002), Recombinant immunoglobulin A: powerful tools for fundamental and applied research. *Trends Biotechnol* 20, 65.

C Janeway, P Travers (1997) *Immunobiology*, 3d Edition, Current Biology, 0-443-05964-0.

Stem cells 128

G Daley (2002), Prospects for stem cell therapeutics: myths and medicines. *Curr Opin Genet Dev* 12, 607.

M Larru (2001), Adult stem cells: an alternative to embryonic stem cells? *Trends Biotechnol* 19, 487.

A Colman, A Kind (2000), Therapeutic cloning: concepts and practicalities *Trends Biotechnol* 18, 192.

JA Thomson, JS Odorico (2000), Human embryonic stem cell and embryonic germ cell lines *Trends Biotechnol* 18, 53.

Tissue Engineering 130

MV Risbud, M Sittinger (2002), Tissue engineering: advances in in vitro cartilage generation. *Trends Biotechnol* 20, 351.

LG Griffith, G Naughton (2002), Tissue engineering – current challenges and expanding opportunities. *Science* 295, 1009.

P Bianco, PG Robey (2001), Stem cells in tissue engineering. *Nature* 414, 118.

Interferons 132

G Wetzel (1999) *Medical Applications of Recombinant Proteins in Humans and Animals* in Biotechnology Second, Completely Revised Edition, 125. Vol. 5a, U Ney, D Schomburg (ed.). VCH Verlagsgesellschaft, ISBN 3-527-28315-3

S Wong, J Xing (1995) *Immunotherapeutic Agents* in Kirk-Othmer Encyclopedia of Chemical Technology 4th Edition 14, 64. Interscience-Wiley, ISBN 0-471-52683-5

T Nagabhushan, P Trotta (1989) *Interferons* in Ullmann's Encyclopedia of Industrial Chemistry 5th edition A14, 365. Wiley-VCH, ISBN 3-527-20114-9

Interleukins 134

K Friehs, KF Reardon (1993) Parameters influencing the productivity of recombinant E. coli cultivations. Adv Biochem Eng Biotechnol. 48, 53

Erythropoietin and other growth factors 136

AS Lubiniecki, JH Lupker (1994) Purified protein products of rDNA technology expressed in animal cell culture. Biologicals. 22, 161

Other therapeutic proteins 138

R Schiffmann, RO Brady RO (2002) New prospects for the treatment of lysosomal storage diseases. Drugs 62, 733

CE Kearney, CE Wallis (2000) Deoxyribonuclease for cystic fibrosis Cochrane Database Syst Rev. CD001127

Vaccines 140

C Hsieh, M Ritchey (1997) *Vaccine Technology* in Kirk-Othmer Encyclopedia of Chemical Technology 4th Edition 24, 727. Interscience-Wiley, ISBN 0-471-52693-2

S Cryz, M Granstrom, B Gottstein, L Perrin, A Cross, J Larrick (1989) *Immunotherpay and Vaccines* in Ullmann's Encyclopedia of Industrial Chemistry 5th edition A14, 49. Wiley-VCH, ISBN 3-527-20114-9

Recombinant vaccines 142

FX Berthet, T Coche, C Vinals (2001), Applied genome research in the field of human vaccines *J Biotechnol* 85, 213.

C Olive, I Toth, D Jackson (2001) Technological advances in antigen delivery and synthetic peptide vaccine developmental strategies. Mini Rev Med Chem. 1, 429

AM Walmsley, CJ Arntzen (2000), Plants for delivery of edible vaccines *Curr Opin Biotechnol* 11, 126.

Antibodies 144

K Rajewsky (1996), Clonal selection and learning in the antibody system. *Nature* 381, 751.

CA Janeway, Jr. (1993), How the immune system recognizes invaders. *Sci Am* 269, 72.

Monoclonal antibodies 146

G Galfre, D Secher, P Crawley (1990) *Monoclonal Antibodies* in Ullmann's Encyclopedia of Industrial Chemistry 5th edition A16, 699. Wiley-VCH, ISBN 3-527-20116-5

Recombinant and catalytic antibodies 148

KD Wittrup (2001), Protein engineering by cell-surface display *Curr Opin Biotechnol* 12, 395.

HE Chadd, SM Chamow (2001), Therapeutic antibody expression technology *Curr Opin Biotechnol* 12, 188.

G Blackburn, A Garcon (2000) *Catalytic Antibodies* in Biotechnology 2nd Edition, 491. Vol. 8b, D Kelly (ed.). VCH-Wiley, ISBN 3-527-28324-2

J Adair (1999) *Antibody Engineering and Expression* in Biotechnology Second, Completely Revised Edition, 219. Vol. 5a, U Ney, D Schomburg (ed.). VCH Verlagsgesellschaft, ISBN 3-527-28315-3

A Racher, J Tong, J Bonnerjea (1999) *Manufacture of Therapeutic Antibodies* in Biotechnology Second, Completely Revised Edition, 247. Vol. 5a, U Ney, D Schomburg (ed.). VCH Verlagsgesellschaft, ISBN 3-527-28315-3

E Driggers, PGSchultz (1996), Catalytic Antibodies *Adv. Protein Chem.* 49, 261

Immunoanalysis 150

J Miller, R Niessner (1994) *Enzyme and Immunoassays* in Ullmann's Encyclopedia of Industrial Chemistry 5th edition B5, 129. Wiley-VCH, ISBN 3-527-20135-1

Biosensors 152

FW Scheller, U Wollenberger, A Warsinke, F Lisdat (2001), Research and development in biosensors *Curr Opin Biotechnol* 12, 35.

RL Rich, DG Myszka (2000), Advances in surface plasmon resonance biosensor analysis *Curr Opin Biotechnol* 11, 54.

K Cammann, B Ross, W Hasse, C Dumschat, A Katerkamp, J Reinbold, G Steinhage, B Gründig, R Renneberg, N Buschmann (1994) *Chemical and Biochemical Sensors* in Ullmann's Encyclopedia of Industrial Chemistry 5th edition B6, 121. Wiley-VCH, ISBN 3-527-20136-X

B Mattiasson (1993) *Biosensors* in Biotechnology Second, Completely Revised Edition, 5. Vol. 4, K Schügerl (ed.). VCH Verlagsgesellschaft, ISBN 3-527-28314-5

W Pietro (1992) *Biosensors* in Kirk-Othmer Encyclopedia of Chemical Technology 4th Edition 4, 208. Interscience-Wiley, ISBN 0-471-52672-X

Animal breeding 154

G Bulfield (2000), Farm animal biotechnology *Trends Biotechnol* 18, 10.

Embryo transfer, cloned animals 156

E Wolf, V Zakhartchenko, G Brem (1998), Nuclear transfer in mammals: recent developments and future perspectives *J Biotechnol* 65, 99.

Gene maps 158

GH Yue, P Beeckmann, H Bartenschlager,, G Moser, H Geldermann (1999), Rapid ad precise genotyping of porcine microsatellites Electrophoresis 20, 3358

M Lucy, R Collier (1994) *Genetic Engineering, Animals* in Kirk-Othmer Encyclopedia of Chemical Technology 4th Edition 12, 465. Interscience-Wiley, ISBN 0-471-52681-9

G Wricke, H Geldermann, W Weber (1993) *Gene Mapping in Animals and Plants* in Biotechnology Second, Completely Revised Edition, 141. Vol. 2, A Pühler (ed.). VCH Verlagsgesellschaft, ISBN 3-527-28312-9

H Geldermann (1990) *Application of Genome Analysis in Animal Breeding* in Genome Analysis in Domestic Animals Vol. H Geldermann, F Ellendorf (ed.). VCH Verlagsgesellschaft, ISBN 3-527-28097-9

333

Transgenic animals 160

A Trounson (2001) Nuclear transfer in human medicine and animal breeding Reprod Fertil Dev. 13, 31

D Metzger, R Feil (1999), Engineering the mouse genome by site-specific recombination Curr Opin Biotechnol 10, 470.

G Brem (1993) Transgenic Animals in Biotechnology Second, Completely Revised Edition, 745. Vol. 2, A Pühler (ed.). VCH Verlagsgesellschaft, ISBN 3-527-28312-9

Gene farming and xenotransplantation 162

L Brasile, BM Stubenitsky, G Kootstra (2002) Solving the organ shortage: potential strategies and the likelihood of success. ASAIO J. 248, 211

JW Larrick, DW Thomas (2001), Producing proteins in transgenic plants and animals Curr Opin Biotechnol 12, 411.

A Dove (2000), Milking the genome for profit Nature Biotechnology 18, 1045

NS Rudolph (1999), Biopharmaceutical production in transgenic livestock Trends Biotechnol 17, 367

Plant breeding 164

J Huang, C Pray, S Rozelle (2002) Enhancing the crops to feed the poor Nature 418, 678

RP Tengerdy, G Szakacs (1998), Perspectives in agrobiotechnology J Biotechnol 66, 91.

Plant tissue surface culture 166

BW Grout (1999) Meristem-tip culture for propagation and virus elimination. Methods Mol Biol.111,115.

Plant cell suspension culture 168

S Jennewein, R Croteau (2001) Taxol: biosynthesis, molecular genetics, and biotechnological applications Appl Microbiol Biotechnol. 57,13

J Berlin (1997) Secondary Products from Plant Cell Cultures in Biotechnology 2nd Edition, 593. Vol. 7, Hv Döhren, H Kleinkauf (ed.). VCH-Wiley, ISBN 3-527-28310-2

334 SC Roberts, ML Shuler (1997), Large-scale plant cell culture Curr Opin Biotechnol 8, 154.

PM Kieran, PF MacLoughlin, DM Malone (1997), Plant cell suspension cultures: some engineering considerations J Biotechnol 59, 39.

M Petersen, AW Alfermann (1993) Plant Cell Cultures in Biotechnology Second, Completely Revised Edition, 577. Vol. 1, H Sahm (ed.). VCH Verlagsgesellschaft, ISBN 3-527-28337-4

A Tanaka (1987), Large-scale cultivation of plant cells at high density: a review Process Biochemistry 106

Transgenic plants: methods 170

A Vanavichit, S Tragoonrung, T Toojinda (2001) Genomic Mapping and Positional Cloning with Emphasis on Plant Science in Biotechnology 2nd Edition, 165. Vol. 5b, C Sensen (ed.). VCH-Wiley, ISBN 3-527-28328-5

M Hughes (1996) Plant Molecular Genetics, Longman, ISBN 0-582-24730-6

J Edwards, G Kishore, D Stark (1994) Genetic Engineering, Plants in Kirk-Othmer Encyclopedia of Chemical Technology 4th Edition 12, 491. Interscience-Wiley, ISBN 0-471-52681-9

G Kahl, K Weising (1993) Genetic Engineering of Plant Cells in Biotechnology Second, Completely Revised Edition, 547. Vol. 2, A Pühler (ed.). VCH Verlagsgesellschaft, ISBN 3-527-28312-9

Transgenic plants: resistance 172

CT Verrips, MM Warmoeskerken, JA Post (2001), General introduction to the importance of genomics in food biotechnology and nutrition Curr Opin Biotechnol 12, 483.

I Parkin, S Robinson, A Sharpe, K Rozwadowski, D Hegedus, D Lydiate (2001) Agri-Food and Genomics in Biotechnology 2nd Edition, 145. Vol. 5b, C Sensen (ed.). VCH-Wiley, ISBN 3-527-28328-5

J Mol, E Cornish, J Mason, R Koes (1999), Novel coloured flowers Curr Opin Biotechnol 10, 198.

Transgenic plants: products 174

JW Larrick, DW Thomas (2001), Producing proteins in transgenic plants and animals Curr Opin Biotechnol 12, 411.

G Giddings (2001), Transgenic plants as protein factories Curr Opin Biotechnol 12, 450.

SA Merkle, JF Dean (2000), Forest tree biotechnology *Curr Opin Biotechnol* 11, 298.

Y Poirier (1999), Production of new polymeric compounds in plants *Curr Opin Biotechnol* 10, 181.

L Willmitzer (1993) *Transgenic Plants* in Biotechnology Second, Completely Revised Edition, 627. Vol. 2, A Pühler (ed.). VCH Verlagsgesellschaft, ISBN 3-527-28312-9

Viruses 176

P Ahlquist (2002), RNA-dependent RNA polymerases, viruses, and RNA silencing. *Science* 296, 1270.

E Baranowski, CM Ruiz-Jarabo, E Domingo (2001), Evolution of cell recognition by viruses. *Science* 292, 1102.

JL Dangl, JD Jones (2001), Plant pathogens and integrated defence responses to infection. *Nature 411, 826.*

AJ McMichael, SL Rowland-Jones (2001), Cellular immune responses to HIV. *Nature* 410, 980.

RM Zinkernagel (1996), Immunology taught by viruses. *Science* 271, 173.

Bakteriophages 176

H Sandmeier, J Meyer (1993) *Bacteriophages* in Biotechnology Second, Completely Revised Edition, 543. Vol. 1, H Sahm (ed.). VCH Verlagsgesellschaft, ISBN 3-527-28337-4

Microorganisms 180

C Bertoldo, R Grote, G Antranikian (2001) *Biocatalysis under Extereme Conditions* in Biotechnology 2nd Edition, 61. Vol. 10, H Rehm (ed.). VCH-Wiley, ISBN 3-527-28320-X

Amann, BM Fuchs, S Behrens (2001), The identification of microorganisms by fluorescence in situ hybridisation *Curr Opin Biotechnol* 12, 231

HJ Busse, EB Denner, W Lubitz (1996), Classification and identification of bacteria: current approaches to an old problem. Overview of methods used in bacterial systematics *J Biotechnol* 47, 3.

Bacteria 182

H Bahl, P Dürre (1993) *Methylotrophs* in Biotechnology Second, Completely Revised Edition, 285. Vol. 1, H Sahm (ed.).

VCH Verlagsgesellschaft, ISBN 3-527-28337-4

H König (1993) *Methanogens* in Biotechnology Second, Completely Revised Edition, 251. Vol. 1, H Sahm (ed.). VCH Verlagsgesellschaft, ISBN ISBN 3-527-28337-4

L Dijkhuizen (1993) *Methylotrophs* in Biotechnology Second, Completely Revised Edition, 265. Vol. 1, H Sahm (ed.). VCH Verlagsgesellschaft, ISBN 3-527-28337-4

FG Priest (1993) *Bacillus* in Biotechnology Second, Completely Revised Edition, 367. Vol. 1, H Sahm (ed.). VCH Verlagsgesellschaft, ISBN 3-527-28337-4

G Auling (1993) *Pseudomonads* in Biotechnology Second, Completely Revised Edition, 401 Vol. 1, H Sahm (ed.). VCH Verlagsgesellschaft, ISBN 3-527-28337-4

W Piepersberg (1993) *Streptomycetes and Corynebacteria* in Biotechnology Second, Completely Revised Edition, 433. Vol. 1, H Sahm (ed.). VCH Verlagsgesellschaft, ISBN 3-527-28337-4

Some bacteria of importance for biotechnology 184

PJ Punt, N van Biezen, A Conesa, *et al.* (2002), Filamentous fungi as cell factories for heterologous protein production. *Trends Biotechnol* 20, 200

JR Swartz (2001), Advances in Escherichia coli production of therapeutic proteins *Curr Opin Biotechnol* 12, 195.

F Blattner, G Plunkett, C Bloch, N Perna, V Burland, M Riley, J Collado-Vides, J Glasner, C Rode, G Mayhew, J Gregor, N Davis, H Kirkpatrick, M Goeden, D Rose, B Mau, Y Shao (1997), The Complete Genome Sequence of Escherichia coli K-12 *Science* 277, 1453

Fungi 186

F Meinhardt, K Esser (1993) *Filamentous Fungi* in Biotechnology Second, Completely Revised Edition, 515. Vol. 1, H Sahm (ed.). VCH Verlagsgesellschaft, ISBN 3-527-28337-4

Yeasts 188

GP Cereghino, JM Cregg (1999), Applications of yeast in biotechnology: protein production and genetic analysis *Curr Opin Biotechnol* 10, 422.

D Maloney (1998) *Yeasts* in Kirk-Othmer Encyclopedia of Chemical Technology 4th Edition 25, 761. Interscience-Wiley, ISBN 0-471-52694-0

JJ Heinisch, CP Hollenberg (1993) *Yeasts* in Biotechnology Second, Completely Revised Edition, 469. Vol. 1, H Sahm (ed.). VCH Verlagsgesellschaft, ISBN 3-527-28337-4

Microorganisms: isolation, preservation, safety 190

K Frobel, S Metzger (2001) *New Methods of Screening in Biotechnology* in Biotechnology 2nd Edition, 41. Vol. 10, H Rehm (ed.). VCH-Wiley, ISBN 3-527-28320-X

Microorganisms: strain improvement 192

A Crueger (1993) *Mutagenesis* in Biotechnology Second, Completely Revised Edition, 5. Vol. 2, A Pühler (ed.). VCH Verlagsgesellschaft, ISBN 3-527-28312-9

J Engels, B Sprunkel, E Uhlmann (1993) *DNA Synthesis* in Biotechnology Second, Completely Revised Edition, 317. Vol. 2, A Pühler (ed.). VCH Verlagsgesellschaft, ISBN 3-527-28312-9

L Recio (1990) *Mutagenic Agents* in Ullmann's Encyclopedia of Industrial Chemistry 5th edition A16, 755. Wiley-VCH, ISBN 3-527-20116-5

Growing microorganisms 194

E Stoppok, K Buchholz (1996) *Sugar-Based Raw Materials for Fermentation Applications* in Biotechnology 2nd Edition, 5. Vol. 6, M Roehr (ed.). VCH-Wiley, ISBN 3-527-28316-1

JD Troostembergh (1996) *Starch-Based Raw Materials for Fermentation Applications* in Biotechnology 2nd Edition, 31. Vol. 6, M Roehr (ed.). VCH-Wiley, ISBN 3-527-28316-1

F Schneider, H STeinmüller (1996) *Raw Material Strategies - Economical Problems* in Biotechnology 2nd Edition, 47. Vol. 6, M Roehr (ed.). VCH-Wiley, ISBN 3-527-28316-

S Sengha (1994) *Fermentation* in Kirk-Othmer Encyclopedia of Chemical Technology 4th Edition 10, 361. Interscience-Wiley, ISBN 0-471-52678-9

R Greasham (1993) *Media for Microbial Fermentations* in Biotechnology Second, Completely Revised Edition, 127. Vol. 3, G Stephanopoulos (ed.). VCH Verlagsgesellschaft, ISBN 3-527-28313-7

Growth kinetics and product formation 196

CH Posten, CL Cooney (1993) *Growth of Microorganisms* in Biotechnology Second, Completely Revised Edition, 111. Vol. 1, H Sahm (ed.). VCH Verlagsgesellschaft, ISBN 3-527-28337-4

Fed-batch and continuous fermentation 198

T Imanaka (1993) *Strategies for Fermentation with Recombinant Organisms* in Biotechnology Second, Completely Revised Edition, 283. Vol. 3, G Stephanopoulos (ed.). VCH Verlagsgesellschaft, ISBN 3-527-28313-7

T Yamane, S Shimizu (1984) Fed-Batch Techniques in Microbial Processes, *Adv. Biochem. Eng. Biotechnol.* 30, 147

Fermentation technology 200

KH Lee, V Hatzimanikatis (2002), Biochemical Engineering. *Curr Opin Biotechnol* 13, 85.

R Katzen, GT Tsao (2000), A view of the history of biochemical engineering. *Adv Biochem* Eng Biotechnol 70, 77.

K Schugerl (2000), Development of bioreaction engineering. *Adv Biochem Eng Biotechnol* 70, 41.

R Kleijntjens, K Luyben (2000) *Bioreactors* in Biotechnology 2nd Edition, 329. Vol. 11b, J Klein (ed.). VCH-Wiley, ISBN 3-527-28323-4

G Larsson, SB Jorgensen, MN Pons, B Sonnleitner, A Tijsterman, N Titchener-Hooker (1997), Biochemical engineering science *J Biotechnol* 59, 3.

B Tarmy (1996) *Reactor Technology* in Kirk-Othmer Encyclopedia of Chemical Technology 4th Edition 20, 1006. Interscience-Wiley, ISBN 0-471-52689-4

M Reuss (1993) *Oxygen Transfer and Mixing: Scale-Up Implications* in Biotechnology Second, Completely Revised Edition, 185. Vol. 3, G Stephanopoulos (ed.).

VCH Verlagsgesellschaft, ISBN 3-527-28313-7

B Buckland, M Lilly (1993) *Fermentation – an Overview* in Biotechnology Second, Completely Revised Edition, 7. Vol. 3, G Stephanopoulos (ed.). VCH Verlagsgesellschaft, ISBN 3-527-28313-7

J Nielsen, J Villadsen (1993) *Bioreactors: Description and Modelling* in Biotechnology Second, Completely Revised Edition, 77. Vol. 3, G Stephanopoulos (ed.). VCH Verlagsgesellschaft, ISBN 3-527-28313-7

H Voss (1992) *Bioreactors* in Ullmann's Encyclopedia of Industrial Chemistry 5th edition B4, 381. Wiley-VCH, ISBN 3-527-20134-3

Fermentation technology: scale-up 202

B Sonnleitner (2000), Instrumentation of biotechnological processes *Adv Biochem Eng Biotechnol* 66, 1

G Seidel, C Tollnick, M Beyer, K Schugerl (2000), On-line and off-line monitoring of the production of cephalosporin C by Acremonium chrysogenum *Adv Biochem Eng Biotechnol* 66, 115

HJ Henzler (2000), Particle stress in bioreactors *Adv Biochem Eng Biotechnol* 67, 35

DA Mitchell, M Berovic, N Krieger (2000), Biochemical engineering aspects of solid state bioprocessing *Adv Biochem Eng Biotechnol* 68, 61

W Beyeler, E DaPra, K Schneider (2000), Automation of industrial bioprocesses *Adv Biochem Eng Biotechnol* 70, 139

T Chattaway, G Montague, A Morris (1993) *Fermentation Monitoring and Control* in Biotechnology Second, Completely Revised Edition, 319. Vol. 3, G Stephanopoulos (ed.). VCH Verlagsgesellschaft, ISBN 3-527-28313-7

L Erickson, D Fung, P Tuitemwong (1993) *Anaerobic Fermentations* in Biotechnology Second, Completely Revised Edition, 7. Vol. 3, G Stephanopoulos (ed.). VCH Verlagsgesellschaft, ISBN 3-527-28313-7

Cultivation of mammalian cells 204

C Bardouille (2001) *Maintenance of Cell Cultures – with Special Emphasis on Eukaryotic Cells* in Biotechnology 2nd Edition, 27. Vol. 10, H Rehm (ed.). VCH-Wiley, ISBN 3-527-28320-X

F Hesse, R Wagner (2000), Developments and improvements in the manufacturing of human therapeutics with mammalian cell cultures *Trends Biotechnol* 18, 173.

H Hauser, R Wagner (1997) *Mammalian Cell Biotechnology in Protein Production.* Walter de Gruyter, 3-11-013403-9

R Wolfe (1993) *Media for Cell Culture* in Biotechnology Second, Completely Revised Edition, 141. Vol. 3, G Stephanopoulos (ed.). VCH Verlagsgesellschaft, ISBN 3-527-28313-7

M Wirth, H Hauser (1993) *Genetic Engineering of Animal Cells* in Biotechnology Second, Completely Revised Edition, 663. Vol. 2, A Pühler (ed.). VCH Verlagsgesellschaft, ISBN 3-527-28312-9

Mammalian cell bioreactors 206

G Kretzmer (2002), Industrial processes with animal cells. *Appl Microbiol Biotechnol* 59, 135.

L Chu, DK Robinson (2001), Industrial choices for protein production by large-scale cell culture *Curr Opin Biotechnol* 12, 180.

WS Hu, JG Aunins (1997), Large-scale mammalian cell culture *Curr Opin Biotechnol* 8, 148.

B Kelley, T Chiou, M Rosenberg, D Wang (1993) *Industrial Animal Cell Culture* in Biotechnology Second, Completely Revised Edition, 23. Vol. 3, G Stephanopoulos (ed.). VCH Verlagsgesellschaft, ISBN 3-527-28313-7

J Aunins, H Henzler (1993) *Oxygen Transfer in Cell Culture Bioreactors* in Biotechnology Second, Completely Revised Edition, 219. Vol. 3, G Stephanopoulos (ed.). VCH Verlagsgesellschaft, ISBN 3-527-28313-7

A Sambanis, W Hu (1993) *Cell Culture Bioreactors* in Biotechnology Second, Completely Revised Edition, 105. Vol. 3, G Stephanopoulos (ed.). VCH Verlagsgesellschaft, ISBN 3-527-28313-7

R Bliem, K Konopitzky, H Katinger (1991), Industrial animal cell reactor systems: aspects of selection and evaluation *Adv Biochem Eng Biotechnol* 44, 1

Enzyme and cell reactors 208

337

A Bommarius (1993) *Biotransformations and Enzyme Reactors* in Biotechnology

Second, Completely Revised Edition, 7. Vol. 3, G Stephanopoulos (ed.). VCH Verlagsgesellschaft, ISBN 3-527-28313-7

S Furusaki, M Seki (1992), Use and engineering aspects of immobilized cells in biotechnology *Adv Biochem Eng Biotechnol* 46, 161

Recovery of bioproducts 210

R Rudolph, H Lilie, E Schwarz (1999) *In vitro Folding of Inclusion Body Proteins on an Industrial Scale* in Biotechnology Second, Completely Revised Edition, 111. Vol. 5a, U Ney, D Schomburg (ed.). VCH Verlagsgesellschaft, ISBN 3-527-28315-3

A Mukhopadhyay (1997), Inclusion bodies and purification of proteins in biologically active forms *Adv Biochem Eng Biotechnol* 56, 61

R Spears (1993) *Overwiev of Downstream Processing* in Biotechnology Second, Completely Revised Edition, 39. Vol. 3, G Stephanopoulos (ed.). VCH Verlagsgesellschaft, ISBN 3-527-28313-7

J Shaeiwitz, J Henry (1988) *Separation in Biotechnology: Biochemical Separations* in Ullmann's Encyclopedia of Industrial Chemistry 5th edition B3, 11. Wiley-VCH, ISBN 3-527-20133-5

Recovery of bioproducts: chromatography 212

R Burgess, N Thompson (2002), Advances in gentle immunoaffinity chromatography *Curr Opin Biotechnol* 13, 304.

S Imamoglu (2002), Simulated moving bed chromatography (SMB) for application in bioseparation. *Adv Biochem Eng Biotechnol* 76, 211.

F Svec (2002), Capillary electrochromatography: a rapidly emerging separation method. *Adv Biochem Eng Biotechnol* 76, 1

JA Queiroz, CT Tomaz, JM Cabral (2001), Hydrophobic interaction chromatography of proteins *J Biotechnol* 87, 143.

D King (1999) *Use of Antibodies for Immunopurification* in Biotechnology Second, Completely Revised Edition, 275. Vol. 5a, U Ney, D Schomburg (ed.). VCH Verlagsgesellschaft, ISBN 3-527-28315-3

338 M Ladisch (1998) *Bioseparations* in Kirk-Othmer Encyclopedia of Chemical Technology 4th Edition Supplement, 89. Interscience-Wiley, ISBN 0-471-52696-7

N Labrou, YD Clonis (1994), The affinity technology in downstream processing *J Biotechnol* 36, 95.

PM Boyer, JT Hsu (1993), Protein purification by dye-ligand chromatography *Adv Biochem Eng Biotechnol* 49, 1

AD Diamond, JT Hsu (1992), Aqueous two-phase systems for biomolecule separation *Adv Biochem Eng Biotechnol* 47, 89

G Jagschies (1988) *Separation in Biotechnology: Process-Scale Chromatography* in Ullmann's Encyclopedia of Industrial Chemistry 5th edition B3, 10. Wiley-VCH, ISBN 3-527-20133-5

DNA: structure 216

J Rehmann (1996) *Nucleic Acids* in Kirk-Othmer Encyclopedia of Chemical Technology 4th Edition 17, 507. Interscience-Wiley, ISBN 0-471-52686-X

F Götz (1993) *Structure and Function of DNA* in Biotechnology Second, Completely Revised Edition, 191. Vol. 2, A Pühler (ed.). VCH Verlagsgesellschaft, ISBN 3-527-28312-9

Genetic engineering: general steps 220

F Schmidt (1995) *Genetic Engineering, Procedures* in Kirk-Othmer Encyclopedia of Chemical Technology 4th Edition 12, 440. Interscience-Wiley, ISBN 0-471-52681-9

V Nagarajan (1994) *Genetic Engineering, Microbes* in Kirk-Othmer Encyclopedia of Chemical Technology 4th Edition 12, 481. Interscience-Wiley, ISBN 0-471-52681-9

B Holloway (1993) *Genetic Exchange Processes for Prokaryotes* in Biotechnology Second, Completely Revised Edition, 47. Vol. 2, A Pühler (ed.). VCH Verlagsgesellschaft, ISBN 3-527-28312-9

U Stahl, K Esser (1993) *Genetic Exchange Processes in Lower Eukaryotes* in Biotechnology Second, Completely Revised Edition, 73. Vol. 2, A Pühler (ed.). VCH Verlagsgesellschaft, ISBN 3-527-28312-9

PCR: general method 226

YL Ong, A Irvine (2002) Quantitative real-time PCR: a critique of method and practical considerations. Hematology 7, 59

HA Erlich, D Gelfand, JJ Sninsky (1991), Recent advances in the polymerase chain reaction. *Science* 252, 1643.

DNA sequencing 232

L Middendorf, P Humphrery, N Narayanan, S Roemer (2001) *Sequencing Technology* in Biotechnology 2nd Edition, 193. Vol. 5b, C Sensen (ed.). VCH-Wiley, ISBN 3-527-28328-5

G Volckaert, P Verhasselt, M Voet, J Robben (1993) *DNA Sequencing* in Biotechnology Second, Completely Revised Edition, 257. Vol. 2, A Pühler (ed.). VCH Verlagsgesellschaft, ISBN 3-527-28312-9

Transfer of foreign DNA in living cells (transformation) 234

H Schwab (1993) *Principles of Genetic Engineering for Escherichia coli* in Biotechnology Second, Completely Revised Edition, 373. Vol. 2, A Pühler (ed.). VCH Verlagsgesellschaft, ISBN 3-527-28312-9

W Wohlleben, G Muth, J Kalinowski (1993) *Genetic Engineering of Gram-Positive Bacteria* in Biotechnology Second, Completely Revised Edition, 455. Vol. 2, A Pühler (ed.). VCH Verlagsgesellschaft, ISBN ISBN 3-527-28312-9

U Priefer (1993) *Principles of Genetic Engineering of Gram-negative Bacteria* in Biotechnology Second, Completely Revised Edition, 427. Vol. 2, A Pühler (ed.). VCH Verlagsgesellschaft, ISBN 3-527-28312-9

P Sudbery (1993) *Genetic Engineering of Yeast* in Biotechnology Second, Completely Revised Edition, 507. Vol. 2, A Pühler (ed.). VCH Verlagsgesellschaft, ISBN 3-527-28312-9

G Turner (1993) *Genetic Engineering of Filamentous Fungi* in Biotechnology Second, Completely Revised Edition, 529. Vol. 2, A Pühler (ed.). VCH Verlagsgesellschaft, ISBN 3-527-28312-9

Gene expression 236

C Gorman, C Bullock (2000), Site-specific gene targeting for gene expression in eukaryotes *Curr Opin Biotechnol* 11, 455.

F Baneyx (1999), Recombinant protein expression in Escherichia coli *Curr Opin Biotechnol* 10, 411

R Mattes (1993) *Principles of Gene Expression* in Biotechnology Second, Completely Revised Edition, 233. Vol. 2, A Pühler (ed.). VCH Verlagsgesellschaft, ISBN 3-527-28312-9

Gene silencing 240

SW Ding (2000), RNA silencing *Curr Opin Biotechnol* 11, 152.

RNA 242

A Fatica, D Tollervey (2002), Making ribosomes. *Curr Opin Cell Biol* 14, 313.

JA Doudna, TR Cech (2002), The chemical repertoire of natural ribozymes. *Nature* 418, 222.

GF Joyce (2002), The antiquity of RNA-based evolution. *Nature* 418, 214.

BA Sullenger, E Gilboa (2002), Emerging clinical applications of RNA. *Nature* 418, 252.

GJ Hannon (2002), RNA interference. *Nature* 418, 244.

T Hermann, DJ Patel (2000) Adaptive recognition by nucleic acid aptamers. *Science* 287, 820.

M Famulok, G Mayer, M Blind (2000) Nucleic acid aptamers-from selection in vitro to applications in vivo. Acc Chem Res. 33, 591

Gene libraries and gene mapping 244

F Sterky, J Lundeberg (2000), Sequence analysis of genes and genomes *J Biotechnol* 76, 1.

Genomes of prokaryotee and eukaryontes 246, 248

A Pühler, D Jording, J Kalinowski, D Buttgereit, R Renkawitz-Pohl, L Altschmied, A Danchin, H Feldmann, H Kleink, M Kröger (2001) *Genome Projects of Model Organisms* in Biotechnology 2nd Edition, 5. Vol. 5b, C Sensen (ed.). VCH-Wiley, ISBN 3-527-28328-5

The human genome 250

L Tsui, S Scherer (2001) *The Human Genome Project* in Biotechnology 2nd Edition, 41. Vol. 5b, C Sensen (ed.). VCH-Wiley, ISBN 3-527-28328-5

Functional analysis of the human genome 252

R Green (2001) *Genomics and Human Disease* in Biotechnology 2nd Edition, 105. Vol. 5b, C Sensen (ed.). VCH-Wiley, ISBN 3-527-28328-5

G Dellaire (2001) *Genetic Disease* in Biotechnology 2nd Edition, 61. Vol. 5b, C Sensen (ed.). VCH-Wiley, ISBN 3-527-28328-5

Mange, A Mange (1994) *Basic Human Genetics* Sinauer Associates, ISBN 0-87893-495-2.

DNA assays 254

PM Hurley, CH Rodeck (1997) *Prenatal diagnosis* in Molecular Biology in Medicine 299. TM Cox, J Sinclair (ed.). Blackwell Science Ltd., ISBN 0-632-02785-1

DNA and protein arrays 256

DD Shoemaker, PS Linsley (2002), Recent developments in DNA microarrays. *Curr Opin Microbiol* 5, 334.

MF Templin, D Stoll, M Schrenk, *et al.* (2002) Protein microarray technology. *Trends Biotechnol* 20, 160.

H Eickhoff, Z Konthur, A Lueking, *et al.* (2002), Protein array technology: the tool to bridge genomics and proteomics. *Adv Biochem Eng Biotechnol* 77, 103.

DH Blohm, A Guiseppi-Elie (2001), New developments in microarray technology *Curr Opin Biotechnol* 12, 41.

D Tessier, D Thomas, R Brousseau (2001) *A DNA Microarrays Fabrication Strategy* in Biotechnology 2nd Edition, 227. Vol. 5b, C Sensen (ed.). VCH-Wiley, ISBN 3-527-28328-5

ET Fung, V Thulasiraman, SR Weinberger, EA Dalmasso (2001), Protein biochips for differential profiling *Curr Opin Biotechnol* 12, 65.

NL van Hal, O Vorst, AM van Houwelingen, EJ Kok, A Peijnenburg, A Aharoni, AJ van Tunen, J Keijer (2000), The application of DNA microarrays in gene expression analysis *J Biotechnol* 78, 271.

M Schena (2000) *Microarray Technology* Eaton Publishing, ISBN 1-881299-37-6.

M Hagmann (2000), Doing Immunology on a chip. *Science* 290, 82.

A Lueking, M Horn, H Eickhoff, *et al.* (1999), Protein microarrays for gene expression and antibody screening. *Anal Biochem* 270, 103.

Reporter groups 258

R Rigler (1995), Fluorescence correlations, single molecule detection and large number screening. Applications in biotechnology *J Biotechnol* 41, 177.

Protein Design 260

H Zhao, K Chockalingam, Z Chen (2002), Directed evolution of enzymes and pathways for industrial biocatalysis. *Curr Opin Biotechnol* 13, 104.

B van den Burg, V Eijsink (2002), Selection of mutations for increased protein stability. *Curr Opin Biotechnol* 13, 333.

U Heinemann, G Illing, H Oschkinat (2001), High-throughput three-dimensional protein structure determination *Curr Opin Biotechnol* 12, 348.

D Wahler, JL Reymond (2001), High-throughput screening for biocatalysts. *Curr Opin Biotechnol* 12, 535.

Gene therapy 262

A Mountain (2000), Gene therapy: the first decade *Trends Biotechnol* 18, 119

N Wu, MM Ataai (2000), Production of viral vectors for gene therapy applications *Curr Opin Biotechnol* 11, 205

A Mountain (1999) *Overview of Gene Therapy* in Biotechnology Second, Completely Revised Edition, 383. Vol. 5a, U Ney, D Schomburg (ed.). VCH Verlagsgesellschaft, ISBN 3-527-28315-3

B Carter (1999) *Viral Vectors for Gene Therapy* in Biotechnology Second, Completely Revised Edition, 395. Vol. 5a, U Ney, D Schomburg (ed.). VCH Verlagsgesellschaft, ISBN 3-527-28315-3

N Weir (1999) *Non-Viral Vectors for Gene Therapy* in Biotechnology Second, Completely Revised Edition, 427. Vol. 5a, U Ney, D Schomburg (ed.). VCH Verlagsgesellschaft, ISBN 3-527-28315-3

AML Lever (1997) *Gene therapy* in Molecular Biology in Medicine, 284. TM Cox, J Sinclair (ed.). Blackwell Science Ltd., ISBN 0-632-02785-1

Proteomics 264

R Burgess, B Witholt (2002), Protein research, proteomics and applied enzymology. *Curr Opin Biotechnol* 13, 289.

N Dovichi, S Hu, D Michels, Z Zhang, S Krylov (2001) *Proteome Analysis by Capillary Electrophoresis* in Biotechnology 2nd Edition, 269. Vol. 5b, C Sensen (ed.). VCH-Wiley, ISBN 3-527-28328-5

D Figeys (2001) *Two-Dimensional Gel Electrophoresis and Mass Spectrometry for Proteomic Studies: State-of-the-Art* in Biotechnology 2nd Edition, 241. Vol. 5b, C Sensen (ed.). VCH-Wiley, ISBN 3-527-28328-5

YF Leung, CP Pang (2001), Trends in proteomics *Trends Biotechnol* 19, 480

SP Gygi, GL Corthals, Y Zhang, Y Rochon, R Aebersold (2000), Evaluation of two-dimensional gel electrophoresis-based proteome analysis technology. *Proc Natl Acad Sci*

U S A 97, 9390.

PA Haynes, JR Yates, 3rd (2000), Proteome profiling-pitfalls and progress. *Yeast* 17, 81.

MF Lopez (2000), Better approaches to finding the needle in a haystack: optimizing proteome analysis through automation. *Electrophoresis* 21, 1082.

R Kellner, F Lottspeich, H Meyer (1999) *Microcharacterization of Proteins*, 2nd, Wiley VCH, ISBN 3-527-30084-8.

F Lottspeich (1999), Proteome Analysis: A Pathway to the Functional Analysis of Proteins *Angew. Chem. Int. Ed.* 38, 2476

Drug screening 266

T Hansson, C Oostenbrink, W van Gunsteren (2002), Molecular dynamics simulations. *Curr Opin Struct Biol* 12, 190.

LN Kinch, NV Grishin (2002), Evolution of protein structures and functions. *Curr Opin Struct Biol* 12, 400

M Norin, M Sundstrom (2002), Structural proteomics: developments in structure-to-function predictions. *Trends Biotechnol* 20, 79.

J Saven (2002), Combinatorial protein design. *Curr Opin Struct Biol* 12, 453.

T Reiss (2001), Drug discovery of the future: the implications of the human genome project *Trends Biotechnol* 19, 496.

C Ramanathan, D Davison (2001) *Pharmaceutical Bioinformatics and Drug Discovery* in Biotechnology 2nd Edition, 123. Vol. 5b, C Sensen (ed.). VCH-Wiley, ISBN 3-527-28328-5

JN Kyranos, H Cai, D Wei, WK Goetzinger (2001), High-throughput high-performance liquid chromatography/mass spectrometry for modern drug discovery *Curr Opin Biotechnol* 12, 105.

RM Lawn, LA Lasky (2000), Pharmaceutical biotechnology. The genomes are just the beginning *Curr Opin Biotechnol* 11, 579.

J Gregersen (1995) *Biomedicinal Product Development* in Biotechnology 2nd Edition, 213. Vol. 12, D Brauer (ed.). VCH-Wiley, ISBN 3-527-28322-6

Bioinformatics 268

TD Wu (2001), Bioinformatics in the postgenomic era *Trends Biotechnol* 19, 479

P Rice (2001) *Bioinformatics: Tools for DNA Technologies* in Biotechnology 2nd Edition, 61. Vol. 5b, C Sensen (ed.). VCH-Wiley, ISBN 3-527-28328-5

D Wishart (2001) *Bioinformatics: Tools for Protein Technologies* in Biotechnology 2nd Edition, 325. Vol. 5b, C Sensen (ed.). VCH-Wiley, ISBN 3-527-28328-5

E Zdobnov, R Lopez, R Apweiler, T Etzold (2001) *Bioinformatics: Using the Molecular Biology Data* in Biotechnology 2nd Edition, 281. Vol. 5b, C Sensen (ed.). VCH-Wiley, ISBN 3-527-28328-5

M Cygler, A Matte, J Schrag (2001) *Bioinformatics: Structure Information* in Biotechnology 2nd Edition, 345. Vol. 5b, C Sensen (ed.). VCH-Wiley, ISBN 3-527-28328-5

E Poetzsch (1995) *Databases in Biotechnology* in Biotechnology 2nd Edition, 323. Vol. 12, D Brauer (ed.). VCH-Wiley, ISBN 3-527-28322-6

Metabolism 270

G Michal (1999) *Biochemical Pathways* Spektrum Akademischer Verlag, ISBN 3-86025-239-9.

R Krämer, G Sprenger (1993) *Metabolism* in Biotechnology Second, Completely Revised Edition, 50. Vol. 1, H Sahm (ed.). VCH Verlagsgesellschaft, ISBN ISBN 3-527-28337-4

Metabolic Engineering 272

B Christensen, J Nielsen (2000), Metabolic network analysis. A powerful tool in metabolic engineering *Adv Biochem Eng Biotechnol* 66, 209

I Fotheringham (1999) *Engineering Microbial Pathways for Amino Acid Production* in Biotechnology 2nd Edition, 313. Vol. 8a, D Kelly (ed.). VCH-Wiley, ISBN 3-527-30104-6

GN Stephanopoulos, AA Aristidou, J Nielsen (1998) *Metabolic Engineering - Principles and Methodologies*, Academic Press, ISBN 0-12-666260-6.

J Hansen, MC Kielland-Brandt (1996), Modification of biochemical pathways in industrial yeasts *J Biotechnol* 49, 1.

H Sahm (1993) *Metabolic Design* in Biotechnology Second, Completely Revised Edition, 189. Vol. 1, H Sahm (ed.). VCH Verlagsgesellschaft, ISBN 3-527-28337-4

Systems biology 274

H Kitano (2002), Systems biology: a brief overview. *Science* 295, 1662.

LM Loew, JC Schaff (2001), The Virtual Cell: a software environment for computational cell biology. *Trends Biotechnol* 19, 401.

ML Simpson, GS Sayler, JT Fleming, B Applegate (2001), Whole-cell biocomputing. *Trends Biotechnol* 19, 317.

Safety in genetic engineering 276

M Droge, A Puhler, W Selbitschka (1998), Horizontal gene transfer as a biosafety issue: a natural phenomenon of public concern *J Biotechnol* 64, 75.

D Brauer, M Bröker, C Kellermann, E Winnacker (1995) *Biosafety in rDNA Research and Production* in Biotechnology 2nd Edition, 63. Vol. 12, D Brauer (ed.). VCH-Wiley, ISBN 3-527-28322-6

T Medley, S McCammon (1995) *Strategic Regulations for Safe Development of Transgenic Plants* in Biotechnology 2nd Edition, 197. Vol. 12, D Brauer (ed.). VCH-Wiley, ISBN 3-527-28322-6

R Simon, W Frommer (1993) *Safety Aspects in Biotechnology* in Biotechnology Second, Completely Revised Edition, 835. Vol. 2, A Pühler (ed.). VCH Verlagsgesellschaft, ISBN 3-527-28312-9

Regulations of products derived from biotechnology 278

H Hasskarl, R Kretzschmar, K-J Hahn, M Zahn (1991) *Pharmaceuticals, General Survey and Development* in Ullmann's Encyclopedia of Industrial Chemistry 5th edition A19, 273. Wiley-VCH, ISBN 3-527-20119-X

Ethical considerations and acceptance 280

PJ Dale (1999), Public reactions and scientific responses to transgenic crops *Curr Opin Biotechnol* 10, 203.

RE Spier (1998), Animal and plant cell technology: a critical evaluation of the technology/society interface *J Biotechnol* 65, 111.

S Huttner (1995) *Government, Researchers and Activists: The Critical Public Policy Interface* in Biotechnology 2nd Edition, 459. Vol. 12, D Brauer (ed.). VCH-Wiley, ISBN 3-527-28322-6

D Macer (1995) *Biotechnology and Bioethics: What is Ethical Biotechnology?* in Biotechnology 2nd Edition, 115. Vol. 12, D Brauer (ed.). VCH-Wiley, ISBN 3-527-28322-6

E Weise, H Friege, G Altner, P Schmitz, K Klostermaier (1995) *Ethics and Industrial Chemistry* in Ullmann's Encyclopedia of Industrial Chemistry 5th edition B7, 1. Wiley-VCH, ISBN 3-527-20137-8

H Kepplinger, S Ehmig (1995) *Press Coverage of Genetic Engineering in Germany: Facts, Faults and Causes* in Biotechnology 2nd Edition, 495. Vol. 12, D Brauer (ed.). VCH-Wiley, ISBN 3-527-28322-6

Patents in biotechnology 282

RS Crespi (2000), Genomics, proteomics and patents *Trends Biotechnol* 18, 405.

E Szarka (1999), Patenting in biotechnology: a review of the 20th symposium of ECB8 *J Biotechnol* 67, 1.

J Gresens (1996) *Patents and Trade Secrets* in Kirk-Othmer Encyclopedia of Chemical Technology 4th Edition 18, 61. Interscience-Wiley, ISBN 0-471-52687-8

J Straus (1995) *Biotechnology and Intellectual Property* in Biotechnology 2nd Edition, 281. Vol. 12, D Brauer (ed.). VCH-Wiley, ISBN 3-527-28322-6

J Gregersen (1995) *Patent Applications for Biomedicinal Products* in Biotechnology 2nd Edition, 299. Vol. 12, D Brauer (ed.). VCH-Wiley, ISBN 3-527-28322-6

W Hauf (1990) *Patents* in Ullmann's Encyclopedia of Industrial Chemistry 5th edition B1, 13. Wiley-VCH, ISBN 3-527-20131-9

International aspects of biotechnology 284

A Schmid, F Hollmann, J Park, B Buhler (2002), The use of enzymes in the chemical industry in Europe. *Curr Opin Biotechnol* 13, 359.

Ernst & Young (2002) *Beyond Borders – The Global Biotechnology Report 2000. Ernst & Young 2002.* www.ey.com/uk

Ernst & Young (2001) *Integration – Ernst & Young's Eigth Annual European Life Science Report 2001.* www.ey.com/uk

Ernst & Young (2001) *Focus on Fundamentals – The Biotechnology Report.* www.ey.com/uk

Ernst & Young (2000) *Convergence - The Biotechnology Industry Report, Millenium Edition.* www.ey.com/uk

T Beppu (2000), Development of applied microbiology to modern biotechnology in Japan *Adv Biochem Eng Biotechnol* 69, 41

R Schmid, B Chung, A Jones, S Saono, J Scriven, J Tsai (1995) *Biotechnology in the Asian-Pacific Region* in Biotechnology 2nd Edition, 369. Vol. 12, D Brauer (ed.). VCH-Wiley, ISBN 3-527-28322-6

Origin of materials

Material from the following publications inspired some of the graphs in this pocket guide:

p 3 R Renneberg, (1990). *Biohorizonte* Urania Publisher, ISBN 3-332-00320-8

pp 12, 101, 191 HG Schlegel (1992). *Allgemeine Mikrobiologie* Thieme Publisher, ISBN 3-13-444607-3

p 63 H Heslot, Biochimie (1998). 80, 19

p 71 K Faber (2000). *Biotransformations in Organic Chemistry* Springer Publisher ISBN 3-540-66334-7

pp 87, 91, 97 H Uhlig, *Enzyme arbeiten für uns* (1991). Hanser Publisher, ISBN 3-446-15702-6

p 103 P Präve et al. (1994). *Handbuch der Biotechnologie* R Oldenbourg Publisher, ISBN 3-486-26223-8

p 121 EGD Tuddenheim (1997). in *Molecular Biology in Medicine*, TM Cox, J Sinclair (ed.). Blackwell Science Publisher, ISBN 0-632-02785-1

p 123 W Bode, H Renatur (1997). Current Opinions in Structural Biology 6, 865

pp 127, 135 C Janeway, P Travers (1997) *Immunobiology*. Current Biology Publisher, ISBN 0-443-05964-0

p 131 S Thude, Fraunhofer Institute for Bioprocess and Membrane Technology, Stuttgart, personal communication

p 139 EJ Mange, AP Mange (1992). *Basic Human Genetics*. Sinauer Associates, ISBN 0-87893-495-2

pp 159, 163 J Schenkel (1995). *Transgene Tiere*. Spektrum Akademischer Verlag ISBN 3-86025-269-0

pp 177, 181, 183, 187 J Black (1999). *Microbiology – principles and explorations* Prentice Hall ISBN 0-13-920711-2

p 209 K Buchholz, V Kasche (1997). *Biokatalysatoren und Enzymtechnologie* VCH Publishers ISBN 3-527-2831-7

p 217, 219, 227 R Knippers (2001). *Molekulare Genetik* Thieme Publishers ISBN 3-1347-7008-3

p 231, 237 T Brown (1996) *Gentechnologie für Einsteiger* Spektrum Akademischer Verlag, ISBN 3-8274-0059-7

p 247, 249, 251, 253, 255 TA Brown (1999). *Genomes* John Wiley Sons ISBN 1-85996-201-7 (p. 251: the genetic map of Drosophila is from Arthur Sturtevant) (p. 251: data from the following publications have been used: Oliver et al. (1992). Nature 357, 38 San Miguel et al. (1996). Science 274, 765 Blattner et al. (1997). Science 277, 1453)

p 255, 257 SB Primrose (1966). *Genomanalyse* Spektrum Akademischer Verlag ISBN 3-8274-0116-X

p 267 F Lottspeich (1999). Angewandte Chemie Int. Ed. 38, 2476

p 271, 273 GN Stephanopoulos et al. (1998). *Metabolic Engineering – principles and Methodologies* Academic Press ISBN 0-12-666260-6

p 281 D Schulte (2000). *Nachr. aus der Chemie* 626

p 285 Ernst & Young Reports (2002) *Beyond Borders – The Global Biotechnology Report*; (2001) *Integration*, (2001) *Focus on Fundamentals;* (2001) *Convergence*

Copyright permission was requested for the following photographs:

p 137 CSIRO, Clayton, Australia, cited from Science 281, p 511 (1998)

p 139 T. Cox et al., Molecular Biology in Medicine, plate 25 (1997)

p 161 Nature 300 (5893) (1982) title page, Macmillan Magazines Ltd.

p 173 AJ Büchting, Biologie in unserer Zeit 28, p 16 (1998)

p 173 G Krczal, Transgene Pflanzen in der landwirtschaftlichen Produktion, Nachr Chem Tech Lab 45, p 867 (1997)

p 175 Susan Colburn/Monsanto, cited from Science 282, p 2178 (1998)

p 187 J Black (1999). *Microbiology – principles and explorations* Prentice Hall ISBN 0-13-920 711-2

p 205 B Atkinson et al., Biochemical Engineering and Biotechnology Handbook (1992), p. 459, plate C

DEMCO